Artificial Intelligence in Tribology

Artificial Intelligence in Tribology presents the use of machine learning, AI, neural networks, and optimization algorithms in study of the friction, wear, and lubrication. It explores the selection of materials, modeling of the tribo-system, prediction of tribological behavior, and assessment and optimization of performance across numerous systems.

Addressing tribological problems from macroscale to nanoparticles, this book discusses AI-driven methods to improve productivity, detect disasters, and support sustainable engineering. Case studies from various industries, such as automotive, aerospace, renewable energy, and marine engineering, highlight the transformative impact of AI in real-world scenarios.

This book will interest researchers and graduate students studying AI integration and advances in surface engineering, tribology, and materials science.

Dr. Jashanpreet Singh is an Assistant Professor at Chandigarh University, Mohali, India. In 2019, he earned his Ph.D. from Thapar Institute of Engineering and Technology, Patiala, India. He has several years of both teaching experience and industry experience. His research interests include tribology, solid particle erosion, power plant engineering, ash conveying, thermal spray coatings, CFD simulation, composites, and materials characterization. He has authored over a hundred research publications indexed in the Scopus database. He has authored and edited five books with CRC Press. He serves as a reviewer for various journals, and he is a member of AICTSD, IAENG, and NITTSD.

Dr. Hitesh Vasudev is a Professor in the School of Mechanical Engineering at Lovely Professional University, Phagwara, Punjab, India. He earned his Ph.D. in 2018 from Guru Nanak Dev Engineering College, Ludhiana, India. His primary research interests are surface engineering, with particular emphasis on thermal spray technologies; development of nanostructured, multimodal, and high-entropy alloy coatings for high-temperature oxidation and corrosion resistance; thermal barrier coatings (TBCs); and microwave processing of materials. He has made significant scholarly contributions to the field of thermal spray coatings through numerous publications in reputed international journals.

Advances in Surface Engineering and Tribology

Series Editors: Hitesh Vasudev, Jashanpreet Singh, and Jeyaprakash Natarajan

Advances in Surface Engineering and Tribology explores the interdisciplinary aspects of surface science and engineering, as well as tribology discussions pertaining to the examination of friction, wear, and lubrication. It covers topics ranging from surface modification techniques, nanomaterials and their applications in nanotechnology, catalytic and photocatalytic processes, synthesis of materials, colloids and surfaces, thin film deposition processes, methods for characterizing surfaces, and surface engineering for diverse industrial purposes. The volumes comprehensively explore the complex mechanisms that regulate surface phenomena, presenting both established principles and emerging developments in the field of surface science and engineering. The series offers a forum for professionals, engineers, and researchers in the energy, automotive, aerospace, and biomedical sectors to delve into innovative developments in these interrelated disciplines.

Should you have an idea for this series, please get in contact with Kyra Lindholm (kyra.lindholm@taylorandfrancis.com).

Thermal Spray Coatings for High-Temperature Conditions
Edited by Gaurav Prashar, Hitesh Vasudev, and Lalit Thakur

Microwave Claddings: Engineering Applications
Edited by Sarbjeet Kaushal

Fabrication of Wear and Corrosion Resistant Materials: Surface Processing Techniques
Edited by Gurbhej Singh and Hitesh Vasudev

Metal Oxides: Advances in Sustainability and Nanotechnology
Edited by Virat Khanna, Rohit Jasrotia, Tejeet Singh, and Gulab Singh

Multifunctional and Smart Materials: Advances in Sustainable Engineering Applications
Edited by Hitesh Vasudev and Gurbhej Singh

Microwave Processing of Metallic Materials: Revolutionizing Surface Engineering for Green Manufacturing
Edited by Hitesh Vasudev, Gurbhej Singh, and Amit Bansal

Slurry Erosion: Flow Phenomenon, Complexities, and Protection Methods
Edited by Jashanpreet Singh and Satish Kumar

Artificial Intelligence in Tribology

Edited by
Jashanpreet Singh and Hitesh Vasudev

CRC Press
Taylor & Francis Group
Boca Raton London New York

CRC Press is an imprint of the
Taylor & Francis Group, an **informa** business

Designed cover image: Shutterstock

First edition published 2027
by CRC Press
2385 NW Executive Center Drive, Suite 320, Boca Raton FL 33431

and by CRC Press
4 Park Square, Milton Park, Abingdon, Oxon, OX14 4RN

CRC Press is an imprint of Taylor & Francis Group, LLC

ISBN: 978-1-041-03528-2 (hbk)
ISBN: 978-1-041-03529-9 (pbk)
ISBN: 978-1-003-62416-5 (ebk)

DOI: 10.1201/9781003624165

Typeset in Times
by codeMantra

This book is dedicated to our families, friends, research supervisors and teachers.

This book is dedicated to all families, Filipino

Contents

About the Editors

Jashanpreet Singh is currently working as an Assistant Professor in the Department of Mechanical Engineering, University Centre of Research and Development, Chandigarh University, Mohali, India. He earned his Ph.D. in Slurry Erosion and Thermal Spray Coatings from Thapar Institute of Engineering and Technology, Patiala, in 2019. His research interests include tribology, solid particle erosion, power plant engineering, ash conveying, thermal spray coatings, CFD simulation, composites, biotechnology, additive manufacturing, and materials characterization. He has teaching experience of more than 5 years and industrial experience of 2 years. He is capable of handling the academic duties and research activities. He is familiar with CAD software, artificial intelligence, design of experiments, and materials characterization techniques. He has contributed comprehensively to thermal spray coatings and tribology in reputed journals under the flagship of various publishers. He is serving as a reviewer for various Scopus- and Web of Science-indexed journals. He is a lifetime member of AICTSD, IAENG, and NITTSD. He has written over 25 book chapters in different edited books. As per the Scopus database, his h-index is 33, and he has over 115 Scopus-indexed publications with over 2,000 citations. He is the Book Series Editor with different publishers. He is holding editorial positions in several journals. He had received Trusted Reviewer Status from the Institute of Physics (IOP) in 2020. He has been awarded an Outstanding Reviewer Award 2021 by the Institute of Physics (IOP) for *Surface Topography: Metrology and Properties*.

Hitesh Vasudev is working as a Professor in the School of Mechanical Engineering, Lovely Professional University, Phagwara, Punjab, India. He earned his Ph.D. from Guru Nanak Dev Engineering College, Ludhiana, India, in 2018 under the guidance of Prof. (Dr.) Harmeet Singh, GNDEC, Ludhiana, and Dr. Lalit Thakur, NIT, Kurukshetra. His research areas include surface engineering and thermal spraying (HVOF, flame spray, cold and plasma spray, and microwave claddings). He is currently working on the development of nanostructured, multimodal, and high-entropy alloy coatings for high-temperature oxidation and corrosion resistance, thermal barrier coatings (TBCs), and microwave processing of materials. He has contributed extensively in thermal spray coatings in repute journals under the flagship of various publication groups. He has been awarded a "Top Cited Paper Awards India 2022" by @IOP Publishing, United Kingdom, in the Review Category. He has won the Research Excellence Award – 2019, 2020, 2021, and 2023 at Lovely Professional University. He has 50+ SCI-indexed papers, 200 Scopus-indexed papers, 15 conference papers, 12 books, and 25 book chapters. He has granted a patent titled "High-Temperature Oxidation and Erosion Resistant Alloy-718/ Al_2O_3 Composite Coatings". He has organized a national conference and international conferences. He has received SEED MONEY for a project titled "Fabrication of coatings for bio-medical applications", from Lovely Professional University, India. He has also received a grant from the Scientific Engineering and Research Board (SERB), New Delhi, Government of India, to conduct a conference in 2023.

Contributors

Mohamed Aboughaly
Faculty of Energy Systems and Nuclear Science
University of Ontario Institute of Technology
Oshawa, ON, Canada

Jyoti Bhattacharjee
Department of Chemical Engineering
University of Calcutta
Kolkata, India

Pushpendra Gupta
Department of Data Science and Artificial Intelligence
BITS Pilani Digital, Birla Institute of Technology & Science
Bangalore, Karnataka, India

Abdul Wahab Hashmi
School of Mechanical Engineering
Shandong University of Technology
Zibo, China

Hyun Suk Jung
Department of Advanced Materials Science & Engineering
Sungkyunkwan University (SKKU)
Republic of Korea

Shivam Kumar
WA School of Mines: Minerals, Energy and Chemical Engineering
Curtin University
Perth, Australia

Virinder Kumar
Department of Mechanical Engineering
University Centre for Research and Development
Chandigarh University
Mohali, Punjab, India

Amrinder Mehta
Research and Development Cell
Lovely Professional University
Phagwara, Punjab, India

Brijesh Prasad
Department of Mechanical Engineering
Graphic Era (Deemed to be University)
Dehradun, Uttarakhand, India

Subhasis Roy
Department of Chemical Engineering
University of Calcutta
Kolkata, India

Vikas Sharma
Department of Mechanical Engineering
University Center for Research and Development
Chandigarh University
Mohali, Punjab, India

Adityabir Singh
Department of Mechanical Engineering
University Centre for Research and Development
Chandigarh University
Mohali, Punjab, India

Jashanpreet Singh
Department of Mechanical Engineering
University Centre for Research and Development
Chandigarh University
Mohali, Punjab, India

Parminder Singh
Quantum Materials & Devices Unit
Institute of Nano Science and Technology
Mohali, Punjab, India

Surinder Singh
LaserBond Limited
New South Wales, Australia

Manpreet Singh
Department of Mechanical Engineering
College of Engineering and Information Technology
University of Maryland
Baltimore, Maryland

Miroslaw Szala
Department of Materials Engineering
Lublin University of Technology
Lublin, Poland

Hitesh Vasudev
School of Mechanical Engineering
Research and Development Cell
Lovely Professional University
Phagwara, Punjab, India

Praveen Kumar Verma
Research and Development Cell
Lovely Professional University
Phagwara, Punjab, India

Sunirmit Verma
Department of Mechanical Engineering
University Centre for Research and Development
Chandigarh University
Mohali, Punjab, India

Jibrin Muhammad Yelwa
Department of Scientific and Industrial Research
National Research Institute for Chemical Technology
Zaria, Kaduna State, Nigeria

Preface

Tribology, or the study of friction, wear, and lubrication, plays a pivotal role in a wide range of applications in different industries such as automotive, aerospace, renewable energy, and biomedicine. The tribological principles greatly affect the efficiency, durability, and sustainability of mechanical systems. At the same time, artificial intelligence (AI) has emerged as a breakthrough in science and engineering and revolutionized the approach to solving problems. However, it is a significant combination in order to adequately address the increasing demands of modern technology.

This book, *Artificial Intelligence in Tribology*, explores the intersection between the field of tribology and AI in a comprehensive manner. It combines the theoretical foundations with practical insights and emphasizes the real-world applications where AI-driven methods improve productivity, detect disasters, and support sustainable engineering. Previously, the sciences of tribology were used to improve the lubrication design and predict the wear behavior or design advanced coating. The introduction of AI into the field of tribology has brought a new dimension to its practice.

This book is structured into different chapters to target diverse audience. It will be useful to researchers and academicians as a reference point to study the novelties in the existing approaches and future developments. Many engineers and industry specialists will find real-world methods that they could then use to apply AI to the tribological process. Students and young practitioners will find the material basics and case descriptions useful for understanding the gap between the academic and practical approaches.

In this book, the readers will explore a range of diverse topics such as machine learning models for wear analysis and optimization procedures for lubrication and surface engineering. Various case studies from different industries, such as automotive, aerospace and renewable, biomedical, and marine, will be presented to highlight the transformative impact of AI in real-world scenarios. Moreover, this book refers to difficulties and constraints, which makes it possible to acquire a critical view and invent solutions by questioning.

At present day, tribology is facing various challenges regarding sustainability and efficiency. Therefore, the implication of AI in tribology shapes new horizons. This book is intended to help readers understand and accept this intersection, as well as use AI to advance tribology and create a better future. It is expected that this book will act as a compass and a resource for those interested in the future of AI in the profound and demanding domain of tribology.

1 Introduction to Tribology and Artificial Intelligence

Jashanpreet Singh

1.1 INTRODUCTION TO TRIBOLOGY

Tribology is an interdisciplinary science that deals with the principles of friction, wear, and lubrication that take place when two surfaces move relative to each other [1]. It provides the interface between the science of materials, mechanical engineering, surface chemistry, and physics so as to improve efficiency and longevity of the mechanical systems. Knowledge of tribology is important in improving efficiency, minimizing energy loss, and prolonging the lifetime of parts in various applications in the automotive engine market, manufacturing equipment, biomedical implants, and aircraft mechanisms [2]. In this segment, the history, scope, and importance of tribology in engineering systems today are described. Tribology was named after tribos, a Greek word meaning rubbing, and was coined officially in the 1966 Jost Report in the United Kingdom [3]. It was characterized as the science and technology of interacting surfaces in relative motion and associated topics and activities. Tribology not only includes information about mechanical contact but also surrounds the issue of surface contacts that are dependent on materials, environmental factors, and working conditions. Friction and wear processes are comprehended and controlled using surface engineering, nanotechnology, and computational modelling that are joined together in the modern tribology [4]. The field, therefore, helps in conservation of energy and in environmental sustainability and creation of high-quality materials with better surface characteristics. Tribology has its roots in the oldest civilizations, when early engineers used tribological ideas (although they were not aware of how) in transport and construction. As an example, the ancient Egyptians employed the use of rollers lubricated with wood to move heavy stones. Leonardo da Vinci (15th century) was the first to investigate the topic of friction systematically, coming up with empirical laws that were later formalized by Amontons and Coulomb [5]. The Industrial Revolution was an important event because the need to have a dependable machine prompted the systematic examination of the subject of lubrication and bearing design. The scientific field of tribology was recognized as a discipline in its own right with the release of the Jost Report in 1966, with tribology of particular interest to economics and industry [6]. The combination of surface characterization tools, computational models, and data analytics via AI has changed the field over the past few decades in a way that is capable of predictive maintenance and design of smart tribological systems.

Tribology is important in streamlining mechanical systems through loss reduction in friction, wear, and efficient lubrication [7]. Tribological performance in

DOI: 10.1201/9781003624165-1

engineering practice has a direct effect on the energy efficiency, component reliability, and maintenance cost. As just an example, better lubrication regimes can be used in engines and engines to consume less fuel and emit less carbon in the environment and good wear control can be used in manufacturing tools and improves product quality and equipment life. The tribological principles are also important in biomedical devices, micro-electromechanical systems and renewable energy technologies, where surface interactions are critical to functionality and safety. As the need to be more sustainable and digitalized is increasing, the field of tribology is now expanding into smart materials, coating, and AI-based diagnostics, which is one of the pillars of the innovative engineering of the next generation.

1.2 FUNDAMENTAL CONCEPTS OF FRICTION, WEAR, AND LUBRICATION

The above three interrelated phenomena of friction, wear, and lubrication are the fundamental pillars of tribology since they determine the operation and stability of mechanical systems with regard to surface interactions. This knowledge is important in the design of materials, components, and systems with the ability to work efficiently under different conditions in the environment and load [8]. Friction opposes motion, wear defines material losses and deterioration, and lubrication lessens the two by creating a shielding layer between interacting surfaces [9]. The combination of these ideas is the scientific foundation of tribology and has many industrial implications.

1.2.1 Friction and Its Mechanisms

Friction is the force that opposes some relative motion between two contacting surfaces. It is a result of the microscopic asperities (peaks of surface roughness) which interlock or deform during contact [10]. Friction may be of two types: static (opposes the establishment of motion) and kinetic (acts in motion) [11]. Classical laws of friction of Amontons and Coulomb posit that friction is linear to the normal load, and does not depend on the apparent contact area, although recent research has indicated that surface chemistry, roughness, and environmental influence can have considerable effects on it. Mechanically, friction has three primary causes: adhesion, which is the force of molecular attraction between surfaces, creating a resistance; deformation, which is the plowing or deformation of a softer material; and third-body effects, which is a contact mediated by some substance [12]. Friction control is a vital area of engineering systems, not only to reduce the energy losses in engines but also to provide the correct grip in braking systems and medical prosthetics.

1.2.2 Types and Mechanisms of Wear

Wear is a gradual solid material loss when subjected to mechanical contact. It is a natural, yet manageable process which defines service life and reliability of components [13]. The main categories of wear are:

- Adhesive wear is due to localized bonding between surfaces and then the material transfer [14].
- Abrasive wear takes place when a tougher surface is cut or scratched by harder particles or asperities [15].
- Corrosive (chemical) wear also have relationship with tribology in which chemical reactions destroy surface integrity [16].
- Fatigue wear is caused by the repetitive cycles of stress, localized on surfaces causing cracking and delamination [17].
- Erosive wear is due to the action of high velocity particles or fluids [18]. The mechanisms of wear are based on the material hardness, surface finish, environment, and operating conditions that include load, speed, and temperature.

In contemporary tribology, surface engineering techniques, such as coating, texturing, and alloying, are used to predictively model the wear and also to prolong the life of a component.

1.2.3 Lubrication Principles and Regimes

Lubrication involves the addition of a substance (liquid, solid, or gas) between moving surfaces in order to support a reduction in friction and wear [19]. The lubricant forms a film that serves as a separator between the surfaces, and this lowers direct contact and associated energy losses. There are three dominant lubrication regimes which are denoted in the Stribeck curve:

- Boundary lubrication is a type of lubrication in which the contact between surfaces and the lubricant is asperity contact with the lubricant acting mainly via chemical films or lubricant additives.
- Mixed lubrication considers some fluid film support and asperity contact, where the load is borne by a complete film of fluid (hydrodynamic and elastohydrodynamic lubrication) as always happens in bearings and gears.
- Lubricants may also have other properties such as cooling, corrosion control, and debris removal. Among other things, solid lubricants (graphite, MoS_2) and nanolubricants are modern innovations that use nanoparticles to increase thermal stability and load-bearing capacity.

1.2.4 Applications of Tribology in Industry

The industrial significance of tribology is extremely broad in industries such as automotive, aerospace, manufacturing, energy, and biomedical engineering. Tribological interfaces in engines, brakes, and transmissions in the automotive industry are optimized to be more efficient and low-emission [20]. Bearings, wear resistant coats, and lubricants are employed in manufacturing to enhance the wear life of tools and surface finish. High-performance lubrication is needed in the extreme conditions of the aerospace industry, whereas the integrity of artificial joints and prosthetic devices is ensured in the biomedical tribology [21]. Moreover, tribological design is also important for renewable energy systems such as wind turbines and solar trackers because it ensures sustainability by reducing maintenance and downtime. As digital technologies and artificial intelligence (AI) continue to develop, tribological data are currently being processed in predicting system failure, in optimizing lubrication times, and in generating adaptive materials, becoming intelligent tribological systems.

1.3 CHALLENGES IN TRIBOLOGY

Tribology, therefore, having long-established roots, is still evolving with the new engineering challenges arising with the use of new, advanced materials, extreme operational environments, and sustainable manufacturing systems. The need for high efficiency, low energy losses, and long component life has increased pressure for new knowledge and innovation in the field of friction, wear, and lubrication science [22]. Recent research directions are centred on mixture of material design, surface engineering, and data-driven modelling to resolve the rich interactions within tribological interfaces. In this part, the author presents some of the key issues and emerging trends that affect tribology in the 21st century focusing on the choice of materials, new testing methods, and predictive computerized techniques. Figure 1.1 represents the flowchart of challenges in tribology.

1.3.1 Material Selection and Surface Engineering

Any tribological system largely depends on the materials employed as well as the design of the surfaces to perform. Selecting proper materials for contact pairs is a challenge because of the conflicting requirements of hardness, toughness, and compatibility under varied conditions of operation [23]. Each of the metallic materials, ceramics, composite polymers, and composites possesses its own tribological

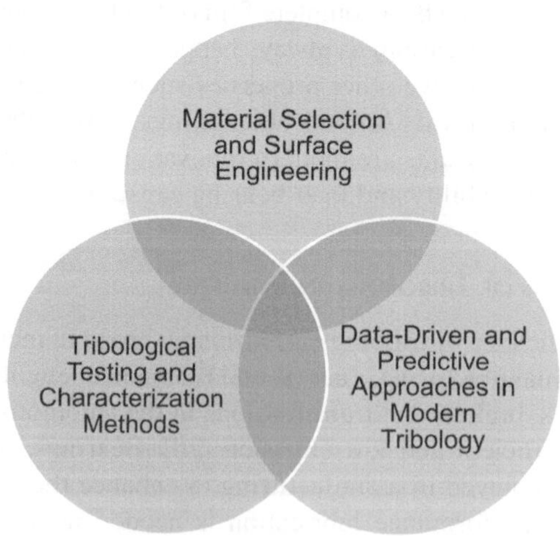

FIGURE 1.1 Flowchart of challenges in the tribology.

behaviour that varies with the surface roughness, temperature, and environment. Surface engineering has become an important method of improving wear resistance and frictional performance through methods that do not change the bulk material properties. Techniques such as coatings (Physical Vapor Deposition (PVD), Chemical Vapor Deposition (CVD), and thermal spray), surface texturing, ion implantation, and laser surface modification techniques are widely used in order to tailor surface energy, hardness, and lubrication response [24]. Additionally, nanostructured coatings and solid lubricants are being developed to achieve a superior tribological stability in a high load and temperature situation. This has remained a challenge to maximize these techniques to be economically viable, environmentally friendly, and enduring in operating systems.

1.3.2 Tribological Testing and Characterization Methods

The tribological performance should be assessed with proper evaluation procedures and testing of surfaces. The traditional techniques such as pin-on-disk, four-ball testers, and reciprocating tribometers remain significant to investigate the friction and wear behaviour under controlled environments [25]. The needs of real-world applications, however, demand multi-scale high precision analysis, which is able to replicate complex loading, temperature, and lubrication conditions. In situ observation, atomic force microscopy, scanning electron microscopy (SEM), and 3D surface profilometry are all used in modern tribology to comprehend the wear mechanism at the micro and nano scale [26]. Surface energy measurements, nano indentation, and Raman spectroscopy are used in relating mechanical and chemical properties to tribological results. Moreover, real-time monitoring of frictional interfaces can be applied using non-contact and high-speed sensing technologies, which is useful in predictive maintenance [27]. Experimental data, along with computational modelling, have offered new avenues in the context of simulating the contact behaviour and testing the efficiency of materials before their use.

1.3.3 Data-Driven and Predictive Approaches in Modern Tribology

Tribological interactions are governed by a large number of variables, including load, speed, temperature, and material properties, which make analytical prediction difficult. The annual research and development of AI, machine learning (ML), and computational modelling has fundamentally altered the principles by which tribological systems are analysed and optimized. The methods based on data allow the researcher to reveal the hidden correlations, in addition to predicting the friction and wear patterns and optimization of lubrication regimes without necessarily undertaking massive experimentation.

Artificial neural networks (ANNs), support vector machine (SVMs), genetic algorithms, and other types of analytical methods are becoming more common in modelling the tribological behaviour as well as predictive system performance under different conditions [28]. Since digital twins - virtual models of tribological systems can be developed with the help of their tools along with finite element analysis and multiphysics simulations, they can be utilized to monitor and optimize real time.

AI-assisted tribology is a significant research trend that will eventually lead to adaptive self-learning surfaces and self-controlling lubrication. However, data acquisition, standardization, and model interpretability still remain the bottleneck to industrial adoption. The combination of tribology and data science is, therefore, a major frontier in the path toward intelligent, sustainable, and high-performance engineering systems.

1.4 BASICS OF ARTIFICIAL INTELLIGENCE (AI)

AI is now a revolutionary technology in almost all branches of science and industry, and the systems can execute functions that previously needed the human mind to execute [29]. In the framework of tribology and engineering, AI offers effective methods to analyse the data of high complexity, identify patterns, and predict the work of systems with maximum accuracy. In this section, the history of AI, its development, its main principles – including ML and neural networks – and its increasing use in the engineering and manufacturing industries are presented. This implies that these concepts are important in order to understand the overlap of AI and tribology in the current literature and practice.

The AI field's ML approaches are having far-reaching effects on many facets of Industrial Revolution 4.0, from the scientific community and manufacturing to our daily lives. ML techniques can sift through mountains of data in novel ways, identify patterns, generate predictions, and draw conclusions based on evidence. As a result, new AI applications will be more likely to be developed, which will help current companies continue to use AI [30,31]. ML provides a fresh perspective on dataset data-driven approaches that may be used for physics-based simulation. A big win for AI and ML techniques is the development of these new spectacles. An advantage of ML is that it can evaluate massive data sets and determine their structure in a manner that can't be handled by a human [30]. Figure 1.2 is a schematic representation of the shared characteristics and areas of application across several AI subfields [31].

Knowledge discovery in databases and data mining (using convolutional neural networks (CNNs) and similar algorithms) are similar to ML. In the same vein as pattern recognition and other subfields of statistics (such as regression models) (like failure analysis) contribute significantly to data-driven analysis, particularly in applications like failure analysis. A simplified representation of a typical ML application in thermal spray is shown in Figure 1.3. Building a dataset by actual testing, modelling, or mining is the first step in using an ML approach. The dataset takes composition, density, hardness, and other variables as individual inputs. One may therefore access data held in a database, including characteristics to learn about the glass's composition, manufacturing process, and similar details (components or substances). Once correlations within the dataset have been identified, ML is used to build a prediction model [30]. Supervised and unsupervised ML algorithms are the two most common types. We will explore supervised and unsupervised ML techniques in greater detail in the sections that follow. While supervised learning makes use of models for classification and regression, unsupervised learning relies on clustering models for prediction. Following a number of best practices is essential for accurately predicting outcomes with AI/ML algorithms. Data partitioning, problem

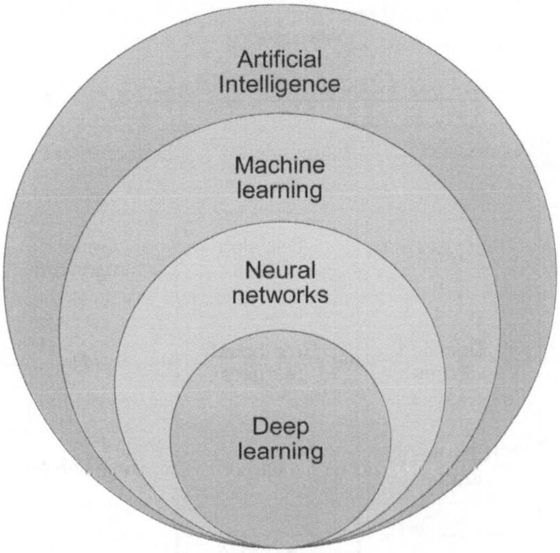

FIGURE 1.2 Illustration demonstrating the general scope of AI.

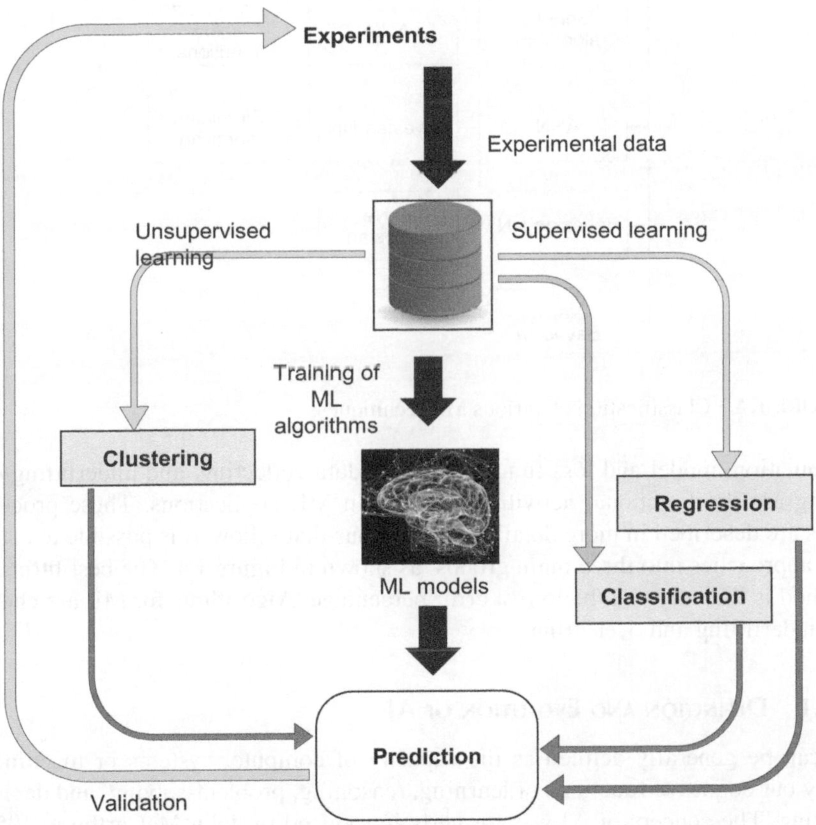

FIGURE 1.3 Overview of machine learning (ML) applications.

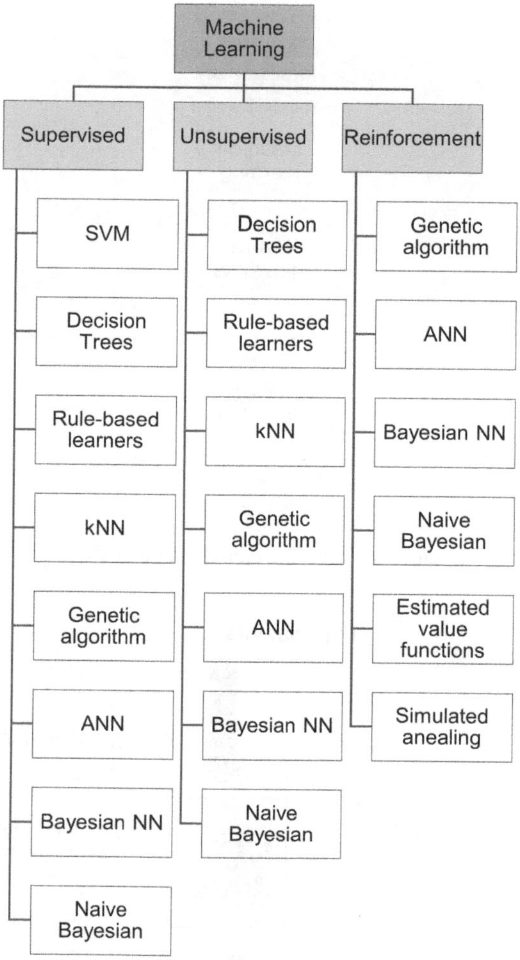

FIGURE 1.4 Classification of various ML techniques.

formulation, model and loss function tuning, data collecting, and underfitting/overfitting are the functional activities involved in ML applications. These procedure steps are described in more detail in the sections that follow. It is possible to classify ML approaches into three main groups, as shown in Figure 1.4. The best-fitting ML method is the one with the lowest error percentage. Algorithms for ML are checked for underfitting and overfitting.

1.4.1 Definition and Evolution of AI

AI can be generally defined as the capacity of computer systems or machines to carry out cognitive functions of learning, reasoning, problem-solving, and decision-making. The concept of AI was formerly introduced by John McCarthy in 1956 as the Dartmouth Conference, and this is where AI as a research discipline first came into being [32]. To begin with, AI was directed at rule-based systems and symbolic reasoning, in which knowledge was coded into machines by humans. These early systems were, however, restricted by the fact that they could not adjust to new data or circumstances. The history of AI can be divided into a number of waves. The initial (1950s–1970s) wave was focused on symbolic AI and logical thinking. The second wave (1980s–1990s) has introduced expert systems and fuzzy logic applications to the control system. The third wave (2000s–present) has been typified by data-driven learning, which was enabled by the development of computational power, high volumes of data, and better algorithms. The contemporary AI generation has seen the introduction of ML, deep learning (DL), and neural networks, enabling machines to learn and experience on their own. These technologies are becoming widely used in tribology to simulate complicated surface interactions, predict wear behaviour, and develop optimized lubrication strategies.

AI is a 60-year-old discipline that organizes theories and techniques from many areas, such as computational neuroscience, mathematical logic, probability, and computer science. AI is a rapidly expanding branch of computer science that aims to produce programmes that can do intelligent calculations on par with humans. Computer programmes, hardware, and networks that mimic the way the human brain learns and solves problems are collectively known as AI [33]. In Figure 1.5, we can see a short timeline of AI. The modern field of AI was formally launched by John McCarthy, who coined the term at a 1956 conference at Dartmouth College [34]. This was the birth of the field of AI, and it was at this time when simple, innocent algorithms were the major emphasis. Having said that, this algorithm works best with relatively short sentences. After 1960, there was an incredible surge in development. For the purpose of solving algebraic problems and proving mathematical theorems, several academics and scientists have focused on automated reasoning and used AI. Many AI pioneers believed that completely intelligent computers would soon be realized after these accomplishments, which gave them unrestrained optimism. But they soon discovered that there was still a long way to go before computers showed signs of intelligence comparable to humans. Many nontrivial problems were intractable to the logic-based programmes. Another challenge was the accessibility of computing resources to address more complicated problems. Companies and investors stopped putting money into these AI initiatives because they weren't successful.

A kind of AI system called "Expert systems" was created in the 1980s by a number of universities and research centres. It aims to help non-experts make judgements by summarizing a number of basic concepts from expert knowledge. The XCON, developed at Carnegie Mellon University, and the MYCIN, developed at Stanford University, are two such examples. First was the application of logic rules

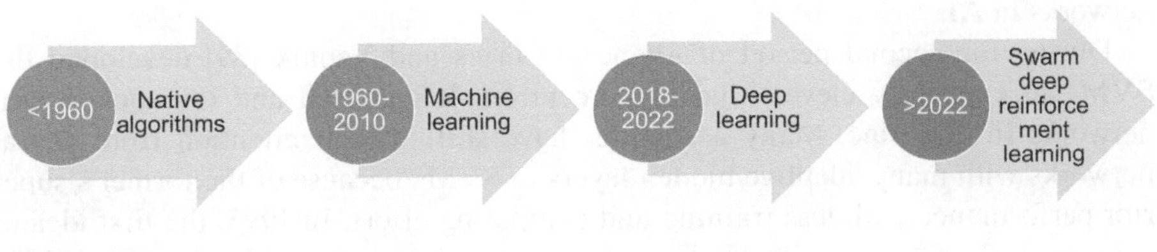

FIGURE 1.5 Schematic flow chart representing the history of AI.

derived from expert knowledge by an expert system to real-world problems. During this period, AI research was based on the notion that made computers "smarter". Privacy technologies, limited adaptability, low flexibility, high-maintenance costs, etc. were among the numerous downsides that the expert system did, however, bring out gradually.

Meanwhile, despite receiving substantial support from the Japanese government, the 5th generation computer project failed to achieve its original goals. AI development funding was once again slashed, bringing the sector to its second-lowest rank ever. In 2006, Geoffrey Hinton and colleagues [35,36] made a substantial contribution to AI by proposing a technique for creating deep neural networks. Since this is the case, DL techniques have become a hotspot for AI studies. DL employs representation learning and several layers of neural networks; it is a subset of ML [37]. On the other hand, ML is a part of AI that lets computers and programmes learn and adapt to new situations without any human intervention. More and more novel neural network topologies and training methods are being developed to improve DL's representational learning capabilities and expand its use to broader domains. Via the application of ML approaches to high-throughput data, new insights may be uncovered through classification, forecasting, and evidence-based decision-making. Therefore, AI technologies have been very successful in all areas of life. Furthermore, they have shown themselves to be valuable as the bedrock of scientific reasoning and real-world implementations. As an alternative, ML is an AI component that allows computers or programmes to learn and become smarter without human intervention. An increasing amount of novel neural network designs and training methods have been created to expand DL's representational learning capabilities and deepen its use in a wider range of applications. ML methods may be used to high-throughput data in order to categorize, predict, and arrive at new reasoning-based judgements. AI has therefore been immensely successful in all areas of life. Aside from that, they have shown themselves to be valuable as the bedrock of scientific reasoning and real-world applications.

However, AI with reactive machinery, self-awareness, limited memory, and theory of mind is distinct. Nonetheless, AI development encompasses perceptual, cognitive, and decision-making intelligence. The area of AI has seen two main kinds of advancements. One area of study focuses on software and methodologies that aim to simulate human intelligence via rule-based deduction. One such tool is the expert system. Systems that mimic brain activity, such as ANNs, make up the second group [33]. ANNs were first developed in 1943 by McCulloch and Pitts [38], who primarily used threshold logic methods to build a computational model for neural networks. This paradigm shift cleared the way for research to be divided into two streams, one concentrating on biological processes and the other focusing on using neural networks in AI.

During the second period of idleness, Cortes and Vapnik [39] developed the SVM, an extremely clever kind of perceptron. It persisted and overtook neural networks in due time. Many academics have shifted their attention from neural networks with many adaptive hidden layers to SVMs because of the former's superior performance with less training and computing effort. In 1993, the first idea of using neural computing for TS procedures was presented by Einerson et al. [40].

This research used an ANN to explain the connection between certain processing parameters and the in-flight particle properties.

1.4.2 Core Concepts: Machine Learning, Deep Learning, and Neural Networks

ML and DL are the subfields of AI that are the most applicable in terms of engineering and tribological practice. ML can be described as the process of training computers to learn by observing data and make predictions or decisions without their programmed implementation. It can be divided into three broad categories: first, supervised learning in which models are trained using labelled data to perform predictions (e.g., wear rate prediction); second, unsupervised learning where algorithms discover concealed patterns or groups in unclassified data (e.g., assembling materials in tribologically useful groupings) [41]; and third, reinforcement learning where systems can be taught on trial and error to maximize performance amid volatile environments. DL is a branch of ML deploying multi-layered ANNs in order to handle non-linear relationships that are not easy to handle [42]. Neural networks are inspired by human brain structure and are composed of interconnected nodes (neurons) in layers- input, hidden, and output layers. These networks can be trained on complex relationships in large data sets and are therefore applicable to tribological analysis with a combination of several dependent parameters, including load and speed, among others. DL is based on neural networks.

The goal of ANNs is to simulate the way the human brain learns. Neural connections are established via edges. These neurons and edges are given different relative weights as learning advances. A simple ANN with a single hidden layer is shown in Figure 1.6a. Everything from the input to the hidden to the output layers is there. The next step is to compare the calculated output to the actual production and adjust the weights accordingly. Conversely, many different problems may be solved with ANNs with several layers. As the number of hidden layers increases, the difficulties also become more complicated. Problems that might normally be too complex are made easier using this network topology. Figure 1.6b displays a multi-layered ANN.

Other implementation variants, such as CNNs and recurrent neural networks, can be applied to image recognition, time-series prediction, and fault detection. With tribology, CNNs can be applied to wear images on a surface of the wear or pattern on a lubrication film, whereas recurrent neural networks can be applied to model wear behaviour in time under cyclic loading. These AI concepts combined allow intelligent systems that can be predicted, optimized, and controlled based on the adaptive characteristics, which are significant in the development of modern tribological systems.

1.5 INTERSECTION OF AI AND TRIBOLOGY

The introduction of AI to tribology is a new phase of smart surface engineering, proactive maintenance, and intelligent production. By nature, tribological systems are very complex systems that are controlled by a variety of interacting variables, including an extension of load and speed, temperature, material properties, and lubrication

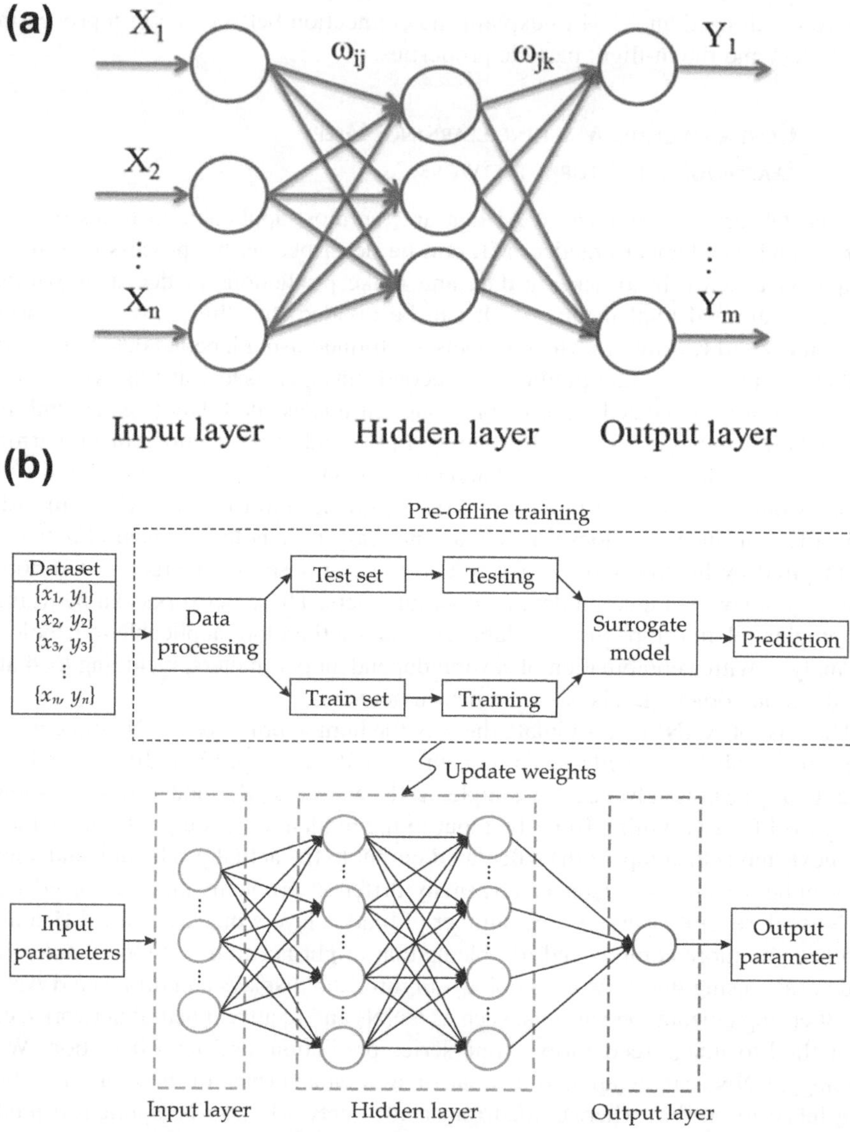

FIGURE 1.6 (a) A single hidden layer in an ANN model [43], and (b) many hidden layers, i.e., deep learning [44]. (Creatives Commons Attribution 4.0.)

regimes [4]. These non-linear, interdependent phenomena are generally difficult to describe using traditional analytical and empirical models. The alternative to this is presented in AI, which learns through experimental and operating data to foresee, streamline, and even regulate tribological conduct in the present time. The flowchart of integration of AI into tribology discusses the integration of AI in tribological analysis, as shown in Figure 1.7.

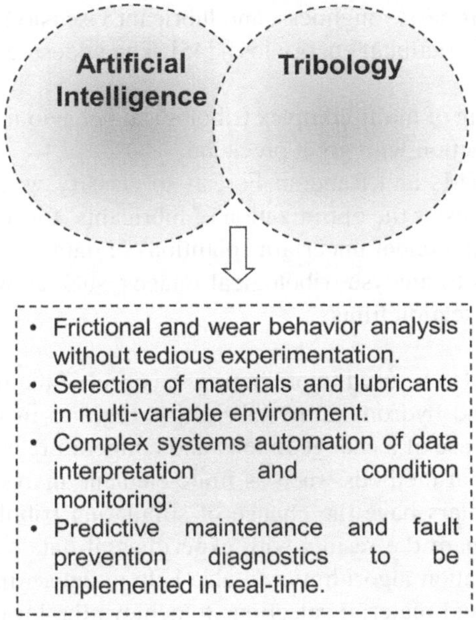

FIGURE 1.7 Flowchart of integration of artificial intelligence (AI) in to tribology.

1.5.1 Motivation for AI-Driven Tribological Analysis

The desire to use AI in tribology is predetermined by the necessity to overcome the limitations of simple models, the increased supply of tribological information, and the challenge of increased efficiency and sustainability in engineering systems. Conventional ways of investigating the friction and wear processes are dependent on physical experimentation and semi-empirical correlations that can be time-consuming, expensive, and limited in their focus. However, in contrast, AI methods, in particular, ML and DL, have the ability to process massive datasets generated by simulations, sensors, and experiments to identify subconscious patterns and causality.

In addition, the world trend in the direction of smart manufacturing and Industry 5.0 implies the focus on intelligent, adaptive systems that learn and develop themselves, as well as correct themselves. Tribological design and operation by incorporating AI would allow engineers to make an important contribution to energy efficiency, reliability of components, and environmental sustainability.

1.5.2 Role of AI in Predicting Friction, Wear, and Lubrication Performance

AI is significant in coming up with predictive models that can be used to estimate tribological parameters, which include friction coefficient, wear rate, the temperature of the surface, and the thickness of the lubrication film at various operating conditions. ML algorithms could be used to investigate experimental and simulation data sets on a large scale to develop non-linear correlations between input parameters (e.g., load, speed, hardness, roughness, and lubricant viscosity) and output responses (e.g., wear depth and frictional energy loss) [45]. The widespread AI techniques are:

- ANN modelling of multi-complex tribological behaviour and the prediction of wear and friction with great precision.
- The use of SVMs and Random Forests to classify wear modes that take place on surfaces or the optimization of lubricants' formulations.
- Decision-making under uncertain conditions of data.
- DL and CNNs to analyse tribological images, such as wear scars, surface texture, or lubrication films.

The AI models can also be used in predicting the onset of the transition between the boundary, mixed, and hydrodynamic lubrication regimes in lubrication studies to eliminate failure in case of a change in load and temperature conditions. Also, using AI with computational methods, such as finite element analysis and computational fluid dynamics, scholars have the chance of simulating tribological systems more effectively and compare the results with experimental data [28]. The combination of AI-based optimization algorithms can also help to determine the optimal topographies of surfaces and material selections to achieve the lowest possible losses on friction and prolong the life of components.

1.6 CONCLUSIONS

This chapter presented the principles of tribology and AI and defined their relationship as the ground-breaking change in present-day engineering. Tribology, which deals with friction, wear, and lubrication, has been at the core of enhancing the performance, reliability, and sustainability of mechanical systems. The fact that it has gradually developed over time into a data-driven, multidisciplinary science demonstrates the increasing demand for such a science to be more precise, efficient, and environmentally conscious in industrial practices. At the same time, AI has become a breakthrough that can solve the complexity of tribological systems. AI can be used by engineers to process large datasets and identify latent relationships and predict surface behaviour in a wide range of operating conditions through ML, DL, and predictive modelling. The adoption of AI in tribology has created the need to reorient the emphasis of trial and error experimentation to intelligent design of tribology, real-time and predictive maintenance. The two areas of interest combine to bring about exhilarating prospects of AI-enabled tribology to develop self-adaptable materials, intelligent lubrication systems, and autonomous condition-based systems. This synergy is in accordance with the idea of Industry 5.0, which is centred on sustainability, human innovation, and technology resilience.

REFERENCES

1. Stolarski TA. *Tribology in Machine Design*. Industrial Press Inc., London; 1990.
2. Reeber RR. Research and Technology Transfer Ion Implantation Technology for Specialty Materials: Proceedings of a Joint Workshop Held in Knoxville, Tennessee on 26–27 October 1989; 1991.

3. Gohar R, Rahnejat H. *Fundamentals of Tribology.* Imperial College Press, London; 2008.
4. Mang T, Bobzin K, Bartels T. *Industrial Tribology: Tribosystems, Friction, Wear and Surface Engineering, Lubrication.* John Wiley & Sons, Berlin; 2011.
5. Pitenis AA, Dowson D, Gregory Sawyer W. Leonardo da Vinci's friction experiments: An old story acknowledged and repeated. *Tribol Lett.* 2014;56:509–515.
6. Myers CJ. Tribology: Science in motion. *Phys Technol.* 1975;6:54.
7. Yu X, Lin Y, Wang P, et al. Analysis of lubrication characteristics of dynamic-static pressure hybrid thrust bearing considering key factors under eccentric loads. *Tribol Int.* 2024;194:109471.
8. Chong KP, Garboczi EJ. Smart and designer structural material systems. *Prog Struct Eng Mater.* 2002;4:417–430.
9. Al-Samarai RA, Al-Douri Y. *The Wear: Friction and Wear in Metals.* Springer, Berlin; 2013. pp. 1–31.
10. Zhang S, Li D, Liu Y. Friction behavior of rough surfaces on the basis of contact mechanics: A review and prospects. *Micromachines.* 2022;13:1907.
11. Persson BNJ, Albohr O, Mancosu F, et al. On the nature of the static friction, kinetic friction and creep. *Wear.* 2003;254:835–851.
12. Faghihnejad A, Zeng H. Fundamentals of surface adhesion, friction, and lubrication. *Polym Adhes Frict Lubr.* 2013;35:1–57.
13. Sikorska JZ, Hodkiewicz M, Ma L. Prognostic modelling options for remaining useful life estimation by industry. *Mech Syst Signal Process.* 2011;25:1803–1836.
14. Aghababaei R, Warner DH, Molinari J-F. Critical length scale controls adhesive wear mechanisms. *Nat Commun.* 2016;7:11816.
15. Singh S, Garg J, Singh P, et al. Effect of hard faced Cr-alloy on abrasive wear of low carbon rotavator blades using design of experiments. *Mater Today Proc.* 2018;5:3390–3395. Available from: https://doi.org/10.1016/j.matpr.2017.11.583.
16. Aiming F, Jinming L, Ziyun T. Failure analysis of the impeller of a slurry pump subjected to corrosive wear. *Wear.* 1995;183:876–882.
17. Potzies C, Kainer KU. Fatigue of magnesium alloys. *Adv Eng Mater.* 2004;6:281–289.
18. Singh J, Kumar S, Vasudev H, et al. Erosion prediction gaussian process regression algorithm for alumina and chromia reinforced nickel-based high-velocity oxy-fuel coatings. *Recent Patents Mech Eng.* 2025;18:1–12. Available from: https://www.eurekaselect.com/article/139361.
19. Menezes PL, Reeves CJ, Lovell MR. Fundamentals of lubrication. In Menezes PL (Ed) *Tribology for Scientists and Engineers.* Springer, Berlin; 2013. pp. 295–340.
20. Hemanth G, Suresha B. Hybrid and electric vehicle tribology: A review. *Surf Topogr Metrol Prop.* 2021;9:43001.
21. Qin C, Yang H, Lu Y, et al. Tribology in nature: Inspirations for advanced lubrication materials. *Adv Mater.* 2025;37:2420626.
22. Tung SC, McMillan ML. Automotive tribology overview of current advances and challenges for the future. *Tribol Int.* 2004;37:517–536.
23. Zhai W, Bai L, Zhou R, et al. Recent progress on wear-resistant materials: Designs, properties, and applications. *Adv Sci.* 2021;8:2003739.
24. Singh SK, Sharma D, Alam N, et al. Enhancing the tribological properties of surfaces through various surface modification and coating techniques. *Mach Tribol Adv Mater From Coatings, Lubr Surf Treat to Model Simul.* 2025;12:107.
25. Pan C, Gu Y, Chang J, et al. Recent patents on friction and wear tester. *Recent Patents Eng.* 2023;17:86–102.
26. Sadeghi B, Cavaliere P, Shabani A, et al. Nano-scale wear: A critical review on its measuring methods and parameters affecting nano-tribology. *Proc Inst Mech Eng Part J J Eng Tribol.* 2024;238:125–155.

27. Yedurkar DP, Schlech T, Sause MGR. A systematic review on smart and predictive maintenance in tool condition monitoring. *IEEE Access.* 2025;13:106246–106286.
28. Marian M, Tremmel S. Current trends and applications of machine learning in tribology—A review. *Lubricants.* 2021;9:86.
29. Shabbir J, Anwer T. Artificial intelligence and its role in near future. arXiv Prepr arXiv180401396. 2018.
30. Bulgarevich DS, Tsukamoto S, Kasuya T, et al. Pattern recognition with machine learning on optical microscopy images of typical metallurgical microstructures. *Sci Rep.* 2018;8:3–9. Available from: https://doi.org/10.1038/s41598-018-20438-6.
31. Shobha G, Rangaswamy S. Machine learning. *Handb Stat.* 2018;38:197–228. Available from: https://doi.org/10.1016/bs.host.2018.07.004.
32. Cordeschi R. AI turns fifty: Revisiting its origins. *Appl Artif Intell.* 2007;21:259–279.
33. Agatonovic-Kustrin S, Beresford R. Basic concepts of artificial neural network (ANN) modeling and its application in pharmaceutical research. *J Pharm Biomed Anal.* 2000;22:717–727.
34. Xu Y, Liu X, Cao X, et al. Artificial intelligence: A powerful paradigm for scientific research. *Innovation.* 2021;2:100179.
35. Hinton GE, Osindero S, Teh Y-W. A fast learning algorithm for deep belief nets. *Neural Comput.* 2006;18:1527–1554.
36. Hinton GE, Salakhutdinov RR. Reducing the dimensionality of data with neural networks. *Science.* 2006;313:504–507.
37. LeCun Y, Bengio Y, Hinton G. Deep learning. *Nature.* 2015;521:436–444.
38. McCulloch WS, Pitts W. A logical calculus of the ideas immanent in nervous activity. *Bull Math Biol.* 1990;52:99–115.
39. Cortes C, Vapnik V. Support-vector networks. *Mach Learn.* 1995;20:273–297.
40. Einerson CJ, Clark DE, Detering B. Intelligent control strategies for the plasma spray process. *Therm Spray Coat Res, Des Appl, Proc Natl Spray Conf.* 1993;11:205–211.
41. Singh J, Kumar S, Kumar R, et al. Machine learning regression tools for erosion prediction of WC-10Co4Cr thermal spray coating. *Results Surf Interf.* 2023;13:100156. Available from: https://www.sciencedirect.com/science/article/pii/S2666845923000636.
42. Singh J, Singh S. A review on machine learning aspect in physics and mechanics of glasses. *Mater Sci Eng B.* 2022;284:115858. Available from: https://doi.org/10.1016/j.mseb.2022.115858.
43. Zhong K, Meng Q, Sun M, et al. Artificial neural network (ANN) modeling for predicting performance of SBS modified asphalt. *Materials.* 2022;15:8695.
44. Avila Saccol G, Bui V-H, Su W. A deep-neural-network-based surrogate model for DC/AC converter topology selection using multi-domain simulations. *Energies.* 2024;17:6467.
45. Paturi UMR, Palakurthy ST, Reddy NS. The role of machine learning in tribology: A systematic review. *Arch Comput Methods Eng.* 2023;30:1345–1397.

2 AI Foundations Relevant to Tribology

Amrinder Mehta, Brijesh Prasad, Hitesh Vasudev, and Mohamed Aboughaly

2.1 INTRODUCTION

Artificial intelligence (AI) has transformed tribology and introduced potent tools for analyzing and predicting complex tribological processes. Neural networks and support machine learning (ML) systems have been successfully applied to predict the coefficient of friction (COF) and wear rates under different operating conditions. Such AI-based models can also handle large volumes of experimental data and simulations, including the complex interactions between material behavior, surface topography, and the environment that affect the tribological behavior [1]. Physics-informed AI methods have also increased the precision and explainability of tribological predictions using fundamental physical principles in ML. Moreover, AI has facilitated the development of new materials with custom tribological behaviors, including generative design and materials informatics. Such developments hold tremendous potential in automotive, aerospace, and manufacturing fields, where the optimization of tribological systems can result in enormous gains in energy efficiency, component lifespan, and performance [2].

Figure 2.1 traces the progressive integration of AI into tribology, highlighting milestones from the foundational era of AI in 1956 to the rise of ML in 1980, deep learning (DL) in 2006, and recent foundation models in 2021. The initial application of computer-aided design facilitated simple tribological choices based on computer databases. Later, neural networks added data-driven predictions of tribological properties, and more traditional ML algorithms, such as random forests, were used to better predict tool wear and to optimize systems. Recent developments have been made in DL architecture, that is, time-series data and image data, such as recurrent neural networks (RNNs), convolutional neural networks (CNNs), and transformer-based models of the complete system. Transfer learning and self-supervision are now available in foundational models. In tribology, AI has been used to aid in the analysis of mechanisms, assessment of the multisource status, optimization of system parameters, and prediction of its behavior using both physical and data-driven models. These AI developments have converged in the emerging field of tribo-informatics, where interdisciplinary approaches to lubrication, friction, and wear research that take advantage of multimodal, large-scale data and multifeature algorithm models are enabled [3].

DOI: 10.1201/9781003624165-2

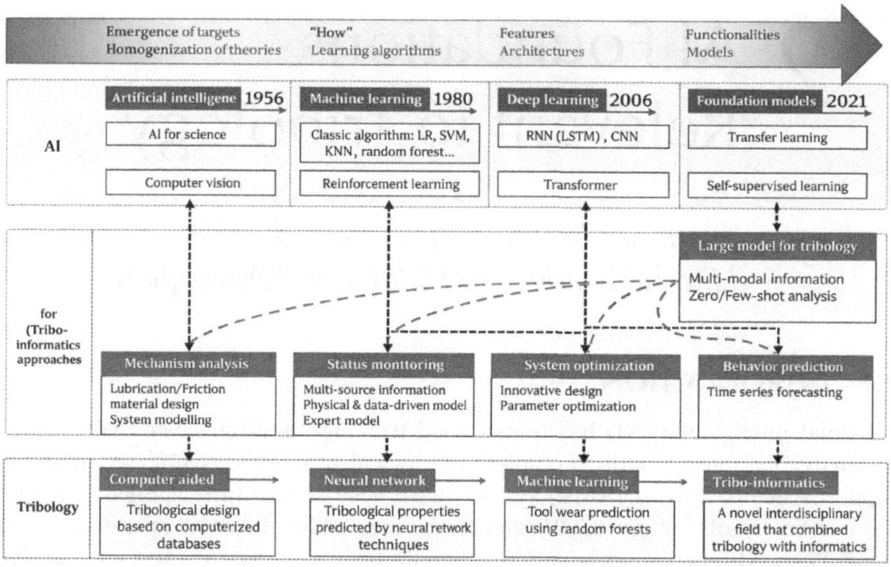

FIGURE 2.1 Development of AI in tribology [3]. (Under CC 4.0 Attribution.)

The use of AI in tribology has resulted in the emergence of new interdisciplinary areas, such as tribo-informatics, which integrates informatics and tribological information to promote better data-based decision-making and system optimization. Such a combination of AI and tribology can change the face of many industries, such as automotive, aerospace, and manufacturing, by improving the design and maintenance of mechanical systems [4].

Some important developments involve artificial neural networks (ANNs), support vector machines, decision trees, and physics-informed neural networks (PINN) to model and predict tribological behaviors, which are usually much more accurate and efficient than traditional models. These ML methods have enabled scientists to study multivariate nonlinear tribological systems. Through big data and computing, AI-based models can help characterize complex wear processes and friction effects that are challenging to measure [5].

Figure 2.2 shows the framework of physics-informed ML (PIML), where neural networks are trained not only on observational data but also by embedding known physical laws, such as partial differential equations (PDEs), into the learning process. The loss functional combines all terms involving fidelity to the governing equations, data accuracy, and compliance with the initial and boundary conditions and balances them against weighted coefficients. This compound loss penalizes both data-driven and physically consistent predictions and drives the network toward devising solutions that are consistent with the laws of science. It is used to predict lubricated contact, wear, and damage and allows a robust analysis of small-data regimes using the knowledge of the domain and available measurements [6].

However, there are still difficulties such as the need for high-quality, standardized datasets and the management of all uncertainties in experimental tribological data.

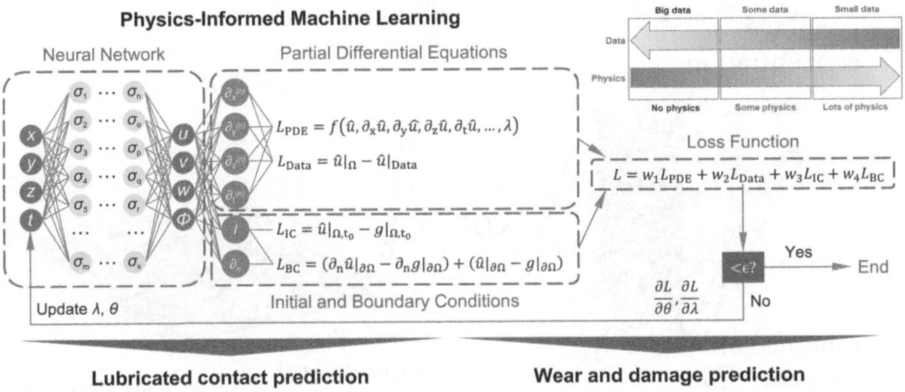

FIGURE 2.2 Physics-informed machine learning (ML) [6]. (Under CC 4.0 Attribution.)

Future studies are needed to create standardized datasets that describe a broad scope of tribological conditions and materials. Moreover, ML models should incorporate more sophisticated statistical tools and methods for quantifying uncertainty to be more representative of the variability and complexity of tribological phenomena [7].

This analysis compiles key concepts, state-of-the-art applications, and new trends in the application of AI in tribology. This review identifies important applications, such as predictive maintenance, quality coating evaluation, adaptive lubrication control, and tribological system multi-objective optimization. It further examines data integration strategies, which include experimental measurements, sensor streams, imaging data, and the laws of physics [8].

2.2 AI FOUNDATIONS AND THEORETICAL INTEGRATION IN TRIBOLOGY

Tribological AI is based on ML, DL, and physics-informed models to study and predict complex tribological phenomena. Initial research has determined the use of ANNs, decision trees, random forests, and support vector machines as a means of modeling friction, wear, and lubrication processes [9].

Recently, advances have been made in combining CNNs and RNNs for tribology image and time-series analyses. PINNs have become a potent instrument for integrating domain knowledge into AI models, improving their accuracy and interpretability in tribological applications. Such AI-driven methods make it possible to more accurately predict material behavior under diverse conditions, formulate lubricants optimally, and build tribological systems with smart capabilities to monitor and control in real time [10].

Figure 2.3 demonstrates the application of ANNs in tribological prediction, specifically for analyzing the COF behavior with varying concentrations of vegetable oil. In the multilayer neural network architecture depicted in panel (a), the input layers receive data on the concentrations of sunflower and rapeseed oils, several hidden layers contain densely connected neurons to achieve the recognition of multilayered patterns, and the output layer predicts the COF. The interconnectedness of the

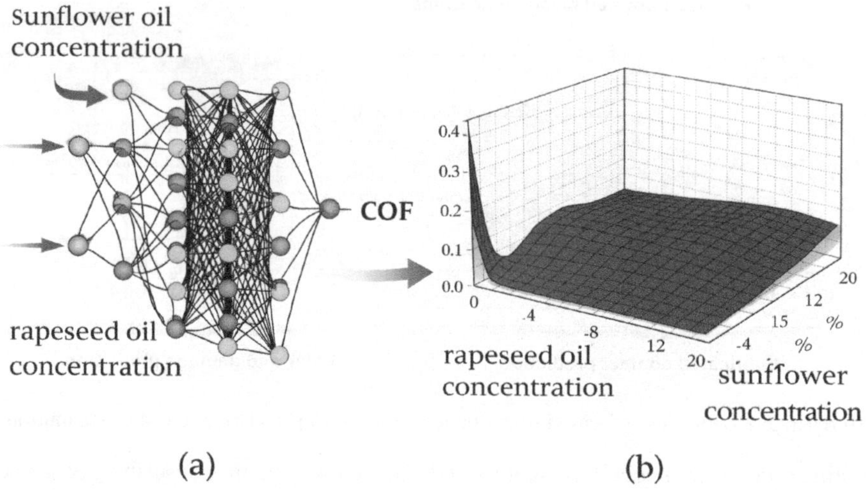

FIGURE 2.3 (a) Artificial neural network (ANN) architecture and (b) coefficient of friction for different vegetable oil concentrations [11]. (Under CC 4.0 Attribution.)

network enables the nonlinear modeling of tribological relationships between oil compositions and friction characteristics. In panel (b), a three-dimensional surface plot is used to show the response of the COF at varying concentrations of rapeseed oil (0%–20%) and sunflower oil (0%–20%). Its surface exhibits low friction coefficients (0.0–0.4 range) with smooth transition, which is observed to display synergistic effects between the two vegetable oils. The ANN successfully modeled the complicated tribological processes, demonstrating how blended bio-lubricant compounds can be used to maximize the friction performance. With this method, the tribological behavior of bio-lubricants can be predicted accurately, making it possible to design bio-lubricants with specific friction properties for use in eco-friendly applications in unique environments [11].

AI and ML have been effectively used to predict tribological quantities (such as friction coefficients, wear rates, and oil film thickness), create microstructured surfaces by design, and optimize lubricated systems. These sophisticated methods have enabled researchers to develop better models for predicting tribological behavior under different operating conditions. Using big data and sophisticated algorithms, AI and ML can detect patterns and relationships that may not be easily discernible using conventional analytical techniques. This increased predictive power has resulted in dramatic improvements in the design and optimization of tribological systems, leading to less friction, wear, and energy consumption in a broad spectrum of applications [12].

Figure 2.4 depicts a schematic of a materials informatics workflow combining Atomistic Green's function (AGF) calculations with Bayesian optimization to discover interfacial structures with enhanced thermal conductance. The AGF module on the left splits a model into three parts: the left lead, interfacial region, and right lead and calculates the phonon transmission by atomic-scale mappings of material arrangements. The interfacial zone is defined by all candidate structures characterized

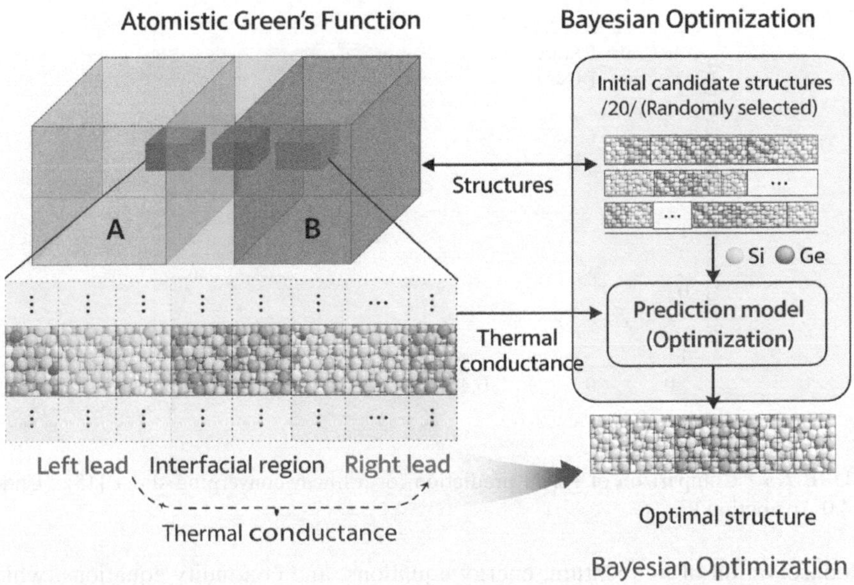

FIGURE 2.4 Schematic of the materials informatics approach integrating the atomistic Green function (AGF) with Bayesian optimization [13]. (Under CC 4.0 Attribution.)

by different Si/Ge atomic arrangements, whose thermal conductance is determined by AGF analysis. A set of randomly chosen candidate interfaces provides the input to the Bayesian optimization loop on the right of these structure–property pairs. The surrogate prediction model is trained using the AGF results and suggests new structures through exploration and exploitation. Following the thermal conductance measurements of each AGF analysis on the proposed designs, the surrogate model was refined [13].

PIML models include PINNs, which enhance the accuracy and interpretability of models by applying physical laws during learning. These methods have proven especially successful in solving complex PDEs and modeling dynamical systems. PIML techniques can use domain knowledge to require less data for training and better predict unseen situations [14].

Figure 2.5 shows that the graph demonstrates exceptional agreement between the PINN predictions and exact analytical solutions for the dimensionless pressure distribution in a linear converging slider bearing. The PINN model accurately captured the characteristic pressure profile, showing an initial linear increase, peak pressure at approximately $x = 0.6$, and a subsequent decline toward the exit. The almost perfect overlap confirms the ability of the PINN to accurately compute Reynolds equation-based lubrication problems [15].

The main benefit of PINNs is that they are used to find solutions to complex boundary value problems on elastic deformation under Hertzian contact, hydrodynamic lubrication with cavitation modeling, and Reynolds equations without requiring significant simulation data. PINNs solve tribological systems (e.g., inverse problems and parameter identification) by constraining neural networks with physical laws, such as

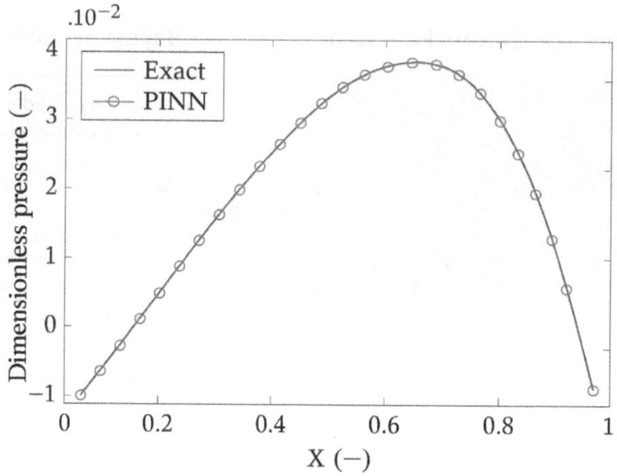

FIGURE 2.5 Comparison of PINN prediction for a linear converging slider [15]. (Under CC 4.0 Attribution.)

the conservation of momentum, energy equations, and continuity equations, which provide physically consistent predictions [16].

AI-based design has also enabled the engineering of new materials and surfaces with custom tribological characteristics. By examining large volumes of material compositions and surface arrangements, ML algorithms can predict the best material combinations to meet the given wear and friction conditions. These AI-based design tools enable engineers to run virtual tests and quickly iterate on designs without the need to physically fabricate the prototype, significantly shortening the design cycle. By using AI to experiment with previously unheard-of combinations of materials and surface geometries, scientists have discovered new tribological solutions that could not have been realized using conventional experimentation techniques alone [17].

2.3 DATA CHALLENGES AND STANDARDIZATION

The heterogeneity and nonstandardization of experimental data serve as one of the biggest difficulties in implementing AI in tribology. The ontological approach seeks to simplify the work of semantic annotation and enhance comparability and data-sharing across studies by formalizing tribological knowledge. Standardized data formats and metadata schemas must support the aggregation and analysis of large volumes of data. Training ML models using various high-quality tribological datasets may provide new insights and predictive behavior. Nevertheless, a substantial collaboration among the tribology community is required to establish the necessary data infrastructure and address existing shortcomings [18].

Figure 2.6 presents a comprehensive diagram illustrating the critical concepts of underfitting, optimal fitting, and overfitting in ML applications for tribology. In tribological problems requiring classification, such as identifying the type of lubricant or modes of failure, underfitting yields overly simplistic decision boundaries, balanced

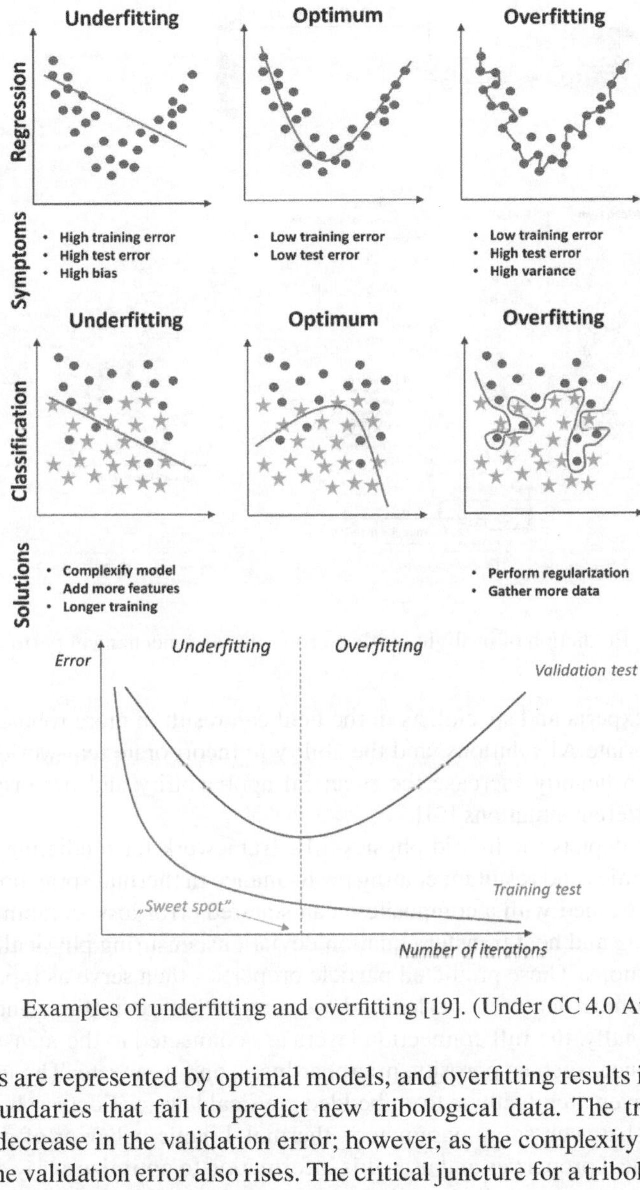

FIGURE 2.6 Examples of underfitting and overfitting [19]. (Under CC 4.0 Attribution.)

complexities are represented by optimal models, and overfitting results in convoluted decision boundaries that fail to predict new tribological data. The training curve indicates a decrease in the validation error; however, as the complexity of the model increases, the validation error also rises. The critical juncture for a tribological model occurs when it exhibits performance characteristics. [19].

The development of such datasets can be laborious and resource-consuming. This issue is especially critical in specific areas where data annotation and validation require expert knowledge. Consequently, various researchers and organizations are investigating new methods for gathering and maintaining data, such as crowdsourcing, artificial data generation, and data-sharing projects [20]. Measuring and controlling uncertainties in AI predictions are required to increase the validity and credibility of AI systems in critical use. Moreover, interdisciplinary cooperation

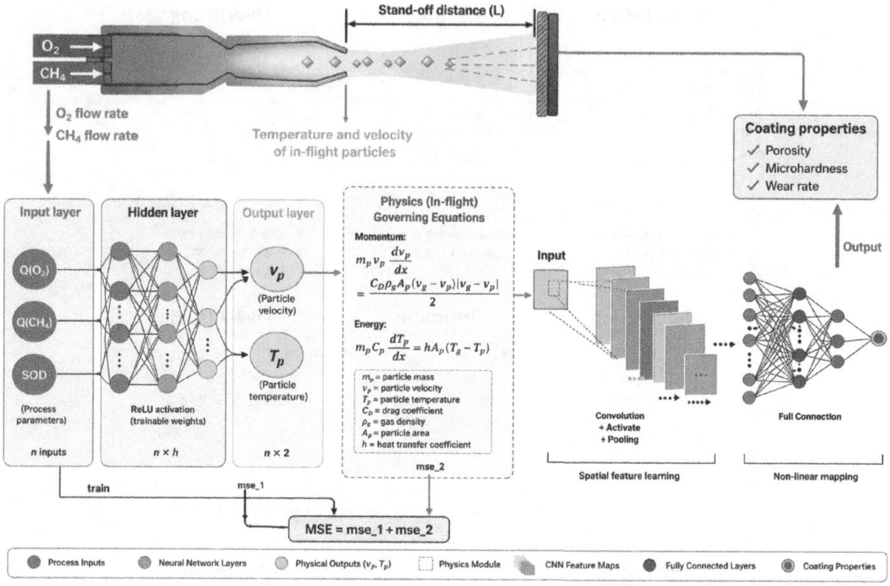

FIGURE 2.7 Prediction of in-flight particle properties and mechanical performance [22].

between AI experts and specialists in the field can result in more robust and contextually appropriate AI solutions, and the ability to incorporate real-world operational data may significantly increase the practical applicability and performance of AI models in different situations [21].

Figure 2.7 depicts the hybrid physics–ML framework for predicting the in-flight particle dynamics and resultant coating performance in thermal spray processes. The network was trained with a composite mean squared error loss, combining data and governing drag and heat transfer equation deviations, ensuring physically consistent model predictions. These predicted particle properties then serve as input to a CNN, where hierarchical features are learned by convolution, activation, and pooling of inputs, and finally, the full connection layers are connected to the measures of coating performance, such as porosity, microhardness, and wear rate. The model, which links data-driven architectures to embedded physical laws, validates a broad range of relationships between spray parameters, thermal–kinetic behavior of particles, and final mechanical properties and uses this information to optimize spray conditions in real time to achieve controlled coating properties [22].

There is a general agreement within the scientific community that AI must operate as an assistant and not as a substitute for traditional tribological research methods. AI is an effective augmentation device that boosts discovery efforts and improves experimental parameters, as well as offers more advanced decision support systems to tribologists. The integration maintains human knowledge and intuition as critical parts of the research process and expands analytical functions beyond conventional boundaries [23].

2.4 FUTURE DIRECTIONS

Future advancements in AI-driven tribology will follow various progressive developments that are set to transform the research approaches and applications in this field. All these platforms will adhere to FAIR (Findable, Accessible, Interoperable, and Reusable) data principles, making tribological data findable, accessible, interoperable, and reusable within various research communities [24].

Figure 2.8 illustrates the four foundational AI network architectures. A multilayer perceptron is a type of neural network with fully connected input, hidden, and output layers, and each neuron is configured to sum its weights through nonlinear activation functions, which allows it to perform a set of general-purpose regression and classification tasks. CNNs use layers of convolutional and pooling that extract grid-based spatial hierarchies of features (such as images) using learnable filters and dimensionality reduction. Generative adversarial networks consist of two competing networks: a generator that creates data outputs starting with noise and a discriminator that learns how to differentiate between real and generated data, pushing the generator to produce more realistic outputs. Collectively, these four architectures form the basis of a wide array of AI applications, including predictive modeling, image analysis, realistic content synthesis, and unsupervised learning [25].

Figure 2.9 shows an outline of a materials informatics-driven workflow for engineering periodic multilayered metamaterials with selective thermal emission. Starting with Bayesian optimization, the prediction model suggests sequences of candidate layers based on electromagnetic simulation results of emissivity spectra using tungsten, germanium,

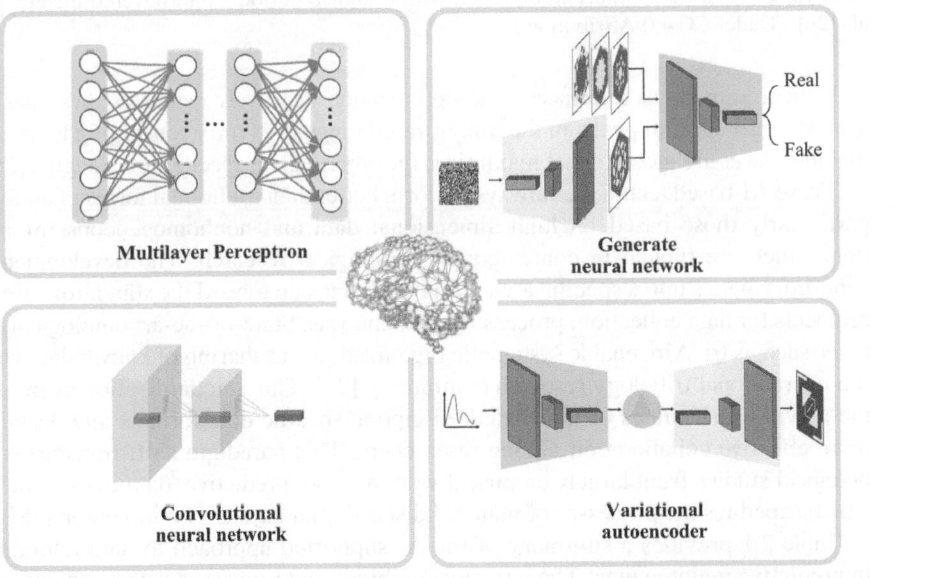

FIGURE 2.8 Illustrations of four basic AI network architectures: multilayer perceptron (MLP), CNN, generative adversarial networks (GAN), and Variational Autoencoder (VAE) [25]. (Under CC 4.0 Attribution.)

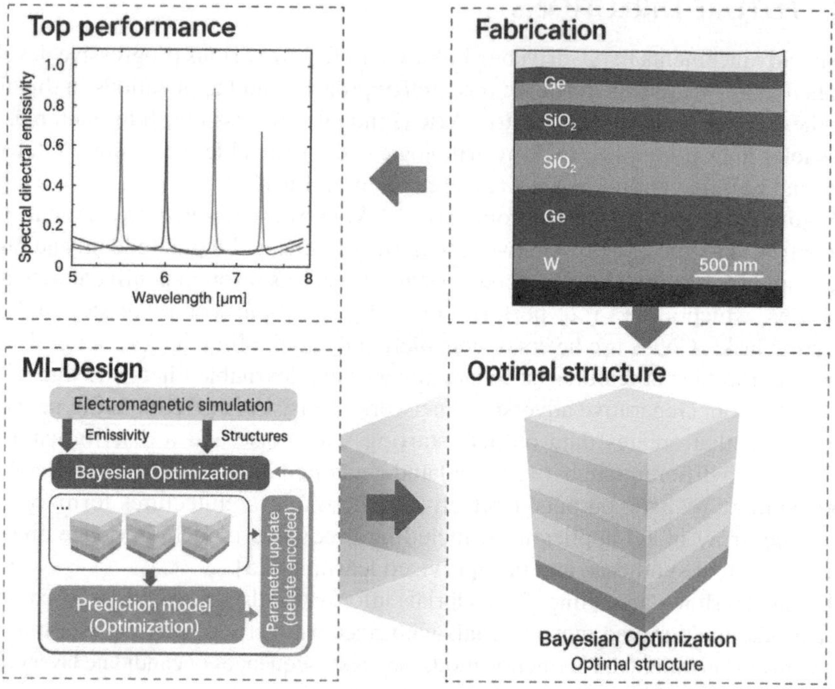

FIGURE 2.9 Selective thermal emission with designed periodic multilayered metamaterials [26]. (Under CC 4.0 Attribution.)

and silicon dioxide as interfaces. The optimization iteratively evaluates these designs against target spectral peaks in the mid-infrared range (5–8 μm), converging to an optimal multilayer architecture that maximizes the emissivity at specific wavelengths [26].

These AI-based techniques always perform better than traditional forms of analysis, particularly those based on high-dimensional data and nonhomogeneous information, which are typical in contemporary tribological research. The development of tribo-informatics into a specific area is an important step toward the standardization of protocols for data collection, processing, and analysis. State-of-the-art ontological systems, such as tribAIn, enable systematic organization and sharing of knowledge within the international tribology research community [27]. This standardization overcomes the previous difficulties of uncollated, discipline-specific data formats and permits a more effective collaboration among researchers. This paradigm shift transforms tribological studies from largely empirical work to more predictive, data-driven studies, which expedites the processes of material discovery and system optimization [28].

Table 2.1 provides a summary of the AI-supported approach to data integration in predictive maintenance. The strategy combines and processes heterogeneous data to increase fault detection and predictive maintenance in renewable energy systems. Wind and solar assets have applications, and the main findings were that the abovementioned strategy has been able to enhance the accuracy, efficiency, and real-time decision support to operational reliability.

TABLE 2.1
AI-Assistance for Methodology Maintenance of the Data Integration Approach

Methodology	Application Area	Key Results	Data Integration Approach
Artificial neural networks (ANN)	Friction coefficient prediction, wear rate analysis, and tool condition monitoring	94%–99% accuracy in friction/wear prediction and 50%–70% reduction in testing time	Experimental data fusion, multisensor integration, and historical datasets [29]
Machine learning (ML) – random forest	Wear prediction, bearing fault diagnosis, and material property estimation	85%–95% accuracy in bearing fault detection and early wear prediction capabilities	Sensor data streams, experimental measurements, and maintenance records [30]
DL – CNN/RNN	Image-based wear detection, surface roughness analysis, and coating quality assessment	96%–98% accuracy in defect detection and automated coating quality assessment	Image/video data, sensor signals, and microscopy data integration [31]
Physics-informed neural networks (PINNs)	Hydrodynamic lubrication modeling, Reynolds equation solving, and contact mechanics	Near-perfect agreement with analytical solutions and physics-constrained predictions	Physical laws + experimental data and governing equations as constraints [32]
Evolutionary neural networks (EvoNN)	Multi-objective tribological optimization and coating properties prediction	94% accuracy in High Entropy Alloy (HEA) tribological behavior and multi-objective optimization	Multisource data fusion and evolutionary optimization with experimental validation [33]
Support vector machine (SVM)	Classification of wear mechanisms and lubricant condition assessment	90%–95% classification accuracy for wear mechanisms and lubricant states	Structured experimental data, feature engineering, and dimensionality reduction [34]
Ensemble methods – gradient boosting	Remaining Useful Life (RUL) prediction, High velocity oxy fuel (HVOF) process optimization, and coating defect detection	96%–99% accuracy in RUL prediction and real-time process optimization	Time-series sensor data, operational parameters, and failure history integration [35]

The integration of AI in tribology is in its early phases, limited by the lack of standardized and high-quality datasets and challenges in integrating various operational data in real-world practices. Additionally, to benefit from all the opportunities of AI, more concerted efforts by tribologists, data scientists, and informaticians are necessary to create joint protocols, interoperable tools, and ontologies tailored to domains [36]. Maintaining traditional processes and incorporating cutting-edge computational AI will enable the field to become more holistic, where human knowledge is complemented by AI-guided understanding to address the new challenges arising in friction, wear, and lubrication.

2.5 CONCLUSION

The optimization of lubrication plans and identification of trends in large volumes of tribological information are now possible using ML algorithms. AI has transformed tribology because it offers potent computational capabilities that can calculate and predict the dynamics of complicated friction, wear, and lubrication operations with unprecedented precision and speed. By harnessing vast amounts of experimental and operational data, AI-driven methods uncover patterns and correlations that were previously hidden in the noise of multivariable datasets. This review explains the growing connection between tribology and new technologies by presenting examples of AI being used to more precisely model and predict friction, wear, and lubrication behavior. It highlights tribo-informatics systems, which combine massive experimental data and informatics systems to streamline surface treatments and lubrication solutions. Beyond prediction, AI empowers the design of novel materials and engineered surfaces with tailored tribological characteristics by rapidly screening millions of candidate compositions and textures. In the future, the field will become more reliant on generative AI models and underlying architectures that can solve creative problems across multiple regions, supported by FAIR (Findable, Accessible, Interoperable, and Reusable) data practices that ensure the findability, accessibility, interoperability, and reuse of tribological data. Future research should explore superlubricity at larger scales and develop environmentally friendly methods.

REFERENCES

1. Paturi UMR, Palakurthy ST, Reddy NS. The role of machine learning in tribology: A systematic review. *Archives of Computational Methods in Engineering*. 2023;30:1345–97.
2. Ji Y, Bao J, Yin Y, Ma C. Applications of artificial intelligence in tribology. *Recent Patents on Mechanical Engineering*. 2016;9:193–205.
3. Yin N, Yang P, Liu S, Pan S, Zhang Z. AI for tribology: Present and future. *Friction*. 2024;12:1060–97.
4. Sose AT, Joshi SY, Kunche LK, Wang F, Deshmukh SA. A review of recent advances and applications of machine learning in tribology. *Physical Chemistry Chemical Physics*. 2023;25:4408–43.
5. Selvaraj SK, Raj A, Dharnidharka M, Chadha U, Sachdeva I, Kapruan C, et al. A Cutting-edge survey of tribological behavior evaluation using artificial and computational intelligence models. *Advances in Materials Science and Engineering*. 2021;2021:9529199.
6. Marian M, Tremmel S. Physics-informed machine learning—An emerging trend in tribology. *Lubricants*. 2023;11:463.
7. Cai M, Zhang P, Xiong Q, Cai Z, Luo S, Gu L, et al. Finite element simulation of fretting wear behaviors under the ball-on-flat contact configuration. *Tribology International*. 2023;177:107930.

8. Deliwala AA, Dubey K, Yerramalli CS. Predicting the erosion rate of uni-directional glass fiber reinforced polymer composites using machine-learning algorithms. *Journal of Tribology*. 2022;144:091707.
9. Maalihan RD, Domalanta MRB, Corrales ACC, Caldona EB. Advances in interfacially engineered surface-functionalized fillers for multifunctional polymer composite coatings. *Polymer Composites*. 2025;46:5857–81.
10. Rahardja U, Sari A, Alsalamy AH, Askar S, Alawadi AHR, Abdullaeva B. Tribological properties assessment of metallic glasses through a genetic algorithm-optimized machine learning model. *Metals and Materials International*. 2024;30:745–55.
11. Humelnicu C, Ciortan S, Amortila V. Artificial neural network-based analysis of the tribological behavior of vegetable oil–diesel fuel mixtures. *Lubricants*. 2019;7:32.
12. Maalihan RD, Domalanta MRB, Ali MRR, Caldona EB. Kelvin probe force microscopy: Uncovering new frontiers in coating and corrosion studies. *Analytical Chemistry*. 2025;97:8133–46.
13. Ju S, Shiga T, Feng L, Hou Z, Tsuda K, Shiomi J. Designing nanostructures for phonon transport via Bayesian optimization. *Physical Review X*. 2017;7:021024.
14. Shah R, Jaramillo R, Thomas G, Rayhan T, Hossain N, Kchaou M, et al. Artificial intelligence and machine learning in tribology: Selected case studies and overall potential. *Advanced Engineering Materials*. 2025;27:2401944.
15. Almqvist A. Fundamentals of physics-informed neural networks applied to solve the Reynolds boundary value problem. *Lubricants*. 2021;9:82.
16. Cuomo S, Di Cola VS, Giampaolo F, Rozza G, Raissi M, Piccialli F. Scientific machine learning through physics–informed neural networks: Where we are and what's next. *Journal of Scientific Computing*. 2022;92:88.
17. Marian M, Almqvist A, Rosenkranz A, Fillon M. Numerical micro-texture optimization for lubricated contacts—A critical discussion. *Friction*. 2022;10:1772–809.
18. Aydin F, Durgut R, Mustu M, Demir B. Prediction of wear performance of ZK60/CeO_2 composites using machine learning models. *Tribology International*. 2023;177:107945.
19. Ledziński Ł, Grześk G. Artificial intelligence technologies in cardiology. *Journal of Cardiovascular Development and Disease*. 2023;10:202.
20. Yin N, Xing Z, He K, Zhang Z. Tribo-informatics approaches in tribology research: A review. *Friction*. 2023;11:1–22.
21. Kumar R, Kariminejad A, Antonov M, Goljandin D, Klimczyk P, Hussainova I. Progress in sustainable recycling and circular economy of tungsten carbide hard metal scraps for industry 5.0 and onwards. *Sustainability*. 2023;15:12249.
22. Gui L, Wang B, Cai R, Yu Z, Liu M, Zhu Q, et al. Prediction of in-flight particle properties and mechanical performances of HVOF-sprayed NiCr–Cr_3C_2 coatings based on a hierarchical neural network. *Materials*. 2023;16:6279.
23. Baş H, Karabacak YE. Machine learning-based prediction of friction torque and friction coefficient in statically loaded radial journal bearings. *Tribology International*. 2023;186:108592.
24. Pasha MB, Rao RN, Ismail S, Gupta M, Prasad PS. Tribo-informatics approach to predict wear and friction coefficient of Mg/Si_3N_4 composites using machine learning techniques. *Tribology International*. 2024;196:109696.
25. Si L, Niu R, Dang C, Bao X, Zhuang Y, Zhu W. Advances in artificial intelligence for artificial metamaterials. *APL Materials*. 2024;12:120602.
26. Sakurai A, Yada K, Simomura T, Ju S, Kashiwagi M, Okada H, et al. Ultranarrow-band wavelength-selective thermal emission with aperiodic multilayered metamaterials designed by Bayesian optimization. *ACS Central Science*. 2019;5:319–26.
27. Marian M, Mursak J, Bartz M, Profito FJ, Rosenkranz A, Wartzack S. Predicting EHL film thickness parameters by machine learning approaches. *Friction*. 2023;11:992–1013.

28. Shandhoosh V, Chakrapani G, Sugumaran V, Ramteke SM, Marian M. Intelligent fault diagnosis for tribo-mechanical systems by machine learning: Multi-feature extraction and ensemble voting methods. *Knowledge-Based Systems.* 2024;305:112694.
29. Neupane D, Bouadjenek MR, Dazeley R, Aryal S. Data-driven machinery fault detection: A comprehensive review. Available at SSRN 4856168. 2024.
30. Palaniappan R. Comparative analysis of support vector machine, random forest and k-nearest neighbor classifiers for predicting remaining usage life of roller bearings. *Informatica.* 2024;48:5726.
31. Ng WL, Goh GL, Goh GD, Ten JSJ, Yeong WY. Progress and opportunities for machine learning in materials and processes of additive manufacturing. *Advanced Materials.* 2024;36:2310006.
32. Zhao W, Tong X, Cai M, Li H, Song T. Physics-informed neural network for solving the coupled Reynolds–energy equations with viscosity–temperature effects in hydrodynamic journal bearings. *Physics of Fluids.* 2025;37:063607.
33. Vashistha S, Mahanta BK, Singh VK, Singh SK. Machine learning assisted optimization of tribological parameters of Al–Co–Cr–Fe–Ni high-entropy alloy. *Materials and Manufacturing Processes.* 2023;38:2093–106.
34. Wakiru J, Pintelon L, Muchiri P, Chemweno P. A data mining approach for lubricant-based fault diagnosis. *Journal of Quality in Maintenance Engineering.* 2021;27:264–91.
35. Sivaraman S, Radhika N, Khan MA. Machine learning-driven prediction of wear rate and phase formation in high entropy alloy coatings for enhanced durability and performance. *IEEE Access.* 2025;13:33956–75.
36. Mahapatra SK, Satapathy A. Development of machine learning models for the prediction of erosion wear of hybrid composites. *Polymer Composites.* 2024;45:7950–66.

3 Machine Learning for Predictive Modeling in Tribology

Jashanpreet Singh and Miroslaw Szala

3.1 INTRODUCTION TO MACHINE LEARNING IN TRIBOLOGY

The interdisciplinary science of the friction, wear, and lubrication between moving surfaces in relative contact is called "tribology," which is derived from a Greek word tribos. It is very important to the performance, reliability, and service life of mechanical components in various industries such as automotive industry, aerospace industry, manufacturing industry, and energy systems. The high demand of energy conservation, sustainability, and state of the art in the material design across the world has raised the significance of tribological studies to a great extent. Although experimental and analytical methods are improved, the prediction of the tribological behavior is still a challenge because of the complexity and nonlinearity of surface interactions. Parameters such as load, speed, temperature, surface roughness, lubrication regime, and material composition are highly complex interacting parameters. These multivariate dependencies motivate the creation of advanced computational tools of machine learning (ML) to make predictions more accurate and conduct intelligent decision-making in tribological systems. Three basic phenomena, namely, friction, wear, and lubrication, are the main aspects of tribology that determine the efficiency and reliability of any mechanical system. The loss of energy is friction, the corrosion of the material is wear, and the ease and stability of the work are the results of lubrication. The main tribological problem is the need to know and model the behavior of these parameters in a variety of operating conditions. Empirical and analytic models of traditional approaches usually have simplifying assumptions that restrict their predictive ability. Although experimental methods are correct, they are time-consuming, expensive, and in certain cases impractical in real-time application or at large scale. Additionally, tribological systems are very sensitive to material microstructures, surface morphologies, as well as environmental conditions; hence, deterministic modeling is challenging. These challenges have led researchers to consider the use of data-driven and intelligent modeling that can learn using experimental data and give outputs with a high level of accuracy.

In tribology, predictive modeling tends to predict important parameters such as friction coefficient, wear rate, and lubrication performance using parameters that are measurable. Proper predictive models will result in an enormous decrease in experimentation, calculated design parameters, and speed up the creation of new materials

DOI: 10.1201/9781003624165-3

and lubricants. Nonlinear, coupled interactions between tribological parameters are not readily described, however, by conventional statistical and theoretical methods.

ML offers an alternative with significant potential by training complicated models wherever possible and system–response associations. The analysis of big experimental or simulation datasets enables ML algorithms to read the hidden patterns and most influential factors, and make reliable predictions for previously unobserved conditions. This feature makes ML-based predictive modeling indispensable to contemporary tribology studies and practice. ML has become a disruptive technology in tribology, which presents new opportunities in the analysis of data, pattern recognition, and predictive analytics. Friction and wear behavior have been effectively predicted using supervised learning methods including regression, decision trees, support systems, and neural networks. Unsupervised learning includes the following: surface texture or real-time monitoring. Deep learning (DL), a subfield of unsupervised learning, involves feature extraction and clustering of experimental data. Experts can use unsupervised learning to cluster and extract features in complex, high-dimensional datasets like surface texture. Not only can ML models improve the predictive accuracy, but they also allow interpreting the tribological mechanisms with the help of feature importance analysis and correlation mapping. Moreover, it can be integrated with the Internet of Things and digital twin technologies to ensure constant monitoring, adaptive regulation, and predictions of tribological systems, which involves a transition to intelligent and self-optimal mechanical systems.

The main goal of this chapter is to provide an overall review of ML use in predictive modeling in the field of tribology. This chapter talks about the basic concepts of tribology, the limitations of traditional prediction methods, and the benefits of ML-based predictors in improving prediction accuracy and the comprehension of the system.

3.2 FUNDAMENTALS OF TRIBOLOGY

3.2.1 Friction, Wear, and Lubrication

The fundamental issues of tribology are three related phenomena: friction, wear, and lubrication that combine to determine the efficiency, durability, and reliability of interacting surfaces during relative motion. The resistance to motion between surfaces in contact is termed as friction. It occurs mainly because of asperity contacts, adhesion joints, and ploughing. Although the classical laws of Amontons and Coulomb provide more or less an approximate description, the frictional behavior is very sensitive to contact conditions, material properties, and other environmental factors such as temperature and humidity. Wear is a process of gradual removal or deformation of the material on one or both of the contacting surfaces. It can be of adhesive, abrasive, corrosive, and fatigue types, which are governed by different mechanisms. Wear is imperative in the prediction of component life and maintenance periods. Lubrication is a mitigating mechanism which lowers friction and wear by introducing a liquid, solid, or gas film between surfaces. Based on the continuity and thickness of this film, it is possible to have a boundary, mixed, or hydrodynamic lubrication regime. Efficiency of lubrication is influenced by the viscosity of lubricants, surface roughness, as well as operating conditions. These three phenomena form the fundamental basis of tribology science, and due to their interdependence with each other, sophisticated modeling techniques are usually needed to predict them and optimize their behavior.

3.2.2 Material and Surface Characteristics

Surface engineering and the choice of material have a central role in defining tribological characteristics of mechanical components. Friction and wear responses depend directly on the mechanical, chemical, and microstructural properties of materials such as hardness, toughness, elastic modulus, and surface energy. The real area of contact and asperity deformation are determined by surface topography, which is defined by such measures as roughness, waviness, and texture orientation. The modern techniques of surface characterization developed through the use of atomic force microscopy (AFM), scanning electron microscopy (SEM), and 3D surface profilometry have allowed high-resolution visualization of surface features, which has broadened our comprehension of contact mechanics. Besides that, surface alterations (e.g., coatings, texturing, and treatments such as nitriding, diamond like carbon (DLC) coatings, or laser structuring) are a common technique to increase wear resistance and decrease friction. The changes affect the surface hardness, chemical reactivity, and energy dissipation. To perform predictive modeling, it is important to accurately identify such surface and material parameters which are important input features for data-driven methods.

3.2.3 Experimental Approaches in Tribological Studies

Tribological research is still based on experimental investigation. Large-scale tests in the laboratory give useful information on frictional behavior, wear mechanism, and lubrication performance in a controlled environment. Typical experimental designs are pin-on-disk and ball-on-disk tests which are used in the assessment of the friction and wear rates during the sliding conditions. Four-ball and reciprocating tribometers are used to determine the performance of lubricants and compatibility of the material. Thrust and block-on-ring tests are used for high-load applications.

Empirical and ML-based models, which are based on experimental data, include the coefficient of friction, depth of wear, surface morphology, and temperature profiles. Tribological experiments are, however, sensitive to noise, environmental variation, and surface preparation, and they need strict calibration and repeatability. Sensors, high-speed imaging, and *in situ* measurement systems have enabled the manufacturing of devices that have enhanced the accuracy of data acquisition, which enables constant monitoring of tribological processes. The developments have increased the quality and quantity of data to be used in predictive modeling and AI applications.

3.2.4 Data Availability and Modeling Limitations

Tribological data can still be scarce, sporadic, and irregular, in spite of the progress made in experimental procedures and computational approaches. The variety of testing criteria, the surface characteristics, and combinations of material make it difficult to come up with universally applicable models. In several instances, the datasets are very small or skewed, and, as a result, these limit the performance of conventional statistical models, as well as some ML algorithms. In addition, tribological systems are nonlinear and stochastic in nature, with a small change in input conditions resulting in high changes in output responses. Nonstandardized data repositories and open-access databases restrict the cross-validation possibilities and transfer learning. Traditional physics-based models, albeit useful in the explanation of underlying mechanisms, tend to be based on simplifying assumptions that fail to consider micro- and nanoscale effects. Conversely, data-driven models need a large and diverse dataset to make effective generalizations. The solution to these gaps and improving the predictive robustness of tribology is bridging gaps with hybrid modes of modeling, that is, a combination of empirical, analytical, and ML methods.

3.3 BASICS OF MACHINE LEARNING IN PREDICTIVE MODELING

In this section, the concepts of ML are elaborated through an introduction to its concepts and principles. ML is a subdivision of AI, wherein computational systems acquire patterns and associations with data without being programmed to do so. In contrast to classical deterministic models, where the equations and related relationships among various variables are predetermined, ML algorithms determine these complex and nonlinear interactions between various variables by using statistical inferences and optimization. ML is an effective predictive modeling tool in the tribology community, which requires many interdependent parameters to determine the behavior of friction, wear, and lubrication. These models have the ability to handle experimental data and simulated data, discover latent correlations, and predict behavior in new operating environments. ML can be used to fill the gap between the theory of modeling and the actual behavior of a system by combining experimental information with computational intelligence. Applications of ML to tribology are generally predictive in nature, using supervised learning, unsupervised learning to cluster data or reduce its dimensionality, and reinforcement learning to act as an adaptive control to a dynamic system. Several classes of ML algorithms might be distinguished by their method of learning and the purpose of use.

3.3.1 Supervised Learning

It is applied when the input and output information are known. Some of the most commonly used algorithms in predicting the friction coefficients, wear rates, and lubrication performance include linear regression, decision trees, Random Forests (RF), support vector machines (SVMs), and artificial neural network (ANN) algorithms.

3.3.2 Unsupervised Learning

It is used to determine latent patterns or clusters in unlabeled groups. The techniques used, including k-means clustering, PCA, and hierarchical clustering, can help in the determination of the prevailing factors that affect tribological behavior and dimensionality reduction of data.

3.3.3 Reinforcement Learning

It includes the sequential process of decision-making through feedback and reward systems. Reinforcement learning is still in its early stage in tribology but has potential in optimizing adaptive lubrication and self-adjusting systems.

3.3.4 Deep Learning

It relies on multilayer neural networks that can represent complicated nonlinear relationships. Conventional neural networks (CNNs) and recurrent neural networks (RNNs) are examples of DL methods that have been utilized to understand surface texture images, wear progression, and real-time monitoring of systems.

3.4 IMPLEMENTATION OF MACHINE LEARNING IN PREDICTIVE MODELING

3.4.1 Data Preparation and Feature Engineering

The quality of data and choice of features are critical to the accuracy and generalization of ML models. The data in tribological research can be acquired through laboratory experiments, finite-element simulation, or sensor-based monitoring. Nevertheless, noise or missing values and outliers can be present in such data, which requires preprocessing. The main data preparation steps are as follows: normalization, removal of the outliers, data balancing, and dimension reduction. In tribology, the choice and conversion of input variables into useful form (i.e., feature engineering) is particularly significant because the material characteristics (e.g., hardness, roughness, and elasticity), operational conditions (e.g., load, speed, and temperature), and environmental conditions interact to affect the results. The most popular techniques are correlation analysis, PCA, and mutual information ranking, which are used to determine the most significant features. The correct selection of features ensures not only a better predictive performance but also better model interpretability to enable the researcher to have an insight into the most effective parameters that influence friction and wear behavior.

AI algorithms are based on prediction models that form the basis of practice [1]. These models gain better performance on particular tasks through continuous training. This chapter discusses supervised learning, in which $x \in R^n$ is a data point of features or independent variables, and targets are values [2,3]. The prediction models are developed using AI algorithms, and they identify relationships among features and targets as a basis and can be optimized through hyperparameter tuning or by using methods such as L2 regularization [4].

Normalization and standardization are some of the steps taken to filter and preprocess data to ensure proper data processing [1–3,5–8]. The datasets are described as being varied, valuable, voluminous, fast-moving, and authentic [1,8], and this parameter determines the extent of big data and large and intricate datasets demanding smart analytical tools. The structured data are organized according to specific formats (e.g., databases), and semi-structured and unstructured data are not strictly

TABLE 3.1
A Typical Structured Data

Properties	Dataset			
	Materials Combination 1	**Materials Combination 2**	**......**	**Materials Combination *N***
Strength	M_{11}	M_{12}	....	M_{1N}
Wear rate	M_{21}	M_{22}	....	M_{2N}
Corrosion	M_{31}	M_{32}	....	M_{3N}
Hardness	M_{41}	M_{42}	....	M_{4N}

structured but still contain attributes that are hierarchical or irregular in nature [1,8,9]. Table 3.1 presents an organized data format.

Furthermore, several observations on the data structure's field-of-use features are addressed in the context of coatings and tribology applications. There are three main categories of data considered in this field: structured, semi-structured, and unstructured. How to utilize quantitative data from photo-based unstructured data for quality and consistency checks is described in the following section. In order to aid with quality control, the structured data that are retrieved from mechanical studies and simulations are utilized to identify patterns and model parameters. According to research, good solutions are the product of a precise and pertinent combination of data and structure [1–3,5,6]. Decisions on experimental strategy features and hardware requirements may then be made based on this. Cost and experimental design issues mean that most statistical datasets only include few observations [10,11]. To successfully apply ML, it becomes necessary to make sure that the nature of the issue at hand, the targeted goals of the endeavor, and the quality and quantity of data used are all carefully considered. The effectiveness of an ML model is greatly determined by the sufficiency, heterogeneity, and the trustworthiness of the training and validation datasets. Knowledge of the underlying structure of the problem such as classification, regression, clustering, or prediction facilitates the selection of appropriate algorithms and feature engineering methods. Also, the appropriate quantity of the data should be representative, unbiased, and complete to allow the model to learn useful patterns and draw relevant generalizations in real-world conditions.

While it's usually not too difficult to handle structured data [7], feature engineering can help create features that are crucial for unstructured data [12]. While modern, highly efficient algorithms can handle data with a large dimensionality, the curse of dimensionality is the complexity that comes with having too many characteristics [1–3]. Applying methods for feature selection and dimensional reduction may resolve the issue. In reality, with the use of assessment techniques like correlation matrix calculation, it is possible to identify important inputs and quantify their characteristics [13]. In addition, each data point may have several input variables; nevertheless, ML models would not benefit from such high-dimensional input spaces due to their inefficiency in computing [14]. When training a model to deal with very high-dimensional data, dimensionality reduction approaches [15] like non-negative factorization [16], linear discriminant analysis [14], and PCA [14] are more beneficial. The idea behind these approaches is to combine data from many sources to generate new, useful inputs, which may then replace any of the original [12,14]. These methods enhance the computational efficiency of a model by reducing the number of input features that a model needs to be represented by without much loss of information. For training purposes, an input needs more than five points if ML is to be successful. Depending on the size of the dataset, dimensionality, or the number of data points in the training set, increases directly with it. In addition, the predicted attribute and the kind of ML technique are associated to this rise.

3.4.2 Model Development and Validation Techniques

The creation of an efficient ML model to predict tribology includes many steps like training, validation, and testing. The training stage enables the algorithm to learn on the basis of the available data, and the validation stage assists in optimizing model hyperparameters (e.g., learning rate, tree depth, and kernel functions) to avoid overfitting. Lastly, the testing stage gauges the model generalization of unknown data. Run-of-the-mill validation methods are k-fold cross-validation, leave-one-out, and bootstrapping. Performance measures such as mean squared error (MSE), coefficient of determination (R^2), mean absolute error (MAE), and root mean square error (RMSE) are used to measure model accuracy. Ensemble algorithms such as bagging, boosting, and stacking are used in order to encourage better predictive stability by using several algorithms to decrease variance and bias. Such methods have been especially effective in tribological modeling, where data are usually small yet highly complex.

3.5 ALGORITHMIC APPROACHES FOR PREDICTIVE TRIBOLOGY

3.5.1 Regression Models

For any given set of inputs and outcomes, regression may provide meaningful results [17]. Figure 3.1 outlines the process according to which regression models can be used to implicate patterns out of a particular set of data; please refer to the mentioned source to understand it further [17] (e.g., linear [18], polynomial [19], or nonlinear functions [20]). Parametric regression is based on given input variables to obtain the response, and nonparametric regression depends on the use of a kernel to give significance to neighboring observations [21]. In order to find the relationship, supervised ML interprets the input characteristics and produces either discrete or continuous output [22].

One of the most common nonparametric regression algorithms is the *k*-nearest neighbors (KNN) algorithm [23,24]. It approximates the output of a particular input by averaging the value of its *k*-nearest data points in the feature space. Conversely, the Gaussian process regression (GPR) approach assumes that the extent of correlation between the data points is a variation of the distance between the points [25]. GPR uses a multivariate normal distribution to characterize the likelihood of the output given observed data on each input [16]. GPR tends to have better prediction reliability than KNN because it measures the uncertainty and reflects the variation in values of the predicted performance [26].

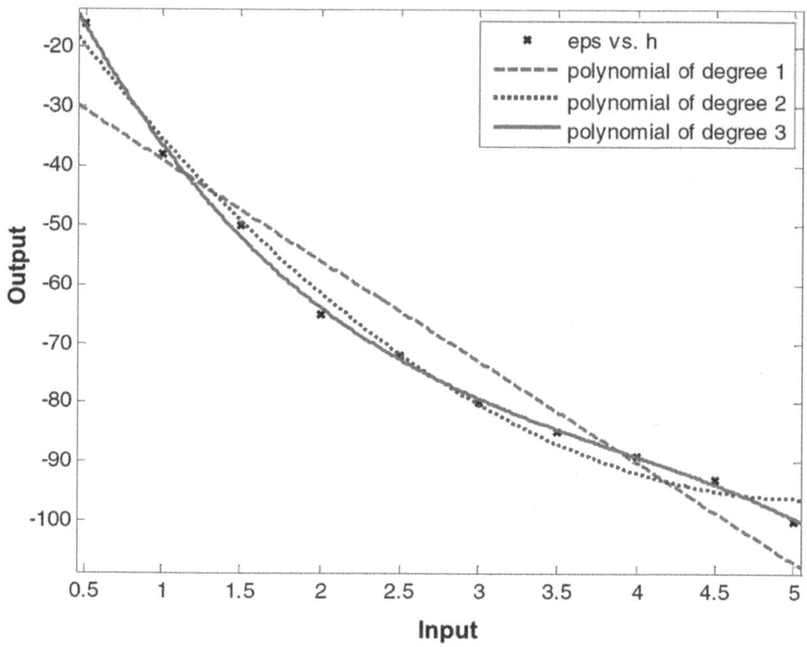

FIGURE 3.1 Sample of a regression model of a polynomial to data points [27]. (CC-BY-NC-ND Attribution.)

In contrast to nonparametric regression, which is primarily concerned with minimizing the objective function (cost), parametric regression is concerned with providing an explicit analytical formulation of the problem that includes the inputs and outputs [17]. This formulation has its components adjusted in accordance with known points. Regression and classification are two areas where more advanced ML techniques may be used.

With the exception of kernel-based approaches [28], most ANNs, SVMs, RF, and Gradient Boosting (GB) algorithms rely on nonlinear parametric formulations [29–31]. Despite their high levels of performance in data interpolation tasks [31], these models do not have high interpretability because of their complex parametric forms [32] and limited hypothesis transparency [33]. Preparation (including configuration and validation) and assessment (the meat and potatoes of model analysis and forecasting) are the two main stages. The complexity of the model, e.g., the maximum degree in a poly-regression, needs to be fine-tuned to obtain the best and reliable predictions [34,35]. The next section gives a step-by-step process of optimizing the complexity of the model, which will guarantee an efficient balance between the underfitting and overfitting while learning the model.

3.5.1.1 Artificial Neural Network (ANN)

ANNs are networks of neurons designed to mimic the way humans learn. Connections between neurons are established by means of edges. As learning progresses, the weights associated with neurons and their connections are updated. Figure 3.2a shows

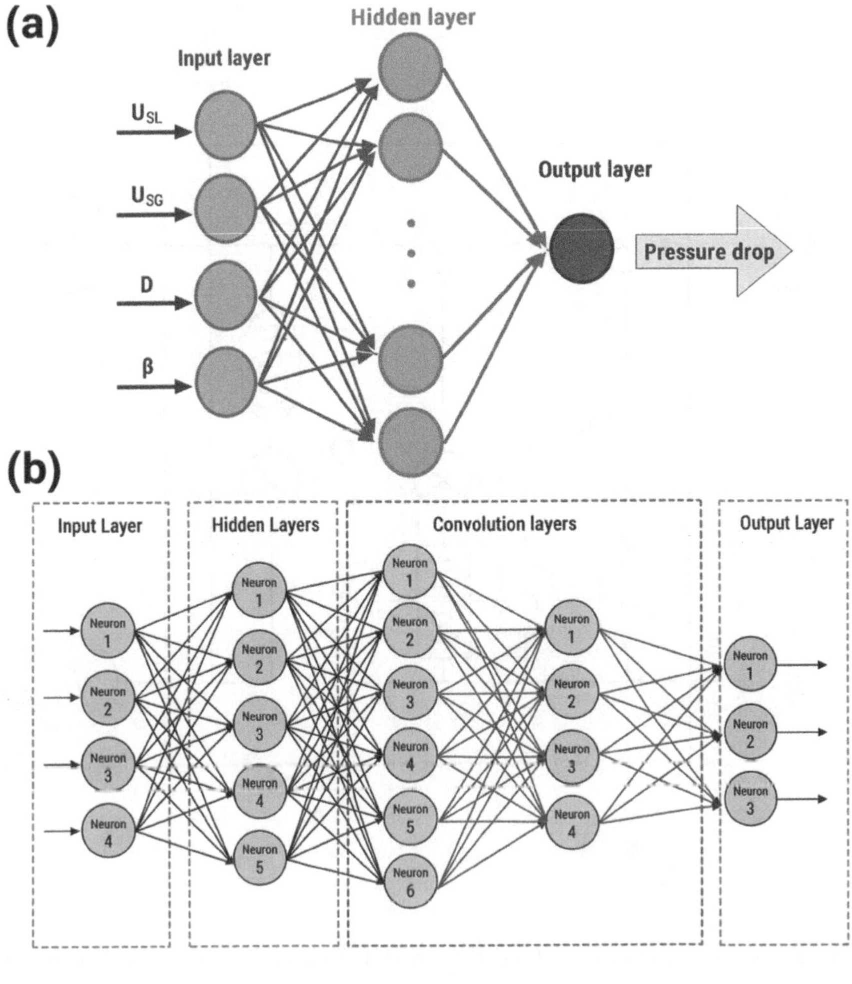

FIGURE 3.2 Structure of an artificial neural network (ANN) model showing (a) a single hidden layer [36] and (b) multiple hidden layers [37]. (Creative Commons Attribution 4.0.)

a basic ANN that just has one hidden layer. The top three layers are input, concealed, and output. The last step in learning is to compare the produced output with the actual output and then alter the weights depending on the difference. However, even ANNs that have many layers are employed to solve many problems. Solving issues becomes increasingly time-consuming as the number of factors increases due to more hidden layers. This network arrangement is used for solving intricate challenges. As shown in Figure 3.2b, an ANN uses many hidden layers.

3.5.1.2 Long Short-Term Memory Network

As illustrated in Figure 3.3, a long short-term memory (LSTM) network is an improved variant of the classic RNN, which is specifically created to address the issue of a vanishing gradient [38]. LSTMs are especially useful for sequential data, thus being well

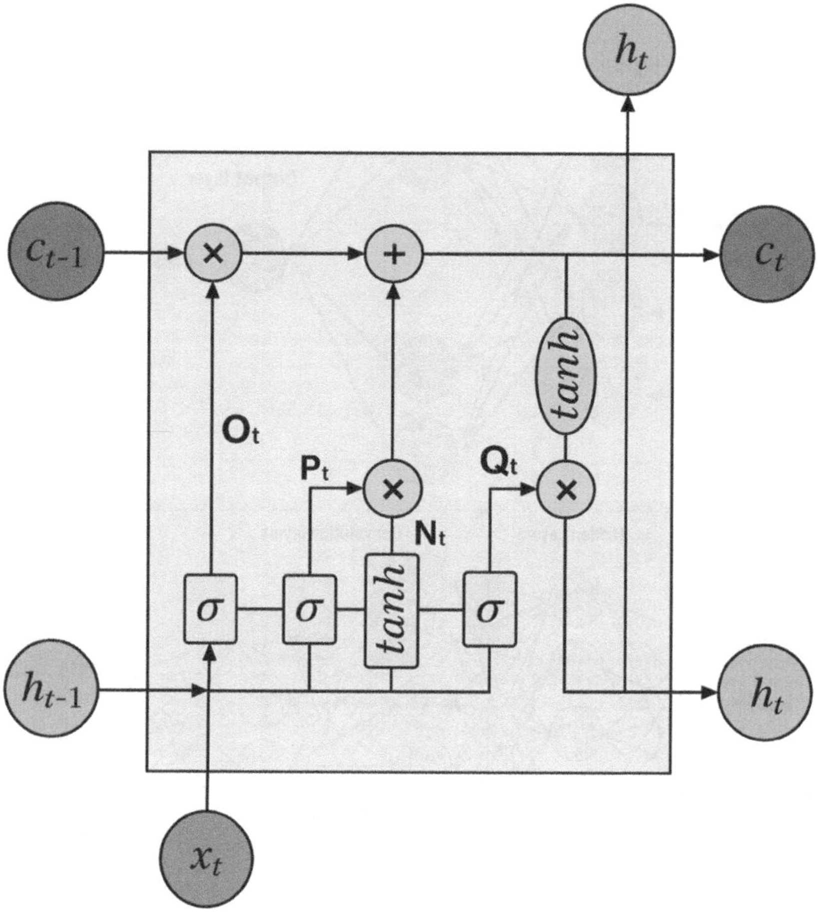

FIGURE 3.3 Architecture of an LSTM model [39]. (Creative Commons Attribution 4.0.)

applicable to tasks like signal processing, stock price prediction, time-series prediction, and speech recognition. In contrast to regular RNNs, LSTMs have a collection of gates, input, forget, and output, that control the information flow and also allow the model to learn both the short-term and the long-term dependencies in an effective way. The forget gate is essential in determining what aspects of the past information are to be remembered or forgotten, therefore preserving the relevant context through long sequences. The functioning of an LSTM can be mathematically expressed as follows:

$$P_t = \sigma(e_P.[U_{t-1}, k_t] + a_P \quad (3.1)$$

$$O_t = \sigma(e_O.[H_{t-1}, x_t] + a_O \quad (3.2)$$

$$N_t = \tanh(e_N.[h_{t-1}, x_t] + a_N \quad (3.3)$$

$$c_t = O_T \circ c_{t-1} + P_t \circ N_t \quad (3.4)$$

$$Q_t = \sigma(e_Q.[U_{t-1}, k_{Pt}] + a_Q \quad (3.5)$$

$$h_t = Q_t \times \tanh(c_t) \quad (3.6)$$

Equations (3.1–3.6) represent the input operation, forget operation, new candidates, cell state, output operation, and next hidden state, respectively. Here, ∘ denotes the Hadamard (element-wise) product, e denotes the weight matrices, a denotes the bias term, and σ and tanh are the respective activation functions. Collectively, these operations enable the LSTM to learn complex temporal patterns and maintain stability during gradient propagation.

3.5.1.3 Gated Recurrent Unit (GRU)

Similar to LSTM, the gated recurrent unit (GRU) is a kind of RNN that performs well with sequential input. There is a distinction based on the quantity of gates. Its novel gate actions—update and reset—reduce the number of gate operations compared to LSTM. Compared to LSTM, the GRU can complete the same work quicker since it uses a less number of operations, as seen in Figure 3.4. The gate does the following things:

$$\mathrm{pd}_t = \sigma\left(e_{\mathrm{pd}}.[U_{t-1}, k_t] + a_{\mathrm{pd}}\right) \quad (3.7)$$

$$\mathrm{es}_t = \sigma\left(e_{\mathrm{es}}.[U_{t-1}, k_t] + a_{\mathrm{es}}\right) \quad (3.8)$$

$$N_t = \tanh(e_N[\mathrm{es}_t \circ [U_{t-1}, k_t] + a_N \quad (3.9)$$

$$U_t = (1 - \mathrm{pd}_t) \circ U_{t-1} + \mathrm{pd}_t \circ N_t \quad (3.10)$$

Equations (3.7–3.10) represent the update operation, reset operation, candidate activation, and next hidden state, respectively, where pd_t and es_t represent the update and reset operation, respectively.

3.5.2 Support Vector Machines and Decision Trees

3.5.2.1 Support Vector Machine (SVM)

Figure 3.5 shows the structure of an SVM model. As shown in Figure 3.5, the linear SVM [29] uses a hyperplane to classify features into categories, whereas a kernel hyperplane [40–42] divides features into subcategories by mapping specific information to them. Like the linear SVM models, it divides the input space into several regions using straight lines. The linear functions are determined by finding the coefficients that maximize the separation or margin on the side of every hyperplane [29].

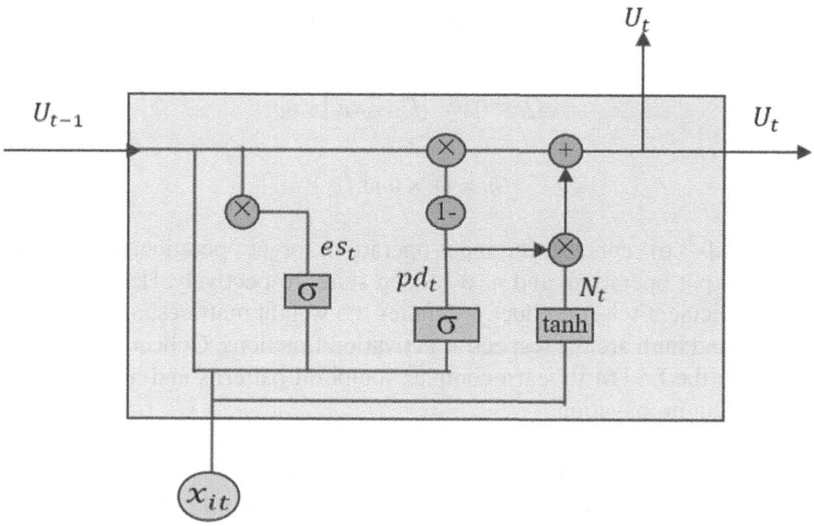

FIGURE 3.4 Architecture of the gated recurrent unit (GRU).

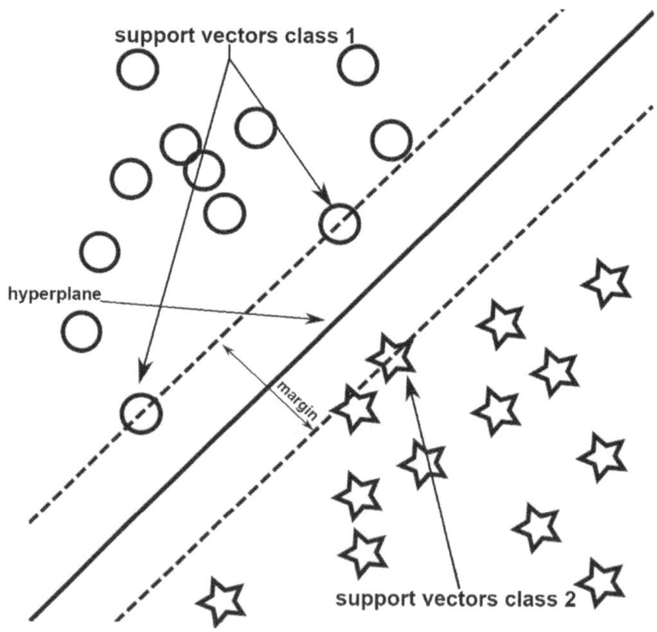

FIGURE 3.5 Architecture of a support vector machine (SVM) [43]. (Creative Commons Attribution 4.0.)

The kernel was used to characterize the connection between the test locations and feedback (i.e., known class). The equation produces nonlinear categorization hyperplanes to achieve the most efficient use of the available space [40–42].

3.5.2.2 Decision Tree

A well-liked variant of decision trees, RF use an ensemble of smaller trees to make decisions [30]. Each decision-making tree represents a prediction, and the model's prediction is the one with the most votes at the end. A decision tree shows considerable potential in cases where multiple models collaborate to get better results than a single prediction model. The level of data collection marks the height of the tree that is defined by the nodes. The prediction error of the validation set can be reduced by maximizing the various requirements and therefore preventing overfitting and underfitting.

3.5.2.3 Ensemble Learning and Random Forests

Ensemble learning is a technique that involves using a combination of various base models to enhance the predictive accuracy and generalization. Ensemble methods, unlike single learning algorithms, combine multiple predictors of data variability, i.e., decision trees and regression models. This strategy alleviates overfitting and improves strength especially where the tribological data are complex with a number of interacting variables including load, speed, material hardness, and temperature. The ensemble techniques are classified into the following types:

- **Bagging (Bootstrap aggregation):** The data are trained on multiple models on diverse subsets, and their predictions are averaged, reducing variance.
- **Boosting:** Boosting is a subset of ensemble learning; it includes two methods: AdaBoost [44] and GB [45,46]. The weak learners are improved in that order in a manner that the weak learner is boosted on the misclassified samples or poorly predicted samples in an attempt to minimize bias. Those inputs on which the learners perform poorly would be used to instruct any bad student in a sequential fashion. The prediction error of the built model after adding each weak classifier determines the weight factor associated with each learner or classifier [31].
- **Stacking:** Uses the predictions of two or more models (meta-learning) in order to create a final high-quality one.

RF is a well-known decision tree ensemble learning algorithm. It builds many trees when training and gives the mode (classification) or the mean (regression) of the predictions made by the trees. RF has many applications in tribological modeling because it can address nonlinearity in relationships, mixed types of data, and high dimensionality. Typical applications of the RF technique are as follows:

- Determining the wear rate and friction coefficient with different mechanical conditions.
- Determining the stability of the lubrication film by using experimental or simulated parameters.
- Prioritizing the significance of influencing factors, which include load, roughness of surfaces, and hardness of materials, in wear performance.

An RF regression model that was trained using the coating data (TiN, DLC, and MoS_2) was able to predict the wear volume in more than 90% of cases, and it outperformed the linear regression and SVM models. It further identified that load and coating hardness are leading factors that affect wear depth, which helps in optimizing materials.

3.5.3 Classification Models

One possible interpretation of regression is as a classification problem [47]. The model is trained in powers of input features and output classes in classification problems, but unlike regression, the output value will always be constant, with each condition corresponding to a different set of classes. Binary categorization has only two categories. The outcomes may be Class 1 or Class 2. In the end, the trained model can classify the characteristics according to the classes it was being trained on. The best separation in the input region can be achieved by defining it in the best way possible [47,48]. A linear nonseparable case classifier and linear one are presented in Figure 3.5. We represent the set of training data (x_i) as the straight lines in Figure 3.5 in the following way:

$$w^T + b = 0 \tag{3.11}$$

This equation takes into consideration information from points with $i = 1, 2, 3, \ldots, n$. The previous equation defines the n-dimensional vector w and the term b. The hyperplane considers two criteria in linear separation [48]. It is demanded that (i) the distances between clustered (class) data be maximum and (ii) the proportion of error in separation data be minimum. In this instance, the classification of the classes is based on the limits of the hyperplane, i.e., the left ($y = 1$) and the right ($y = -1$) sides. The separation is, therefore, controlled as follows:

$$w^T + b \begin{cases} \geq 1\{y_i = 1\} \\ \leq -1\{y_i = -1\} \end{cases} \tag{3.12}$$

Nevertheless, the penalty function (Figure 3.5) may be used to reduce the distance (ξ_i) between the poor classifier and achieve optimum classification.

$$F(\xi) = \sum_{i=1}^{N} \xi_i \tag{3.13}$$

Classification accuracy (CA) is used in order to assess the model's performance on classification tasks [15].

$$CA = \frac{S}{N} \tag{3.14}$$

The total number of samples (N) and the number of samples that were successfully categorized (S) are both represented in the equation above.

3.6 DEEP LEARNING ARCHITECTURES FOR TRIBOLOGICAL PREDICTION

3.6.1 Convolutional Neural Networks (CNNs) for Surface Image Analysis

The modern ANN paradigm, which developed from multilayer perceptrons [49] to sophisticated deep CNNs [42], relies on a single output, a hidden structure per layer of inputs, and a configuration with several layers. When inputted with visual data, a certain kind of neural network known as a CNN may build meaningful connections. In its several layers, which mimic the interconnections of the neurological system, it resembles the human brain. Signals from neurons in one layer are translated into a response signal transmitted to neurons in the layer below using activation functions like the sigmoid function. According to Horne [22], the mapping between inputs and outputs is dynamic and nonlinear. With the right use of filters, CNNs can learn both spatial and temporal relationships. CNNs are executed by applying the appropriate filter to a picture and then multiplying the resulting values by the values of the image and filter that coincide [50]. Direct picture placement of the filters is possible. A CNN is able to learn how to identify pictures [51]. In Figure 3.6, we can see how a CNN is used to identify fractures.

3.6.2 Recurrent Neural Networks (RNNs) for Time-Dependent Friction

The RNN is one type of ANN that accepts data in sequence as the input and tends to establish relevant connections in the time domain. In the case of sequential data, it is most effective when the current inputs depend on the previous ones. Figure 3.7 represents an RNN in its typical form. Although theoretically, RNNs can learn long-term dependencies, the problem owing to the vanishing gradient problem makes it impossible to put them in practice [53].

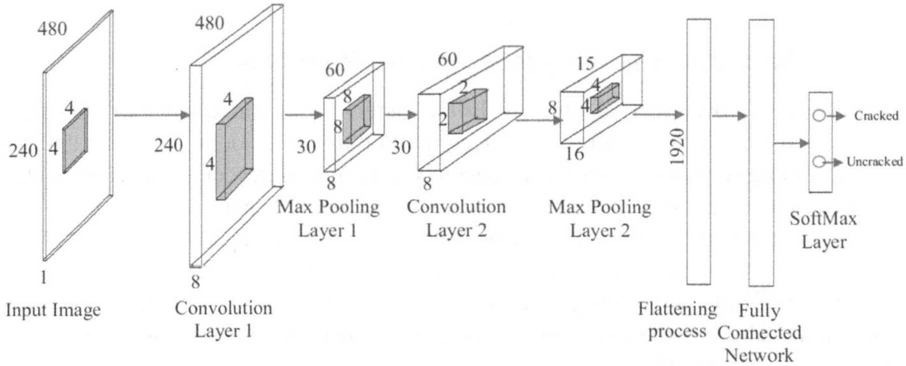

FIGURE 3.6 Illustration of the application of a convolutional neural network (CNN) model [52]. (CC-BY-NC-ND Attribution.)

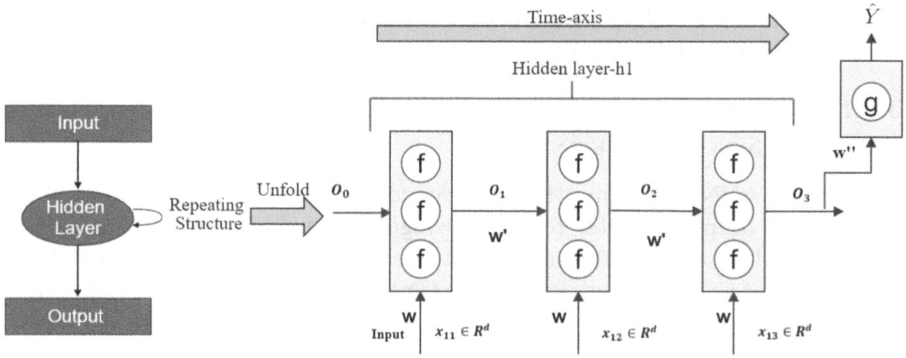

FIGURE 3.7 Typical recurrent neural network (RNN) in its unrolled form.

3.6.3 Generative Adversarial Networks (GAN)

Generative adversarial networks (GANs) are a subset of generative models used in DL. Modeling entails learning from incoming data and then, in an unsupervised manner, producing similar patterns [54]. Training is now more efficient with GANs since they are supervised. The MatGAN architecture for inorganic materials is shown in Figure 3.8a. It utilizes a generator to convert random vectors into samples and a discriminator to identify them [55]. In order to deceive the discriminator, the generator generates synthetic samples that are similar to the actual input data. These samples are then evaluated by the discriminator, and it is either a genuine sample or a generated sample. This antagonistic procedure lets the model learn tricky information outlays. GANs are effective and can be used on small datasets, especially to improve the quality of images, create age-progressed images, and even help in the process of 3D model reconstruction. As shown in Figure 3.8b, the encoder architecture of the generator consists of convolutional and deconvolutional layers. When working with small samples, determining their composition might be a challenge; the encoder eliminates this issue [55].

3.6.4 Unsupervised Techniques for Pattern Recognition

In order to discover several objectively characterized characteristics of the input, unsupervised ML is used. It is a clustering method that may identify unique datasets that have more similarities than differences [47]. In most cases, class-based data clusters with homogeneous samples are the end product of unsupervised DL [56]. Those clusters that emerged from the multivariate analysis are the correct ones. When dealing with two-dimensional input vectors, clustering performs well (Figure 3.9). Utilizing the K-means technique, two clusters were discovered using the two-dimensional data. These data show two different concentrations with and without the use of KNN (Figure 3.9) [57].

Among the many clustering methods available, the *K*-means approach is most used [58–61]. As a basic idea, you should randomly place *K* clusters in the input area's center. During the first round, each data point is assigned a Cluster ID that revolves around the centroid cluster with the highest degree of similarity. Then, based

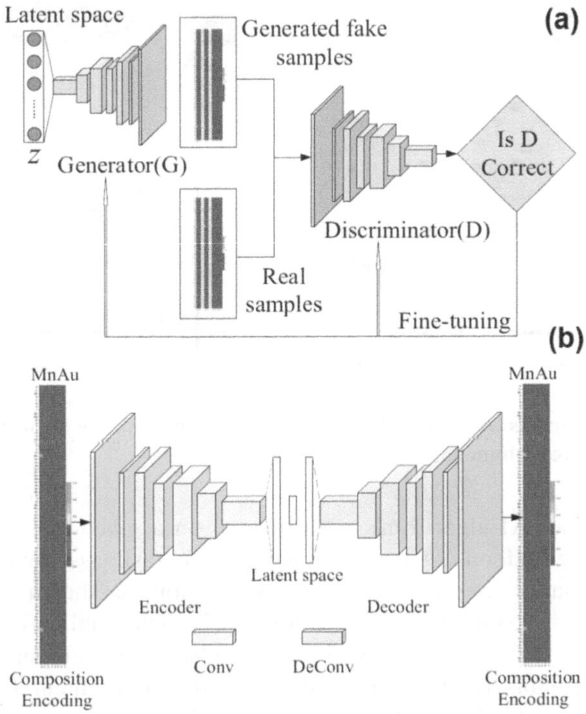

FIGURE 3.8 Schematic representation of (a) the MatGAN framework designed for inorganic material generation and (b) a GAN-based autoencoder architecture illustrating convolution–deconvolution layers [55]. (Creative Commons Attribution 4.0.)

on the average position of the labeled points in the cluster, the centroid locations of each cluster are improved. The process is carried out again and again until the centroid positions match up [58,62,63]. The *K*-means clustering technique determines the number of clusters *K*, which is an input to the model. However, notably, there are a lot of ways to figure out how many *K*-parameters are optimal; for instance, Elbow [58] illustrates this principle. The idea is that you can pick a *K*-value such that adding centroids doesn't significantly reduce cost-related work, like the distances (squared sum) and their related centroid [64,65]. One common issue is that it doesn't get to global minima during optimization but rather lingers in local ones [66]. It is possible to somewhat overcome this limitation by running the calculation numerous times while considering a number of randomly selected beginning positions for the cluster centroids [66,67].

3.7 PERFORMANCE EVALUATION AND MODEL VALIDATION

3.7.1 Model and Loss Function

The assignment function will be denoted by the operator *T* when defining the behavior that an AI model should display when processing data points. Labeling research specimens to determine their loss or intactness is one of the roles of assignment *T*.

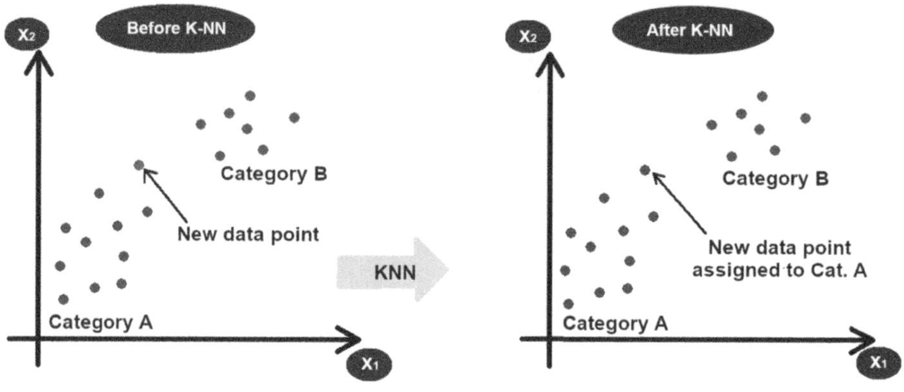

FIGURE 3.9 *K*-means algorithm: two different concentrations with and without use of KNN [43]. (Creative Commons Attribution 4.0.)

In many cases, an AI algorithm's performance on a given task may be reasonably predicted based on its performance on the task assessment measure *P*. Attempts to enhance the accuracy of categorization, as shown in the example, might be seen as "a prospective measure" [5–7]. The application of a formula to defined criteria is comparatively simple to setting the right success measures. Considering the data-driven nature of this research, assignment *T* is restricted to model-based solutions. The performance metric (*P*) is a new mathematical formulation, which gradually improves the model performance by optimization of the numerical parameters. Essentially, *P* is an objective, loss, or cost function as it is typically applied in ML and DL models. Statistically, this study uses the terms of an objective function, which is used to specify whether a quantity is to be maximized or minimized [5,6]. In that regard, within the context of an AI model being trained on a dataset *K*, having adjustable parameters, this notion is denoted by a minimum cost functional (*M*).

$$M = \min_{\theta} L\big(T(K;\theta)\big) \tag{3.15}$$

In this case, *L* is the loss functional, *T* is the assignment functional that maps the dataset *K* using the model parameter theta, and *M* is the optimal cost that can be achieved in the course of optimization. The aim is to find the parameter set θ that minimizes *L*, which enhances model accuracy and generalization.

Table 3.2 summarizes the most commonly used loss functions for model validation. As a fundamental element of any learning algorithm, the choice of loss function plays a crucial role in determining the model's accuracy and overall performance [1,5,6]. A key component of every model, the loss function significantly affects performance. There are always several optimal parameters for different assignments when using different loss functions, but it's hard to tell which one would be best to use before running a simulation.

TABLE 3.2
Summary of Commonly Used Loss Functions and their Applications in Model Validation

S. No	Model Name	Equation	Equation No	References
1	Log loss	$L=-\frac{1}{N}\sum_{j=1}^{N}\log(p_j)$	6	[68]
2	Mean square error (MSE)	$\text{MSE}=\frac{1}{N}\sum_{j=1}^{N}(O_j-p_j)^2$	7	[68]
3	Mean absolute error (MAE)	$\text{MAE}=\frac{1}{N}\sum_{j=1}^{N}\lvert O_j-p_j\rvert$	8	[69]
4	Mean absolute error root mean square error (RMSE)	$\text{RMSE}=\sqrt{\frac{1}{N}\sum_{j=1}^{N}(O_j-p_j)^2}$	9	[69]
5	Coefficient of determination (R^2)	$R^2=1-\frac{E(S)}{T(S)}$	10	[70]
6	Mean absolute percent error (MAPE)	$\text{MAE}=\frac{1}{N}\sum_{j=1}^{N}\frac{\lvert O_j-p_j\rvert}{p_j}$	11	[68]

Note: N denotes the total number of observations, p_n represents the predicted values, O_n represents the original (observed) values, $E(S)$ is the sum of squared errors, and $T(S)$ is the total sum of squares.

3.7.2 Data Splitting

Data splitting involves cleaning and visualizing of data. The goal of tribology researchers is to have a functional ML model with much general-purpose knowledge of the data collected [5,6]. However, the outcomes should be quite similar to what was unknown before, but the final model has to have predictive capabilities. When a model is overfitted to its training data, its generalizability suffers [1–3,5,6]. Bias and variance are two separate but linked issues [1–3]. Other ways of evaluating models are stratified *k*-fold cross-validation (CV) [68], leave-one-out cross-validation [71], *k*-cross-validation [72], and holdout [73].

3.7.3 Training, Testing, and Validation

To prevent overfitting and test model accuracy, a common practice is to split a dataset into three parts (training, validation, and testing) [68]. One way to model anything is to train a model using examples from the training set. The model can't tell the difference between test data and outcomes and training data just yet. The next step is to retrain the model using the validation range, which will increase its accuracy. Figure 3.10 shows that adding more information (a larger *p*) usually improves interpolation but reduces the ability to predict the training set. This is true provided that the model is capable of taking more advanced input information. The ratio of

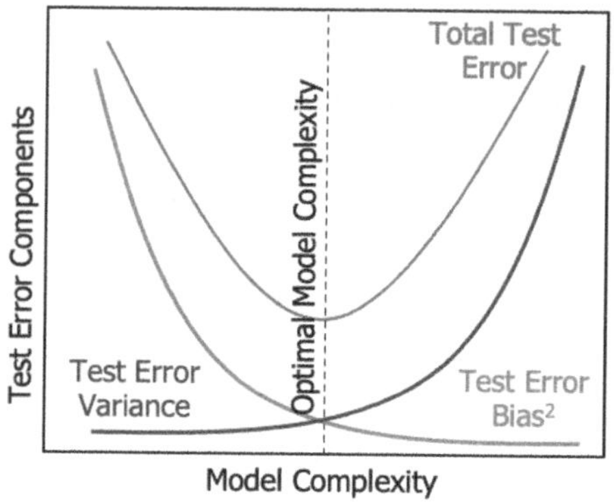

FIGURE 3.10 Relationship between errors and model complexity, where the optimal model achieves the lowest combined bias and variance [75]. (Creative Commons Attribution 4.0.)

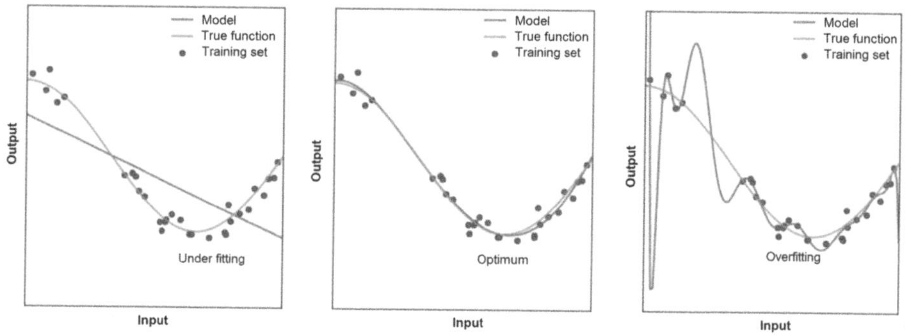

FIGURE 3.11 Illustration showing the concepts of underfitting, optimum fitting, and overfitting in model performance [76]. (Reproduced from open domain by providing citations as per authors' permission.)

difficulty to solutions is optimized when there is the smallest number of possible solutions in the validation set [74]. Lastly, the test set is used to test the generalization capacity of the model based on some of the data that the model is not encountering. This is done to make sure that the model that has been trained obtains the best degree of sophistication and predictability with regard to unseen data.

3.7.4 Underfitting and Overfitting

Underfitting was due to the reduction of model complexity. The model does not give an adequate account as to how the performance of the device is dependent on the input variables. To counteract overfitting, it is helpful to recall dataset "noise" [74]. As a general guideline, while building a model, one should aim to maximize the sum of all non-zero numerical values, total inputs, and word count [12]. Learning causes overfitting and underfitting, as shown in Figure 3.11 (i.e., training collection). As a result, a simple model with one explanatory variable ($p=1$) is not the most suitable one to capture the complexity of the interaction between the predictor and the response variable in the current scenario. Conversely, an empirical model using 15 predictors ($p=15$) may have an acceptable fit to the idiosyncratic observations that comprise the training set but will provide little information about the generalizability required to provide predictions about unobserved outliers, as is the case in the validation set. Although one could remove extraneous stochastic changes in the data, the outcome could be compromised by the loss of statistical power in the analysis due to the reduction in the sample size. However, a 3-degree ($p=3$) poly-regression model still has the ability to capture the overall pattern that can be identified in the empirical findings.

3.7.5 Cross-Validation

In real-world situations, datasets are often rather small, and training models with too little data might be a challenge. Implementing validation procedures will resolve this issue. CV may be used to address hyperparameter adjustment and model selection. CV is a mathematical technique for estimating how applicable the findings are to a broader context [4]. Using a fixed partition of the training set consisting of the test set makes it a one-sided validation scheme that can be applied to any dataset [1,5,6]. The holdout methodology [4] is particularly effective with very large sample sizes, as shown in Figure 3.12. However, three-fold (or three-way) validation also has strong reasons to be considered. Additional preparation datasets are available in the three-way holdout technique [2,3,77]. A normal distribution would indicate that all of the data points in the collection are drawn from the same pool of potential values. A few common data quantification quantities are [1–3].

The problem may be tackled with the help of the *k*-fold cross-validation method (*k*-CV) [72], no matter how intimidating it seems. The data are first partitioned into k subgroups and then tested against each other using this procedure. When evaluating each set of samples, a *k-CV* with a single criterion (*k*) is used. This method is also known as *k*-fold validation. The *k*-fold forward cross-validation (*k*-FCV) and

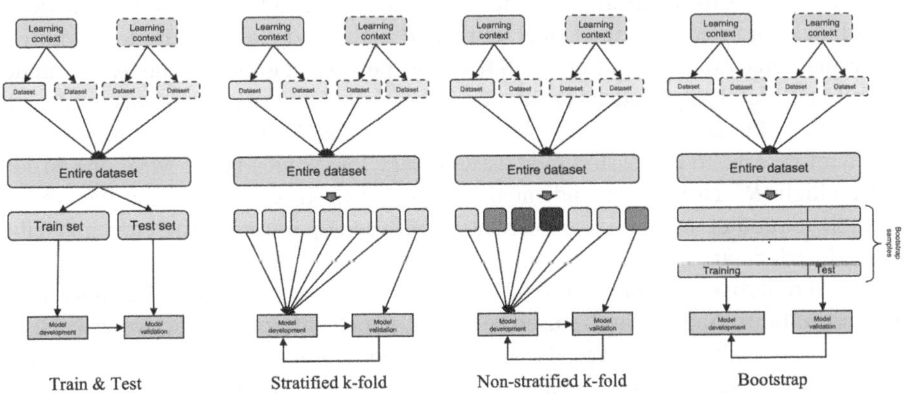

FIGURE 3.12 Common evaluation methods in machine learning [78]. (Creative Commons Attribution 4.0.)

leave-one-out cross-validation are popular versions of the traditional k-CV model. Figure 3.12 shows their schematic models of a total of *k* samples. In each of the *k* cycles of training, *k*–1 folds get training, while the rest perform validation. Iterating through the steps *k* times finishes each validation set. One way to ascertain model accuracy is to average the results from all *k* validation folds.

3.7.6 Methods of Regularization

Regularization techniques [73] like LASSO [79], Ridge [80], or Elastic Net [74,81] may filter out non-important modeling terms, which is another way to minimize a model's complexity [74]. The basic idea of regularization involves two functions: (i) the fact that the model should be able to fit observed data and (ii) a penalty which discourages too complicated models. Lifting the penalty term can have a small effect on removing redundant parameters, which would not significantly improve model accuracy. The determination of the level at which the model complexity can be changed depends on how much the penalty weight has to be adjusted to achieve a good prediction in the valid range [82].

3.8 APPLICATIONS OF MACHINE LEARNING IN TRIBOLOGY

3.8.1 Friction Prediction Models

One of the most developed fields of ML applications in tribology is friction prediction. Friction coefficient (COF) is conditioned by several interacting factors, such as load, velocity of sliding, material hardness, surface roughness, and lubrication condition. The classical models of analysis like the Amontons–Coulomb model and the Bowden–Tabor model provide only approximate descriptions and cannot describe nonlinearities in dependencies in different environments [83]. The real alternative is ML algorithms, which learn to make complex relationships between tribological inputs and frictional outputs. Multilinear regression and support vector regression (SVR) are regression-based models that have been successfully applied to predict COF using various material pairs and lubricants [84]. Friction and wear behavior in a ball on a flat system made of steel lubricated with Jatropha oil with MoS_2 additives were predicted by Mushtaq et al. [84] using multiple regression and ANN model. A full factorial design under varying load, speed, and additive levels was used as the experimental condition. The ANN model demonstrated high accuracy in prediction behavior relative to regression models with very low values in the MSE and MAPE with a high (R^2) (> 0.99), as indicated in Figure 3.13. It was shown that the ANN is suitable to predict the nonlinear correlation between operating parameters and tribological performance.

The predictive accuracy of ANNs and RF has been found to be more accurate since both models can be used to model nonlinear and interactive effects. DL models, especially CNNs, are also used in recent studies to process surface micrographs to predict frictional behaviors based on surface morphology characteristics [85]. These models are able to elicit microtexture-based attributes that are normally challenging to determine manually, which results in enhanced predictive stability.

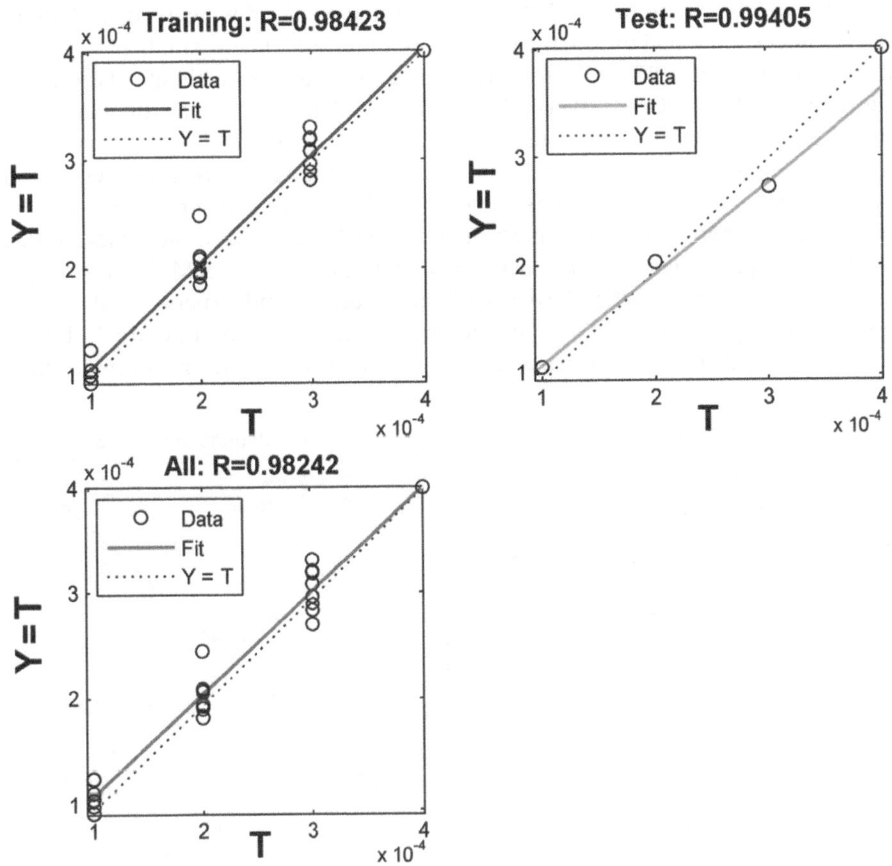

FIGURE 3.13 Output of the artificial neural network (ANN) model showing the predicted wear (weight loss) for training, testing, and combined datasets [84]. (Creative Commons Attribution 4.0.)

The combination of ML and real-time sensor information provides dynamic control of friction and adaptive control, the basis of intelligent tribological systems.

3.8.2 Wear Rate Prediction

Probably, wear rate prediction is one of the most difficult issues of tribological analysis as wear mechanisms are stochastic. Contact geometry, material hardness ratio, load, speed, temperature, and lubrication regime affect the process. Classical wear models like the Archards and the Holms laws give a baseline upon which it does not capture the nonlinear and multivariate nature of interactions in the real world. The decision trees, GB, and the neural networks have been used as ML methods in predicting wear volume and wear rate with better accuracy [86]. Training models on the basis of experimental or simulation data helps researchers determine the prevailing wear mechanisms and predict wear under various operating conditions. In addition, ensemble models, where the prediction of several algorithms is made, contribute to reducing uncertainty and improving robustness, especially during small and/or noisy datasets. Singh et al. [87] utilized ML regression models to estimate the erosion performance of WC-10Co4Cr thermal spray finishes deposited on SS316L metal substances. Based on experimental data of pot tester experiments at 30, 45, and 60 angles, a GPR model was constructed and contrasted with the results of other regression methods such as decision trees, ensemble boosted trees, and linear regression. The GPR model had better predictive performance, and Matern 5/2 kernel (M5/2-GPR) had the best predictive performance (R^2=0.95) and lowest RMSE and MAE. As it could be observed, the closeness of the predicted and experimental erosion rates was the highest in the M5/ 2-GPR model, as demonstrated in Figure 3.14, with the data points falling along the same line, signaling a very good linear relationship.

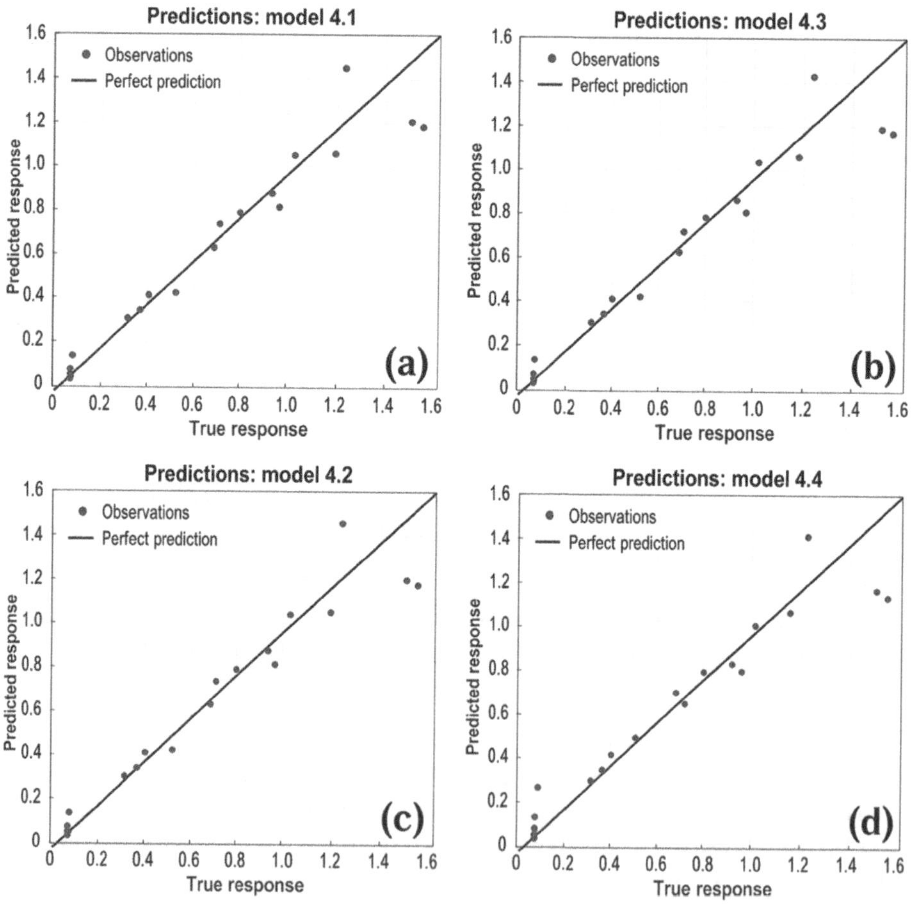

FIGURE 3.14 Comparison between the experimentally measured and model-predicted erosion rates using different Gaussian process regression (GPR) models: (a) Rational quadratic (RQ-GPR), (b) Squared exponential (SE-GPR), (c) Matérn 5/2 (M5/2-GPR), and (d) Exponential (E-GPR) [87]. (Creative Commons Attribution 4.0.)

This proves that the suggested GPR-based model is a good representation of non-linear erosion processes and that it has good predictability in WC-10Co4Cr coating.

However, image-based ML approaches are also included in some studies, where wear track or debris morphology images become the input to CNNs, which also offers a more in-depth insight into wear patterns and their progress over time.

3.8.3 Lubrication Behavior Modeling

Another potential promising field that has seen the application of ML is the prediction of lubrication performance. Tribological efficiency depends on the selection of lubricant, the concentration of additives, film thickness, and viscosity. The dynamics of fluid, thermal, and chemical reactivity are all commonly paramount in predicting the performance of a lubricant—a challenge for existing conventional models of lubricant analysis. Estimation of minimum film thickness, friction coefficient, and power loss in contact between two lubricated surfaces has been done using supervised learning methods, e.g., regression and SVM models. The models based on the neural network can be used to describe the intricate correlation between the properties of the lubricant (e.g., its viscosity index, density, and flash point) and the work parameters to predict the involved transitions in the lubrication regimes. More recently, real-time lubrication control has also been undertaken through reinforcement learning through the optimization of supplying lubricant rate and dynamic real-time adjustment of the conditions to reduce energy loss. The combination of ML with tribo-chemical data and sensor-based systems preconditions the development of smart lubrication strategies, which allow industrial applications to have adaptive and self-optimizing mechanisms.

3.8.4 Surface Roughness and Material Interaction Prediction

The topography and material compatibility of surfaces are the major determinants of tribological behavior. Surface roughness, hardness, and microstructure changes determine the adhesion, deformation, and energy dissipation mechanisms at the contact interface. To make predictions on evolution of surfaces at the condition of loads or sliding, models that are able to work with high-dimensional data are required. Visual analysis of surface textures to define the typical wear patterns and classifying them have been achieved through ML techniques such as PCA, *k*-means clustering, and unsupervised feature learning [88]. DL/ML models are used to process surface images that are acquired through SEM or AFM, which allows automated analysis of the quality of the surface and detection of defects [89]. Moreover, physics-based contact mechanics equations with data-driven learning using hybrid ML models improve predictive performance and at the same time are physically interpretable [90]. Such hybrid structures assist in relating combinations of materials, microstructural variables, and surface features with performance measures including wear rate and friction coefficient.

3.8.5 Hybrid and Ensemble Learning Approaches

Tribological systems have a complex nature, usually making hybrid methods involving the integration of various modeling paradigms. Hybrid ML models combine empirical or analytical relations and ML algorithms to encode both physical and data-driven patterns. As an example, neural networks may be used to enhance the predictions based on the Archard wear model by introducing theoretical constraints into the data-based models. RF, Ada Boost, and XGBoost are ensemble learning methods, which combine several models to enhance prediction stability and generalization [91]. These methods have been found to be especially efficient in the situation of heterogeneous tribological data of various materials, lubricants, or test systems. Also, physics-informed neural networks are becoming an influential instrument in tribology, in which governing equations (e.g., the Reynolds number or energy balance equations) have been written directly into the loss function of the ML model. Such an integration makes predictions physically meaningful, and it has the advantage of utilizing the versatility of DL. Hybrid and ensemble methods can, therefore, be seen as the future of tribological modeling: they combine the advantages of conventional mechanics with the current advances in AI to achieve more accurate, interpretable, and scalable results.

3.8.6 Machine Learning in Thermal Spray

After years of refinement, the thermal spray (TS) method is now standard practice in every major area of engineering. A wide variety of industrial applications may be satisfied by a high-performance and high-quality TS coating. The coating nature of the TS process is highly sensitive to the number of various process parameters, which, in turn, are the primary determinants of the physical state of the particles in their flight and their dynamical behavior. It is challenging for any TS process to create a comprehensive multiphysical model because of the many thermodynamic and chemical processes that occur throughout the deposition process. Among the range of optimization methodologies, the design of experiments methodology by Taguchi is the most popular approach to test the effect of process parameters. It is this type of paradigm that will provide the foundation for identifying the best combinations of parameters and will clearly and undisputably differentiate statistically significant parameters and less influential parameters [92–94]. It may not achieve the exact ideal conditions, however, since the coating quality is usually affected by a combination of several process factors. Numerical modeling and simulation have also been used to regulate and simulate the spray processes [95–97]. These methods typically focus more on studying the spray and combustion process's development. Because the TS process is a complicated multiphysical event, it is challenging for numerical modeling to simulate the real behaviors. Therefore, proper research, prediction, and optimization of the TS coatings are required.

Industrial automation in the previous decade has led to TS's development as a highly automated spraying process, but modern TS applications also need intelligence. It is critical to establish a central automated system that can determine the link between operational settings, in-flight particle behavior, and coating features in order to generate thermal spray coatings that are both durable and of high quality. Because of this need, scientists have developed a trustworthy approach to address the issues with TS-deposited coatings under different working situations. This methodology makes use of new digital tools, such as AI. It is possible to build an intelligent control system by configuring the different TS spray control systems to include models. Smarter, more connected, and more readily integrated spray systems and modern production environments are necessities for the industrial sector. To manage TS operations for improved reliability and resilience using the massive amounts of quantifiable data, new computing-based technologies are needed. AI has the potential to be an integral part in meeting this need.

Using ANNs to simulate the PS process, researchers [98–101] investigated and predicted coating microstructure and attributes including porosity %, unmelted particle content, and different components. Choudhury et al. [102,103] used modular analysis to describe the PS process using multiple ANNs. They then trained the ANN model to predict the in-flight particle state according to several control process elements. A model of the ANN was used by Kanta et al. [104–106] to identify the relationship between the process control parameters and in-flight particle properties through studying the plasma spraying (PS) process. Similarly, Zhang et al. [107] also used ANNs to correlate the states of atmospheric PS, dynamics of in-flight particles, and the influence of the particle temperature on the final coating microstructure. Liu et al. [108] also discussed the relationship between spray gun power, particle temperature, and velocity on a neural network. Among the first uses of ML in TS, the impact of plasma/powder injection conditions on the in-flight characteristics of the particles, namely, average velocity, temperature, and diameter, were measured on an Al_2O_3 feedstock with 13 wt.% TiO_2. The in-flight characteristics predicted by the ML were found to be in good agreement with experimental findings following training, validation, and testing [109]. Furthermore, in a later study, the ANN was enhanced by including coating structural features. Hence, the diametrically opposed patterns in deposition yield and coating porosity [104] and particle melting state [110] were investigated using this model.

The studies that have been mentioned above concentrated on the process of PS mainly. In the case of the high-velocity oxy-fuel (HVOF) method, few studies have used ANN models [111–117]. Cherigui et al. [111] examined the magnetism behavior of HVOF-sprayed FeNb alloys through two ANN models: Model A, associated with the system parameters, which encompassed microstructural features, whereas Model B was associated with the microstructural features with magnetic properties. Zhang et al. [112] used an ANN model to explain structural properties of the NiCrAlY coating deposited through HVOF spraying, which depends on the process parameters, such as oxygen-to-fuel gas ratio and stand-off distance. Kamnis et al. [113] used the ANN model to investigate the airborne acoustic emissions during the HVOF spraying, and it was found that spray distance and powder feed rate have a considerable impact on the microhardness of the coatings. In a similar vein, Mojena et al. [114] applied the ANN to predict the erosion rate of flame-sprayed and HVOF surfaces of the WC-CoCr surface, which revealed porosity to be the most important parameter that dictates erosion, with microhardness and fracture toughness being secondary to it. In order to establish a connection between the process parameters and the mechanical performance of coatings, spraying Cr_3C_2–25NiCr HVOF feedstock was once again studied by Gui et al. [116] for its effects on in-flight particle properties and microstructural coating quality. Gui et al. [116] conducted extensive research on the HVOF spray process using the ANN model. According to the results, the spray distance, oxygen flow rate, and CH_4 flow rate are considered for microhardness and

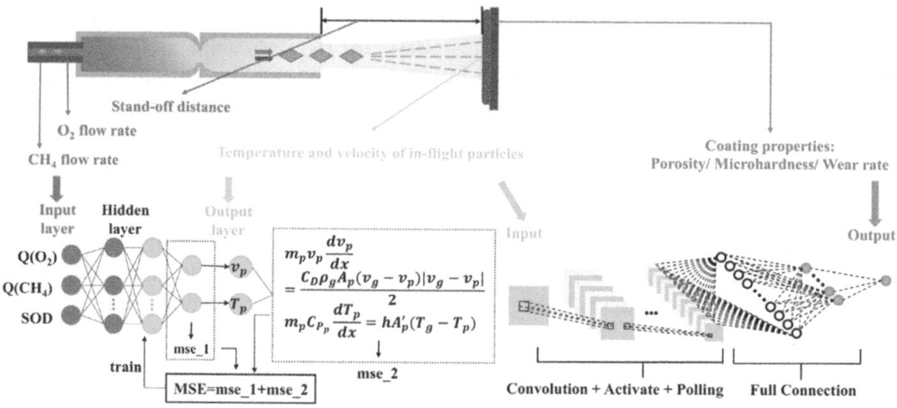

FIGURE 3.15 Representation of the HVOF process modeled using ANNs, illustrating the process parameters and the performance of the resulting coating [116]. (Creative Commons Attribution 4.0.)

porosity in that order. In the case of wear rate, the proper order of consideration is the oxygen flow rate, stand-off distance, and CH_4 flow rate. However, the research did not account for intermediate parameters such as the speed and temperature of particles in flight. Figure 3.15 displays the framework. We picked two models: one that takes as inputs the stand-off distance, gas flow rate, and fuel flow rate and another that uses particle velocity and temperature as goals. Model 2 used particle velocities and temperatures as inputs and coating characteristics as outputs. Improving coating performance controls will be the result of promoting the usage of this state-of-the-art ANN methodology to other thermal spray methods.

3.8.7 ML-Based Wear Prediction in Metal–Metal Contacts

Wear behavior of metallic pairs (steel–steel and steel–aluminum contacts) has been studied extensively to determine the potential of ML in predictive modeling [118]. Experimental studies usually give sets of data that include variables such as normal load, distance of sliding, speed, hardness, temperature, and wear volume or wear rate measured. ANNs were trained with data gathered in pin-on-disk test under both dry and lubricated conditions [119]. The network was able to capture the nonlinear relationships between the operating parameters and wear results with high predictive efficiency compared to conventional regression models. The learned ANN model also indicated the prevalent effect of applied load and sliding velocity on wear behavior. SVR and RF models were also used in another study to predict the wear rate in mild steel alloys. RF proved to be better resistant to experimental noise, and SVR offered less dramatic generalization in the middle parameter space. These findings emphasize the adaptability of ML algorithms in predicting the wear of various metallic systems to provide a predictive model of optimization of material choice and operating environments.

3.8.8 Integration of Experimental and Simulation Data

Tribological studies are becoming more dependent upon experimental studies as well as on computational studies including finite-element analysis (FEA), molecular dynamics (MD), and computational fluid dynamics to understand the behavior of materials and lubrication. ML can also be characterized as the interface between these two fields because it uses the combination of both data to create single predictive systems. As an example, a GPR was trained on experimental wear measurements and MD-simulated contact stresses to predict wear depth under different surface environment conditions. The hybrid dataset increased the extrapolation ability of the model outside the experimental range. Likewise, DL models, such as CNNs, were trained on simulated images of surface roughness generated by FEA and matched them to the measurements of friction trends in experiments. This multi-source learning technique offered more information on the effect of asperity deformation and the distribution of the contact pressure. With this type of integration between experimental and simulation data, it is possible to create digital twins of tribological systems: virtual representations that constantly update themselves with past and real-time data to identify the way that the system will perform, wear, and need maintenance with high accuracy.

3.8.9 Comparative Analysis of Machine Learning Algorithms in Tribological Applications

Various comparative studies have compared performance of various ML algorithms in the task of tribological prediction. Most RF and GB models are effective with small-to-medium datasets because they are only resistant to overfitting and to the ability to detect nonlinear trends. SVMs are useful in case the data are high-dimensional but relatively few in number. Deep architecture neural networks perform better with large and heterogeneous datasets. A comparative study of the models of friction predictions showed that the ensemble methods (RF and GB) were able to predict with an MAE of less than 5%, whereas ANNs needed to be carefully adjusted to the hyperparameters. Conversely, linear regression and *k*-nearest neighbors (kNN) models only showed satisfactory results with low-noise and linear data. The results highlight the fact that there is no single ML model that can be considered effective in all cases, and a model selection process should rely on the size, nature, and quality of the available data. Combination and hybrid approaches that incorporate complementary algorithms provide the most credible direction of strong tribological prediction of a wide range of systems.

3.9 CONCLUSION

ML has proven to be a good actor of the modern tribology, and its predictive functionality as well as flexibility are far much better than those of the previous modeling devices. ML facilitates the quantitative understanding of the processes of friction, wear, and lubrication under diverse operating conditions using the systematic data analysis, algorithm learning, and model optimization. This chapter has shown that supervised, unsupervised, and DL methods could be successfully utilized in tribological prediction and optimization: analysis of surface images with the help of CNNs and time-dependent wear prediction based on RNNs and data augmentation via GANs. Achievement of such techniques, however, depends on the quality of datasets, feature engineering, and the use of strict validation protocols to cushion underfitting and overfitting. As the amount of experimental and computational data increases further, the cooperation between AI and tribology is certain to increase, and new technologies in the surface design, coating technologies, and performance prediction will be discovered. It can be said that the predictive modeling based on ML will form the basis of intelligent tribological systems, allow sustainable production, achieve better reliability, and use less energy across engineering.

REFERENCES

1. Frochte J. *Maschinelles Lernen: Grundlagen und Algorithmen in Python.* Munich, Germany: Carl Hanser Verlag GmbH Co KG; 2020.
2. Kang S, Iwana BK, Uchida S. Complex image processing with less data—Document image binarization by integrating multiple pre-trained U-Net modules. *Pattern Recognit.* 2021;109:107577.
3. Ferreyra-Ramirez A, Aviles-Cruz C, Rodriguez-Martinez E, et al. An improved convolutional neural network architecture for image classification. In: Carrasco JA, Martínez JF, Olvera JA, et al., editors. *11th Mexican Conference on Pattern Recognition, MCPR 2019, held in Querétaro, Mexico.* Cham, Switzerland: Springer Nature; 2020, pp. 89–104.
4. Raschka S. Model evaluation, model selection, and algorithm selection in machine learning. arXiv. 2018.
5. Lecun Y, Bengio Y, Hinton G. Deep learning. *Nature.* 2015;521:436–444.
6. Voulodimos A, Doulamis N, Doulamis A, et al. Deep learning for computer vision: A brief review. *Comput Intell Neurosci.* 2018;2018:1–13.
7. Brownlee J. *Machine Learning Mastery with Python: Understand Your Data, Create Accurate Models, and Work Projects End-to-End. v1.20.* Vermont, Victoria: Machine Learning Mastery; 2021.
8. Vyas S, Jain SS, Choudhary I, et al. Study on use of AI and big data for commercial system. *2019 Amity International Conference on Artificial Intelligence (AICAI)*, Dubai, United Arab Emirates; 2019:737–739.
9. Rusu O, Halcu I, Grigoriu O, et al. Converting unstructured and semi-structured data into knowledge. *2013 11th RoEduNet International Conference*, Sinaia, Romania; 2013.
10. Drass M, Kraus MA. Semi-probabilistic calibration of a partial material safety factor for structural silicone adhesives - part I: Derivation. *Int J Struct Glas Adv Mater Res.* 2020;4:56–68.
11. Drass M, Berthold H, Kraus MA, et al. Semantic segmentation with deep learning: Detection of cracks at the cut edge of glass. *Glas Struct Eng.* 2021;6:21–37. Available from: https://doi.org/10.1007/s40940-020-00133-7.
12. Bulgarevich DS, Tsukamoto S, Kasuya T, et al. Pattern recognition with machine learning on optical microscopy images of typical metallurgical microstructures. *Sci Rep.* 2018;8:3–9. Available from: https://doi.org/10.1038/s41598-018-20438-6.
13. Hancer E, Xue B, Zhang M. A survey on feature selection approaches for clustering. *Artif Intell Rev.* 2020;53:4519–4545. Available from: https://doi.org/10.1007/s10462-019-09800-w.

14. Jordan MI, Mitchell TM. Machine learning: Trends, perspectives, and prospects. *Science*. 2015;349:255–260.
15. Liu Y, Zhao T, Ju W, et al. Materials discovery and design using machine learning. *J Mater*. 2017;3:159–177. Available from: https://doi.org/10.1016/j.jmat.2017.08.002.
16. Efromovich S. *The Multivarient Normal Distribution*. New York: Springer Series in Statistics, Springer-Verlag; 1999.
17. Hunter EA, Brook RJ, Arnold GC. Applied regression analysis and experimental design. *Appl Stat*. 1986;35:77.
18. Montgomery DC, Peck EA, Vining GG. *Introduction to Linear Regression Analysis*, 5th Ed. New York: Wiley series in probability and Statistics; 2012.
19. Nielson GM. Some piecewise polynomial alternatives to splines under tension. *Comput Aided Geom Des*. 1974;5:209–235. Available from: https://doi.org/10.1016/B978-0-12-079050-0.50015-1.
20. Motulsky HJ, Ransnas LA. Fitting curves nonlinear regression: Review a practical. *FASEB J*. 1987;1:365–374.
21. Čížck P, Sadıkoğlu S. Robust nonparametric regression: A review. *Wiley Interdiscip Rev Comput Stat*. 2020;12:1–16.
22. Horne BG. Progress in supervised neural networks. *Signal Process Mag IEEE*. 1993;10:8–39.
23. Altman NS. An introduction to kernel and nearest-neighbor nonparametric regression. *Am Stat*. 1992;46:175–185.
24. Fan D, Zhang X. Short-term traffic flow prediction method based on balanced binary tree and K-nearest neighbor nonparametric regression. *Adv Intell Syst Res*. 2017;132:118–121.
25. Raissi M, Perdikaris P, Karniadakis GE. Machine learning of linear differential equations using Gaussian processes. *J Comput Phys*. 2017;348:683–693. Available from: https://doi.org/10.1016/j.jcp.2017.07.050.
26. Bishnoi S, Singh S, Ravinder R, et al. Predicting Young's modulus of oxide glasses with sparse datasets using machine learning. *J Non Cryst Solids*. 2019;524:119643. Available from: https://doi.org/10.1016/j.jnoncrysol.2019.119643.
27. Ostertagová E. Modelling using polynomial regression. *Procedia Eng*. 2012;48:500–506. Available from: https://doi.org/10.1016/j.proeng.2012.09.545.
28. Rustam Z, Yaurita F. Insolvency prediction in insurance companies using support vector machines and fuzzy kernel C-means. *J Phys Conf Ser*. 2018;1028:012118.
29. Raj JS, Ananthi JV. Recurrent neural networks and LSTM explained. *J Soft Comput Paradig*. 2019;01:33–40. Available from: https://medium.com/@purnasaigudikandula/recurrent-neural-networks-and-lstm-explained-7f51c7f6bbb9.
30. Deng B. Machine learning on density and elastic property of oxide glasses driven by large dataset. *J Non Cryst Solids*. 2020;529:119768. Available from: https://doi.org/10.1016/j.jnoncrysol.2019.119768.
31. Friedman JH. Greedy functional approximation: A gradient boosting machine. *Ann Stat*. 2001;29:1189–1232. Available from: https://marefateadyan.nashriyat.ir/node/150.
32. Villarrubia G, De Paz JF, Chamoso P, et al. Artificial neural networks used in optimization problems. *Neurocomputing*. 2018;272:10–16.
33. Pomerantsev AL. Confidence intervals for nonlinear regression extrapolation. *Chemom Intell Lab Syst*. 1999;49:41–48.
34. Wei Y, Yang F, Wainwright MJ. Early stopping for kernel boosting algorithms: A general analysis with localized complexities. *IEEE Trans Inf Theory*. 2019;65:6685–6703.
35. Cittanti D, Ferraris A, Airale A, et al. Modeling Li-ion batteries for automotive application: A trade-off between accuracy and complexity. *2017 International Conference of Electrical and Electronic Technologies for Automotive*, Turin, Italy; 2017:1–8.

36. Li X, Li N, Lei X, et al. Study on artificial neural network for predicting gas-liquid two-phase pressure drop in pipeline-riser system. *Energies*. 2023;16:1686.
37. Abdolrasol MGM, Hussain SMS, Ustun TS, et al. Artificial neural networks based optimization techniques: A review. *Electronics*. 2021;10:2689.
38. Hochreiter S, Schmidhuber J. Long short-term memory. *Neural Comput*. 1997;9: 1735–1780.
39. He Y, Chen Q. Construction and application of LSTM-based prediction model for tunnel surrounding rock deformation. *Sustainability*. 2023;15:6877.
40. Liu L, Huang W, Wang C. Hyperspectral image classification with kernel-based least-squares support vector machines in sum space. *IEEE J Sel Top Appl Earth Obs Remote Sens*. 2018;11:1144–1157.
41. Liu Y, Wu J, Yang G, et al. Predicting the onset temperature (Tg) of Ge_xSe_{1-x} glass transition: a feature selection based two-stage support vector regression method. *Sci Bull*. 2019;64:1195–1203. Available from: https://doi.org/10.1016/j.scib.2019.06.026.
42. Bai C, Huang L, Pan X, et al. Optimization of deep convolutional neural network for large scale image retrieval. *Neurocomputing*. 2018;303:60–67. Available from: https://doi.org/10.1016/j.neucom.2018.04.034.
43. Kufel J, Bargieł-Łączek K, Kocot S, et al. What is machine learning, artificial neural networks and deep learning?—Examples of practical applications in medicine. *Diagnostics*. 2023;13:2582.
44. Taherkhani A, Cosma G, McGinnity TM. AdaBoost-CNN: An adaptive boosting algorithm for convolutional neural networks to classify multi-class imbalanced datasets using transfer learning. *Neurocomputing*. 2020;404:351–366. Available from: https://doi.org/10.1016/j.neucom.2020.03.064.
45. Benabdellah AC, Benghabrit A, Bouhaddou I. A survey of clustering algorithms for an industrial context. *Procedia Comput Sci*. 2019;148:291–302. Available from: https://doi.org/10.1016/j.procs.2019.01.022.
46. Nerurkar P, Shirke A, Chandane M, et al. Empirical analysis of data clustering algorithms. *Procedia Comput Sci*. 2018;125:770–779. Available from: https://doi.org/10.1016/j.procs.2017.12.099.
47. Wang Z, Joshi S, Savel'Ev S, et al. Fully memristive neural networks for pattern classification with unsupervised learning. *Nat Electron*. 2018;1:137–145. Available from: https://doi.org/10.1038/s41928-018-0023-2.
48. Gholami R, Fakhari N. Support vector machine: Principles, parameters, and applications. *Handb Neural Comput*. 2017;1:515–535. Available from: https://doi.org/10.1016/B978-0-12-811318-9.00027-2.
49. Camacho Olmedo MT, Paegelow M, Mas JF, et al. *Geomatic Approaches for Modeling Land Change Scenarios. An Introduction*. Cham, Switzerland: Springer Nature; 2018.
50. Kuo CCJ. Understanding convolutional neural networks with a mathematical model. *J Vis Commun Image Represent*. 2016;41:406–413.
51. Spanhol FA, Oliveira LS, Petitjean C, et al. Breast cancer histopathological image classification using convolutional neural networks. *International Joint Conference on Neural Networks (IJCNN)*, Vancouver, BC, Canada; 2017:1868–1873.
52. Kouzehgar M, Krishnasamy Tamilselvam Y, Vega Heredia M, et al. Self-reconfigurable façade-cleaning robot equipped with deep-learning-based crack detection based on convolutional neural networks. *Autom Constr*. 2019;108:102959. Available from: https://doi.org/10.1016/j.autcon.2019.102959.
53. Pascanu R, Mikolov T, Bengio Y. On the difficulty of training recurrent neural networks Razvan. *Int Conf Mach Learn*. 2013;28:1310–1318.
54. Goodfellow I, Pouget-Abadie J, Mirza M, et al. Generative adversarial networks. *Commun ACM*. 2020;63:139–144.

55. Dan Y, Zhao Y. Generative adversarial networks (GAN) based ef fi cient sampling of chemical composition space for inverse design of inorganic materials. *Comput Mater.* 2020;6:1–7. Available from: https://doi.org/10.1038/s41524-020-00352-0.
56. Basaran B, Günes F. Data clustering. *Intell Multidimens data Clust Anal.* 2016;31:28–72.
57. Yu J, Huang H, Tian S. Cluster validity and stability of clustering algorithms. *Lect Notes Comput Sci.* 2004;3138:957–965.
58. Varmuza K. Clustering methods. *DATA Min Knowl Discov.* 1980;15:92–96.
59. Omran MGH, Engelbrecht AP, Salman A. An overview of clustering methods. *Intell Data Anal.* 2007;11:583–605.
60. Kumar P, Sirohi D. Comparative analysis of FCM and HCM algorithm on Iris data set. *Int J Comput Appl.* 2010;5:33–37.
61. Jain AK. Data clustering: 50 years beyond K-means. *Pattern Recognit Lett.* 2010;31:651–666. Available from: https://doi.org/10.1016/j.patrec.2009.09.011.
62. Kumar A, Kumar S. Density based initialization method for K-means Clustering algorithm. *Int J Intell Syst Appl.* 2017;9:40–48.
63. Likas A, Vlassis N, Verbeek J. The global k-means clustering algorithm. *Pattern Recognit.* 2003;36:451–461.
64. Saha R, Tariq MT, Hadi M, et al. Pattern recognition using clustering analysis to support transportation system management, operations, and modeling. *J Adv Transp.* 2019;12:417.
65. Aksac A, Özyer T, Alhajj R. CutESC: Cutting edge spatial clustering technique based on proximity graphs. *Pattern Recognit.* 2019;96:106948. Available from: https://doi.org/10.1016/j.patcog.2019.06.014.
66. Qi J, Yu Y, Wang L, et al. An effective and efficient hierarchical K-means clustering algorithm. *Int J Distrib Sens Networks.* 2017;13:1–17.
67. Murugesan VP, Murugesan P. A new initialization and performance measure for the rough k-means clustering. *Soft Comput.* 2020;24:11605–11619. Available from: https://doi.org/10.1007/s00500-019-04625-9.
68. Shobha G, Rangaswamy S. Machine learning. *Handb Stat.* 2018;1.4. Available from: https://doi.org/10.1016/bs.host.2018.07.004.
69. Willmott CJ, Matsuura K. Advantages of the mean absolute error (MAE) over the root mean square error (RMSE) in assessing average model performance. *Clim Res.* 2005;30:79–82.
70. Hagan MT, Menhaj M. Training feedforward networks with the Marquardt algorithm. *IEEE Trans Neural Netw.* 1994;5:989–993.
71. Kuhn M, Johnson K. *Applied Predictive Modeling.* New York: Springer; 2013.
72. Alcobaça E, Mastelini SM, Botari T, et al. Explainable machine learning algorithms for predicting glass transition temperatures. *Acta Mater.* 2020;188:92–100.
73. Lee LC, Liong CY, Jemain AA. Validity of the best practice in splitting data for hold-out validation strategy as performed on the ink strokes in the context of forensic science. *Microchem J.* 2018;139:125–133. Available from: https://doi.org/10.1016/j.microc.2018.02.009.
74. Emmert-Streib F, Dehmer M. Evaluation of regression models: Model assessment, model selection and generalization error. *Mach Learn Knowl Extr.* 2019;1:521–551.
75. Kampezidou SI, Tikayat Ray A, Bhat AP, et al. Fundamental components and principles of supervised machine learning workflows with numerical and categorical data. *Eng.* 2024;5:384–416.
76. Pedregosa F, Varoquaux G, Gramfort A, et al. Scikit-learn: Machine learning in python. *J Mach Learn Res.* 2011;12:2825–2830.
77. Chang M-W, Devlin J, Dragan A, et al. Artificial intelligence, 4th ed. Russell S, Norvig P, editors. *A Modern Approach.* Hoboken, NJ: Pearson Education, Inc.; 2020.

78. Chejara P, Prieto LP, Ruiz-Calleja A, et al. EFAR-MMLA: An evaluation framework to assess and report generalizability of machine learning models in MMLA. *Sensors.* 2021;21:2863.
79. Algamal ZY. Shrinkage parameter selection via modified cross-validation approach for ridge regression model. *Commun Stat Simul Comput.* 2020;49:1922–1930. Available from: https://doi.org/10.1080/03610918.2018.1508704.
80. Wilcox RR. Multicolinearity and ridge regression: Results on type I errors, power and heteroscedasticity. *J Appl Stat.* 2019;46:946–957.
81. Li N, Yang H, Yang J. Nonnegative estimation and variable selection via adaptive elastic-net for high-dimensional data. *Commun Stat - Simul Comput.* 2019;50:1–17. Available from: https://doi.org/10.1080/03610918.2019.1642484.
82. Wang Y, Yang XG, Lu Y. Informative gene selection for microarray classification via adaptive elastic net with conditional mutual information. *Appl Math Model.* 2019;71:286–297. Available from: https://doi.org/10.1016/j.apm.2019.01.044.
83. Assempoor A. *Consideration of Frictional Boundary Modelling in Metal Forming.* Stillwater, OK: Oklahoma State University; 1989.
84. Mushtaq Z, Hanief M, Manroo SA. Prediction of friction and wear during ball-on-flat sliding using multiple regression and ANN: Modeling and experimental validation. *J Tribol.* 2021;28:117–128.
85. Paturi UMR, Palakurthy ST, Reddy NS. The role of machine learning in tribology: A systematic review. *Arch Comput Methods Eng.* 2023;30:1345–1397.
86. Suryawanshi A, Behera N. Application of machine learning for prediction dental material wear. *J Polym Mater.* 2023;40:305–316.
87. Singh J, Kumar S, Kumar R, et al. Machine learning regression tools for erosion prediction of WC-10Co4Cr thermal spray coating. *Results Surf Interf.* 2023;13:100156. Available from: https://www.sciencedirect.com/science/article/pii/S2666845923000636.
88. Cohn R, Holm E. Unsupervised machine learning via transfer learning and k-means clustering to classify materials image data. *Integr Mater Manuf Innov.* 2021;10:231–244.
89. Assafo M, Langendoerfer P. Unsupervised and semisupervised machine learning frameworks for multiclass tool wear recognition. *IEEE Open J Ind Electron Soc.* 2024;5:993–1010.
90. Marian M, Tremmel S. Physics-informed machine learning—an emerging trend in tribology. *Lubricants.* 2023;11:463.
91. Hulipalled P, Algur V, Lokesha V, et al. Interpretable ensemble machine learning framework to predict wear rate of modified ZA-27 alloy. *Tribol Int.* 2023;188:108783.
92. Praveen AS, Sarangan J, Suresh S, et al. Optimization and erosion wear response of NiCrSiB/WC–Co HVOF coating using Taguchi method. *Ceram Int.* 2016; 42:1094–1104.
93. Qiao L, Wu Y, Hong S, et al. Relationships between spray parameters, microstructures and ultrasonic cavitation erosion behavior of HVOF sprayed Fe-based amorphous/nanocrystalline coatings. *Ultrason Sonochem.* 2017;39:39–46.
94. Singh J, Kumar S, Singh G. Taguchi's approach for optimization of tribo-resistance parameters forss304. *Mater Today Proc.* 2018;5:5031–5038. Available from: https://doi.org/10.1016/j.matpr.2017.12.081.
95. Tabbara H, Gu S, McCartney DG. Computational modelling of titanium particles in warm spray. *Comput Fluids.* 2011;44:358–368.
96. Li M, Christofides PD. Modeling and control of high-velocity oxygen-fuel (HVOF) thermal spray: A tutorial review. *J Therm Spray Technol.* 2009;18:753–768.
97. Dongmo E, Wenzelburger M, Gadow R. Analysis and optimization of the HVOF process by combined experimental and numerical approaches. *Surf Coatings Technol.* 2008;202:4470–4478.

98. Guessasma S, Salhi Z, Montavon G, et al. Artificial intelligence implementation in the APS process diagnostic. *Mater Sci Eng B*. 2004;110:285–295.
99. Guessasma S, Montavon G, Coddet C. Modeling of the APS plasma spray process using artificial neural networks: basis, requirements and an example. *Comput Mater Sci*. 2004;29:315–333.
100. Guessasma S, Montavon G, Coddet C. Neural computation to predict in-flight particle characteristic dependences from processing parameters in the APS process. *J Therm Spray Technol*. 2004;13:570–585.
101. Sahraoui T, Guessasma S, Fenineche NE, et al. Friction and wear behaviour prediction of HVOF coatings and electroplated hard chromium using neural computation. *Mater Lett*. 2004;58:654–660.
102. Choudhury TA, Hosseinzadeh N, Berndt CC. Artificial neural network application for predicting in-flight particle characteristics of an atmospheric plasma spray process. *Surf Coatings Technol*. 2011;205:4886–4895.
103. Choudhury TA, Hosseinzadeh N, Berndt CC. Improving the generalization ability of an artificial neural network in predicting in-flight particle characteristics of an atmospheric plasma spray process. *J Therm Spray Technol*. 2012;21:935–949.
104. Kanta A-F, Montavon G, Planche M-P, et al. Artificial neural networks implementation in plasma spray process: Prediction of power parameters and in-flight particle characteristics vs. desired coating structural attributes. *Surf Coatings Technol*. 2009;203:3361–3369.
105. Kanta A-F, Montavon G, Berndt CC, et al. Intelligent system for prediction and control: Application in plasma spray process. *Expert Syst Appl*. 2011;38:260–271.
106. Kanta A-F, Montavon G, Planche M-P, et al. Artificial intelligence computation to establish relationships between aps process parameters and alumina–titania coating properties. *Plasma Chem Plasma Process*. 2008;28:249–262.
107. Zhang C, Kanta A-F, Li C-X, et al. Effect of in-flight particle characteristics on the coating properties of atmospheric plasma-sprayed 8 mol% Y_2O_3–ZrO_2 electrolyte coating studying by artificial neural networks. *Surf Coatings Technol*. 2009;204:463–469.
108. Liu T, Deng S, Planche M-P, et al. Estimating the behavior of particles sprayed by a single-cathode plasma torch based on a nonlinear autoregressive exogenous model. *Surf Coatings Technol*. 2015;268:284–292.
109. Guessasma S, Montavon G, Gougeon P, et al. Designing expert system using neural computation in view of the control of plasma spray processes. *Mater Des*. 2003;24:497–502.
110. Kanta A-F, Planche M-P, Montavon G, et al. In-flight and upon impact particle characteristics modelling in plasma spray process. *Surf Coatings Technol*. 2010;204:1542–1548.
111. Cherigui M, Guessasma S, Fenineche N, et al. Neural computation to correlate HVOF thermal spraying parameters with the magnetic properties of FeNb alloy deposits. *Mater Chem Phys*. 2005;93:181–186.
112. Zhang G, Kanta A-F, Li W-Y, et al. Characterizations of AMT-200 HVOF NiCrAlY coatings. *Mater Des*. 2009;30:622–627.
113. Kamnis S, Malamousi K, Marrs A, et al. Aeroacoustics and artificial neural network modeling of airborne acoustic emissions during high kinetic energy thermal spraying. *J Therm Spray Technol*. 2019;28:946–962.
114. Mojena MAR, Roca AS, Zamora RS, et al. Neural network analysis for erosive wear of hard coatings deposited by thermal spray: Influence of microstructure and mechanical properties. *Wear*. 2017;376:557–565.
115. Liu M, Yu Z, Zhang Y, et al. Prediction and analysis of high velocity oxy fuel (HVOF) sprayed coating using artificial neural network. *Surf Coatings Technol*. 2019;378:124988.
116. Gui L, Wang B, Cai R, et al. Prediction of in-flight particle properties and mechanical performances of HVOF-sprayed NiCr–Cr_3C_2 coatings based on a hierarchical neural network. *Materials*. 2023;16:6279.

117. Liu M, Yu Z, Wu H, et al. Implementation of artificial neural networks for forecasting the HVOF spray process and HVOF sprayed coatings. *J Therm Spray Technol.* 2021;30:1329–1343.
118. Gao G, Fu R, Wang Q, et al. Progress on current-carry friction and wear and prediction models: A review. *Coatings.* 2025;12:1345.
119. Trzepieciński T, Szwajka K, Szewczyk M. Analysis of coefficient of friction of deep-drawing-quality steel sheets using multi-layer neural networks. *Lubricants.* 2024;12:50.

4 AI-Driven Surface Engineering

Parminder Singh, Adityabir Singh, Sunirmit Verma, and Surinder Singh

4.1 INTRODUCTION

Surface engineering has come a long way since its inception as a set of protective treatments, evolving into a highly innovative discipline that has taken center-stage across various industries. In early days, methods such as carburizing, case hardening, and electroplating were employed to increase the durability and resistance to corrosion, thereby prolonging the service life of machines, structures, and tools [1]. Galvanization and chemical passivation were among the common techniques used in mass production during the Industrial Revolution [2]. With time, processes such as thermal spray coating, physical vapor deposition (PVD), chemical vapor deposition (CVD), sol–gel processes, and laser surface engineering were introduced, which help in precise control of surface hardness, adhesion, wear resistance, and texture [3–5]. These developments signify the transition from surface engineering as a purely protective step to a versatile platform where surfaces with certain specific functionalities can be tailored.

In recent years, the development of modern applications has become increasingly complex, giving rise to an increasing demand for highly specialized surface properties of advanced materials. In the aircraft industry, for example, it is necessary to create finishes that will give the products ultra-hard surfaces with superior oxidation resistance in order to survive in harsh operating environments, and in the automotive industry, it is required to create wear- and corrosion-resistant finishes that will enhance fuel efficiency and long life in service [6,7]. Biomedical implants, however, need materials that are both biocompatible and resistant to microbial colonization, whereas the energy sector needs catalytic, conductive, and protective surfaces to enhance the efficiency of batteries, fuel cells, and solar devices [8–10]. Semiconductor and electronics assembly requires nanoscale-accurate surfaces without defects in order to provide the best device performance [11]. A combination of these examples indicates a wider change in the role of surface engineering, where the focus was on improving simple protection and allowing more multifunctional and application-specific design.

Artificial intelligence (AI) is turning materials science into a field of faster innovation and optimization of performance through the introduction of tools. The concept of AI includes systems that perform functions that presuppose human-type intelligence, learning, reasoning, and decision-making [12]. Machine learning (ML) has been used to find patterns in data and make predictions without explicit

DOI: 10.1201/9781003624165-4

programming, while deep learning (DL), a branch of ML, applies multilayered neural networks to find intricate connections in high-dimensional data. The techniques are especially useful in the modeling of structure–property–processing relationships that are hard to solve using traditional techniques. Complementary to experimentation and computation, AI presents an effective new framework for accelerated materials design, characterization, and optimization [13].

In surface engineering, AI is used to solve the interrelated and complicated connections between process parameters and material properties. The characteristics of a coating, such as its hardness, adhesion, and wear resistance, represent a complex issue that varies with numerous factors that include temperature, deposition rate, chemical composition, and microstructural evolution [13]. These interactions are frequently difficult to be fully represented using traditional methods, whereas AI can process a large amount of data and identify missing patterns, predict results, and propose the best conditions. AI can greatly reduce development time and expense when paired with high-throughput experimental and multiscale simulation strategies to enable the creation of the next generation of surfaces with unrivalled precision and customized functionality.

This chapter presents a historical overview of surface engineering; the terminologies of AI, ML, and DL in materials science; and the primary fields of AI revolution in materials science. Through the analysis of the intersection between surface engineering and AI, this discussion demonstrates how smart computational tools can be used with conventional approaches to optimize the processes, enhance the performance of coatings, facilitate predictive modeling, and eventually open new opportunities for next-generation technologies.

4.2 FUNDAMENTALS OF SURFACE ENGINEERING

Surface engineering involves purposeful modification of the surface and near-surface layers of a material in order to enhance its performance without any alterations in its internal structure. The main goal is on properties such as wear resistance, hardness, corrosion resistance, adhesion, and biocompatibility through physical, chemical, mechanical, plasma, ion-based, and hybrid approaches, as well as nano-structured and intelligent surface coatings [3–5,14,15]. These techniques have been summarized in Table 4.1, with their mechanisms, applications, strengths, and limitations highlighted.

Some of the abbreviations used in the above table are high-velocity oxy-fuel spray process (HVOF), microelectromechanical systems (MEMS), and integrated circuits (ICs). Also, Figure 4.1 displays representative schematics of surface modification methods that place a material layer onto the surface, and Figure 4.2 displays schematics of methods that do not place a significant material layer onto the surface.

4.3 FUNDAMENTALS OF AI

AI is a wide and dynamic discipline that integrates concepts from computer science, mathematics, psychology, and engineering to develop systems that have the ability to execute tasks normally associated with human intelligence. Such activities include

TABLE 4.1
Comparison of All Surface Engineering Techniques (Credit statement: Created by authors)

Technique	Mechanism	Applications	Advantages	Limitations
Thermal spray (plasma, HVOF, and cold) [16, 17]	Molten/semi-molten particles propelled at high velocity	Turbines, engines, biomedical, and tools	Thick coatings and versatile materials	Porosity and adhesion challenges
PVD [16]	Vaporized atoms condensed in vacuum	Cutting tools, electronics, and implants	Thin, hard, and environmentally friendly	Vacuum requirement and slower rates
CVD [16]	Chemical reaction of gases forms coatings	Semiconductors, aerospace, and optics	High-purity, conformal films	High temperatures and toxic gases
Laser processing [16]	Focused beam modifies the surface locally	Aerospace, biomedical, and automotive	Precision, no chemicals, and localized	Capital- intensive and directional
Electroplating [16]	Electrochemical deposition	Automotive, jewelery, and electronics	Simple, decorative, and corrosion resistance	Uneven coverage and limited geometries
Electroless plating [16]	Chemical reduction without current	PCBs, aerospace, and complex parts	Uniform coatings and nonconductive substrates	Bath chemistry control critical
Sol–gel [16]	Hydrolysis- condensation films	Optics, sensors, and anti-corrosion	Functional, low-cost, and transparent	Fragile films and long curing
Chemical etching [18]	Selective dissolution with etchants	MEMS, ICs, and jewelery	High precision and stress-free	Hazardous chemicals and waste disposal
Plasma nitriding/carburizing [19]	Glow discharge diffusion	Automotive gears and dies	Controlled case depth and less distortion	Requires vacuum and plasma equipment
Plasma surface activation [20]	Enhances surface adhesion through polar groups	Composites, medical devices, and electronics	Improves adhesion and wettability	Surface effect may fade and controlled plasma setup
Ion implantation [21]	Energetic ion bombardment	Semiconductors and wear parts	Atomic-scale modification and no dimensional change	High cost and shallow depth
Shot peening [22]	Bombardment with spherical media	Aerospace, springs, and gears	Improves fatigue strength	Surface roughness and geometry limits
Ion beam assisted deposition [23]	Ion bombardment modifies coating	Optics, aerospace, and electronics	Dense, adherent, and defect-free	Requires precise process control
Ultrasonic impact treatment [24]	High-frequency impacts refine grains and strengthen surface	Welds, bridges, and machinery	Improves fatigue and strength	Limited to local areas
Weld overlay/hardfacing [25]	Welded resistant alloy layer	Pipelines, turbines, and reactors	Restores life and thick protection	Dilution and thermal distortion

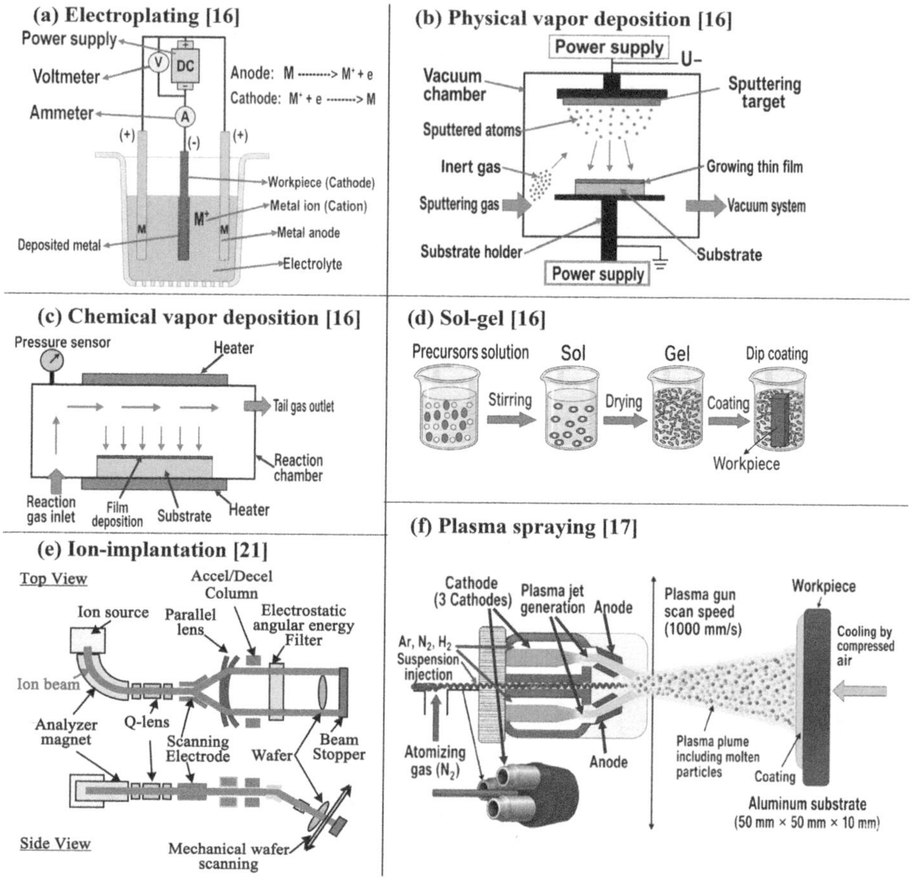

FIGURE 4.1 Schematics of surface modification and treatment techniques involving significant material deposition [16,17,21]. (Credit statement: Creative Commons International Attribution 4.0.)

reasoning, experience-based learning, problem-solving, world perception, and decision-making. During the early years, AI was based on symbolic reasoning and rule-based expert systems or methodologies, which relied on explicitly programmed knowledge. In recent years, however, this has changed to a more data-driven approach, whereby systems are able to learn from large volumes of information. This change has enabled AI to become much more flexible and capable of identifying patterns, making predictions, and gaining experience. Nowadays, AI exists in almost every major industry. In the healthcare sector, it can be applied to identify diseases at an early stage and personalize treatments; in finance, it can identify fraud and monitor investments; in transportation, it can be applied to create autonomous vehicles; and in manufacturing, it can optimize processes through smart automation. Figure 4.3 shows the way in which AI has developed in three major waves, identifying the main areas of interest in each phase and giving clues to the latest trends that promote further development [26].

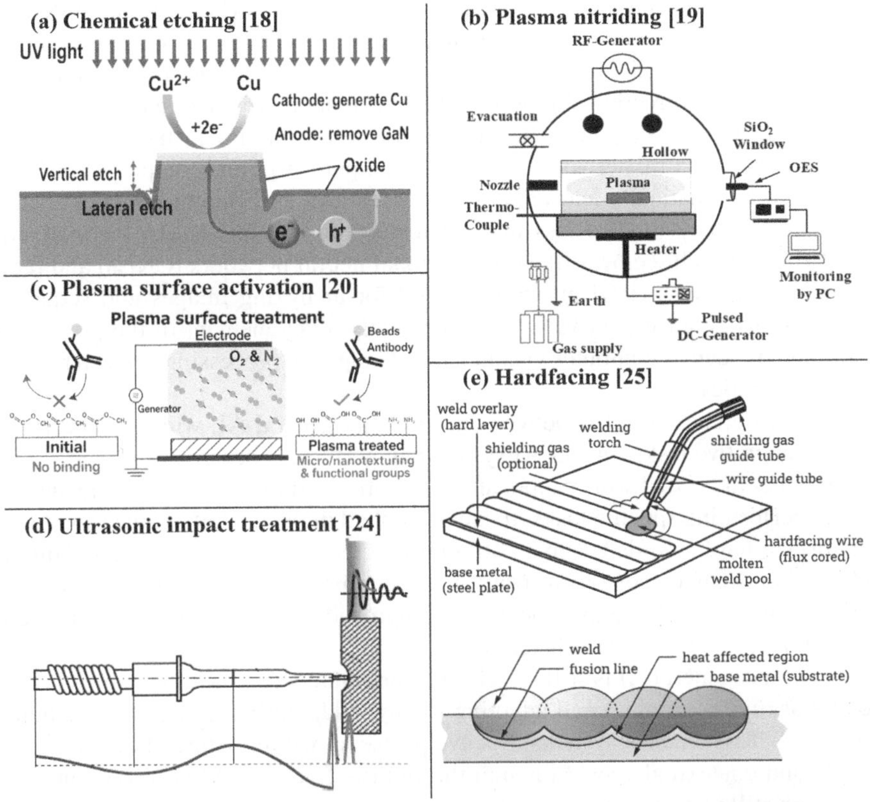

FIGURE 4.2 Schematics of surface modification and treatment techniques that alter surfaces without significant material deposition [18–20,24,25]. (Credit statement: Creative Commons International Attribution 4.0.)

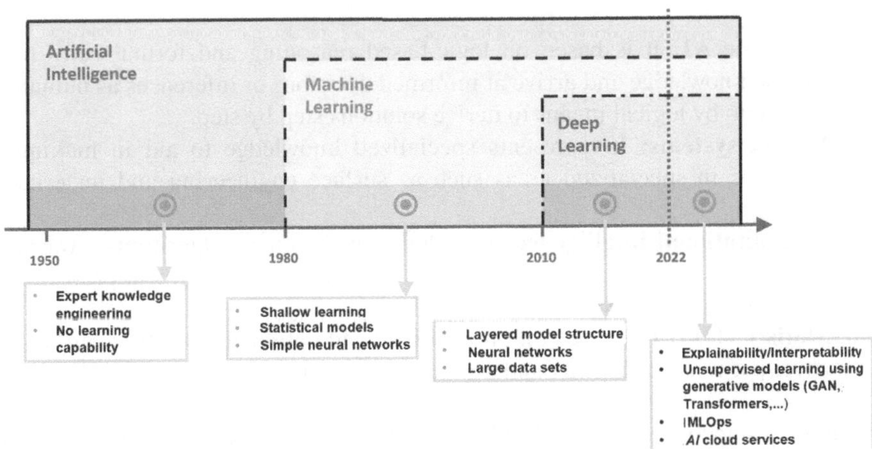

FIGURE 4.3 Evolution of AI: Key waves and emerging future trends [26]. (Credit statement: Creative Commons International Attribution 4.0.)

One of the most significant applications of AI is ML, which allows systems to learn from information and develop without explicit coding. Such systems don't rely on fixed predetermined rules as in traditional programming but instead recognize patterns and learn directly from observable information provided in examples.

This is further developed in DL, a sophisticated form of ML that utilizes multi-layered artificial neural networks (ANNs) to compute large and complex data. These models can automatically learn and enhance features at different levels of abstraction such that they require less manual feature design. DL is also very diverse, with dedicated architectures like convolutional neural networks (CNNs) and recurrent neural networks (RNNs) being applied for analyzing images and sequential or time-series data, respectively. DL is presently a key element in the progress of various fields such as computer vision, natural language processing, speech recognition, and robotics.

A key difference in AI is between narrow and general AI, which quantifies the range of cognitive tasks that an AI system can perform. The current narrow AI that presently pervades real-world applications is optimized to be very accurate and efficient in performing specific tasks. It can be applied in the field of engineering to predict surface hardness, detect microstructural defects, optimize coating conditions, and monitor wear or corrosion. Narrow AI can bring real benefits to specialized work, including quicker experimentation, improved quality control, and an overall improvement in the process efficiency.

In contrast, general AI is still largely goal-orientated, in which the system is capable of learning, reasoning, and adapting to any field, similar to the way the human mind does. This enables systems to solve problems within defined boundaries in a flexible and contextual way. Although the future of general AI is revolutionary, its realization still remains an overwhelming technical and philosophical task. As such, it is safe to assert that narrow AI is still the major driver behind current industrial breakthroughs and research developments.

In general, AI may be classified into a number of interrelated sub-fields, which apply different methodologies:

a. **Symbolic AI:** It is based on logic-based reasoning and formal rules to encode knowledge and arrive at informed decisions or inferences as human beings do by logical means to derive solutions step by step.
b. **Expert systems:** It represents specialized knowledge to aid in making decisions in specialized areas such as surface engineering and material selection.
c. **Computational intelligence:** It includes evolutionary algorithms, swarm intelligence, as well as fuzzy logic to address complex, nonlinear problems based on natural processes.
d. **Statistical AI:** It focuses on probabilistic and data-driven methods to discover patterns and predict and measure uncertainty.

The most compelling model of AI capabilities and development is the cognitive hierarchy model proposed by Ackoff [27] called the data-information-knowledge-insight-wisdom (DIKIW) model. In this framework:

a. Data are raw and unprocessed facts.
b. Information is what gives these facts context and meaning.
c. Knowledge is what brings the information into action.
d. Intelligence allows reasoning and inference.
e. Wisdom is advanced understanding, which enables decision-making.

Scaling the data science technology to this framework largely used data and information levels, which concern collection, structuring, and analysis. On the contrary, AI ventures into new levels such as knowledge, intelligence, and wisdom, aiming to convert structured information into intelligent activities and strategic choices. The crossroad of the two underscores areas of active research in which unprocessed data are increasingly being refined to give useful systems of knowledge.

Figure 4.4 displays the DIKIW hierarchy that depicts the incremental complexity of cognitive treatment of raw data into actionable wisdom. Such progressive enrichment reflects human cognitive development as well as the history of AI development.

This hierarchy can be used in surface engineering to explain how AI can convert raw experimental and simulation data into actionable insights. Surface roughness, hardness, wear, or corrosion measurements are processed into structured information to predict the behavior of materials. This understanding informs the optimization of

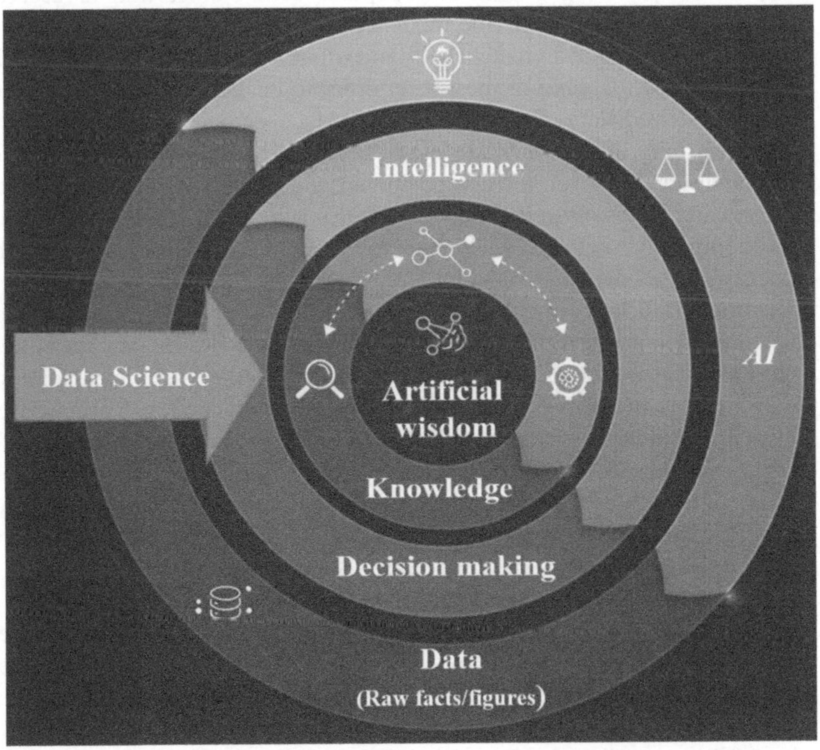

FIGURE 4.4 Cognitive progression from raw data to information, knowledge, and ultimately wisdom. (Credit statement: Created by the authors.)

processes in coatings, additive manufacturing, and surface treatments. At the levels of intelligence and wisdom, AI will assist in strategic decision-making, enhance quality control, and increase the pace of innovation by connecting data-driven predictions to engineering practice.

4.4 ML AND DL IN SURFACE ENGINEERING

The concept of AI in surface engineering has been achieved mostly through the application of ML and its sophisticated version, DL. These computational paradigms provide a strong means of examining complicated data, finding patterns in the microstructure of materials, and eventually allowing integrated forecasting and prescriptive decision-making in engineering.

4.4.1 ML Paradigms

ML is a diverse collection of techniques that differ in the manner in which the model learns using data and the type of supervision or feedback that they depend on. These strategies can be broadly categorized into four major categories, as discussed below:

a. **Supervised learning:** It is the most popular method of learning in research and industry. In this case, the models are trained using labeled datasets, in which each input is matched to an output. The goal is to empirically determine the mapping of the two in order to enable the model to predict the unobserved data. For example, in surface engineering, under supervised learning, it is possible to predict coating hardness, wear resistance, or corrosion rate given the compositional and process-related parameters. Regression, decision trees, support vector machines, and other techniques are commonly used and yield high accuracy in conditions where high-quality labeled data are abundant [28].
b. **Unsupervised learning:** It works without the use of labeled data. It does not predict predetermined results but tries to discover hidden structures, clusters, or associations in the data. This makes it quite useful in exploratory data analysis, anomaly detection, or dimensionality reduction. Unsupervised algorithms including k-means clustering or principal component analysis (PCA) can be used in materials science to classify microstructural patterns, cluster similar surface morphologies, or identify anomalies in surface coatings that can indicate defects. The strength of unsupervised learning is that it identifies patterns that cannot be easily seen by human experts [28].
c. **Semi-supervised learning:** Semi-supervised learning is a paradigm that is intermediate to supervised and unsupervised paradigms and is characterized by the use of both labeled and unlabeled data. As the acquisition of labeled data can be time- and cost-intensive, e.g., annotating high-resolution microscopy images, semi-supervised methods have offered an efficient alternative by taking a small set of such labeled data as inputs and using it to steer the learning process on significantly larger unlabeled datasets. This mixed approach is particularly useful in engineering situations where there are

limited experimental data but lots of unlabeled data. Semi-supervised learning has demonstrated the capability to enhance predictive model strength and lower reliance on large-scale manual labeling [28,29].

d. **Reinforcement learning (RL):** It is a means through which machines can learn via trial and error, as human beings do via experience. An RL agent interacts with the environment, acts on it, and is rewarded or penalized by feedback. In so doing, it works out the optimum plan to maximize rewards in the long run. This renders RL the best choice when adaptive control is required in surface engineering, where the system is constantly changing process parameters to enhance results in complex and dynamic situations [29,30].

4.4.2 DL Architecture

The conventional ML approach in surface engineering uses handcrafted constraints to model surface features but fails to predict the multiscale and multilevel features of surface phenomena. DL, using such architectures as CNNs and RNNs, changes the game by discovering detailed, layer-based representations directly from raw data and automatically learning them. This automated feature extraction enables DL to identify hidden patterns that are not identified by manual processes. CNNs and RNNs improve upon the basic concept of ANNs, which are based on the neurons of the human brain and can create vast interconnected networks to process more complicated information. The combination of all these developments provides an effective platform to explore surface properties better and make more precise predictions to design more advanced materials [31,32].

McCulloch and Pitts [33] presented the first formal model of an artificial neuron, which is the basis of modern computational neuroscience. A neuron is biologically made up of dendrites, soma, axons, and synapses: dendrites receive inputs, the soma processes them, axons send the output, and synapses control the strength of connections. Groups of interconnected neurons together make up subsystems that constitute the human brain, which has close to 100 billion neurons. The behavior of these biological components is simulated in ANNs by using interconnected artificial neurons that work as a unit. The inputs are taken in a similar manner as the dendrites, and the adjustable weights of connections are synaptic strengths that are changed during learning. The summation and activation process reflects the role of the soma, in which the weighted inputs are summed up and converted into an output. These data are then transmitted to other layers through similar connections as axons, allowing more complex signal processing and pattern recognition [32]. ANNs offer an effective way of modeling nonlinearity by harnessing these biological principles in the form of computation, making them an effective tool for producing dependable predictions, as shown in Figure 4.5. This beautiful analogy of biology to computation has opened new avenues in the comprehension of and solutions to complicated issues in several disciplines. It is a miracle that the knowledge provided by nature may lead to great technological developments.

Moreover, CNNs are sophisticated DL models that are reminiscent of multilayer perceptrons (MLPs), which have several convolutional, pooling, and fully connected

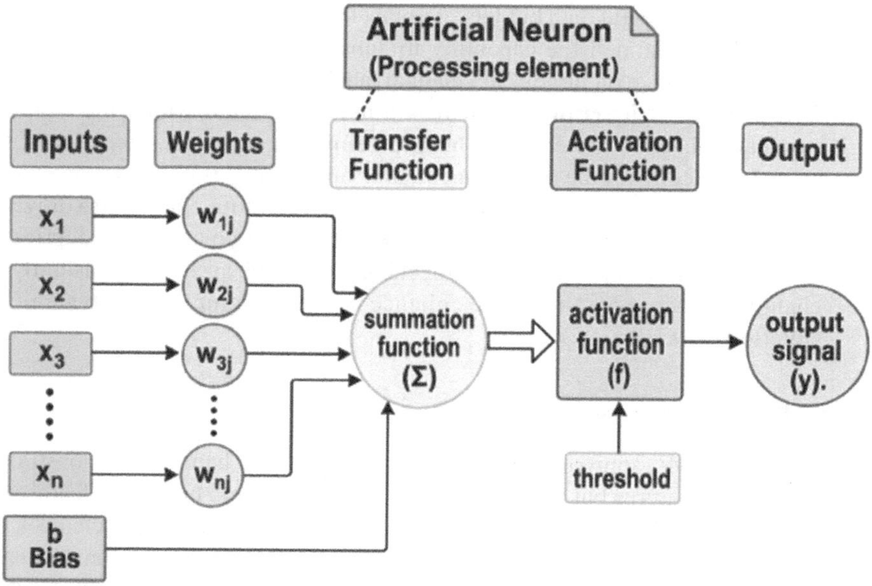

FIGURE 4.5 Schematic of an artificial neuron [32]. (Credit statement: Creative Commons International Attribution 4.0.)

layers. All of the layers accept 3D structured inputs: height, width, and depth (r), with r being the number of channels (e.g., $r=3$ in a red–green–blue image). CNNs are well suited to spatial pattern recognition; hence, they are especially noteworthy in the study of microscopy images as part of surface engineering, as shown in Figure 4.6. Their convolutional layers automatically identify local characteristics using similar weights, lowering computational parameters and preserving high feature extraction rates [32]. These features allow CNNs to be effective in detecting surface defects, measuring grain boundaries, and determining the uniformity of coatings. Visual Geometry Group classification models with 19 layers of weights (VGG19) may also be used to extract features of surface defect datasets, which assist in enhancing automated inspection and quality control. CNNs have proven to be highly effective in the detection of thin cracks in concrete, even in diverse lighting environments, in civil engineering [34]. Besides, electromagnetic imaging is performed using CNN-based DL methods that characterize rough surface profiles to solve complex inverse problems. Outside materials science, CNNs have also been used to recognize the wetness of road surfaces using acoustic signals, which is crucial to autonomous vehicles, and to detect cranial features in 3D format, such as in craniofacial reconstruction and recognition [35].

Additionally, RNNs, and especially long short-term memory (LSTM) models, are best suited to analyzing the temporal characteristics of sequential data, which are essential in analyzing changes in the state of the surface over time. In contrast to CNNs, RNNs have the property of having feedback connections that enable them to perform long-term dependent learning by storing information about past inputs.

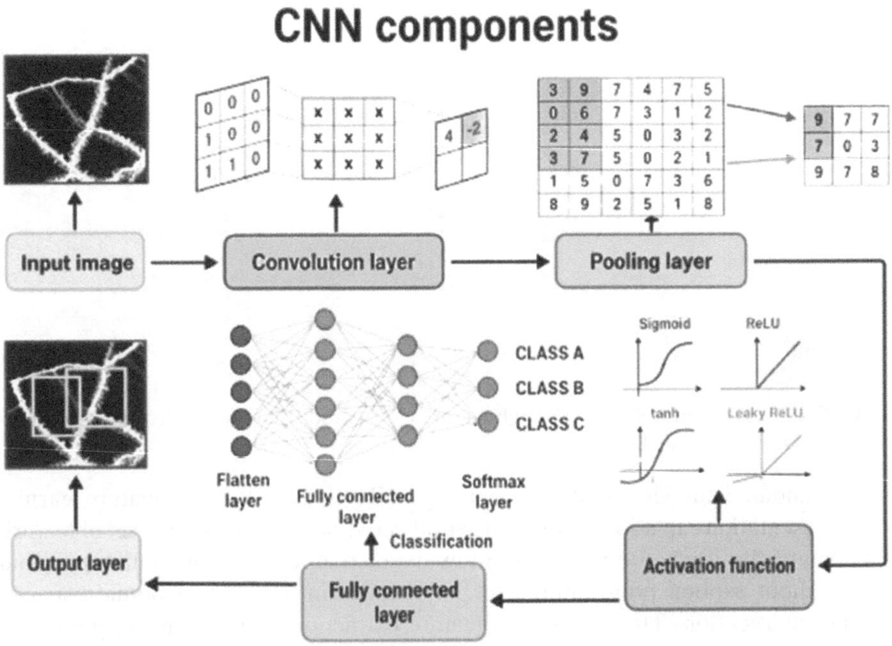

FIGURE 4.6 Key components of a CNN architecture [32]. (Credit statement: Creative Commons International Attribution 4.0.)

This property renders LSTMs very useful in time-dependent degradation prediction such as wear and corrosion. RNNs can predict surface deterioration under varying conditions of operation, enhancing material reliability, particularly in demanding industries such as aerospace and biomedical engineering, by analyzing sequential data. Vibration signals in this study are considered time-series data; therefore, RNNs will be used in the analysis. The information is restructured into two-dimensional data according to time windows in order to preserve the order of the sequence. The RNN takes the present input at every time step and integrates it with the past state by using weighted connections and retains the necessary information over time needed to accurately predict the future [32,36]. Figure 4.7 below presents the RNN structure.

Even further practical uses demonstrate the clear benefits of such DL structures in surface engineering. For example, CNNs are capable of distinguishing finer corrosion signatures in scanning electron microscopy (SEM) scans with the potential to signify initial material decay and replace proactive maintenance [38]. In predictive maintenance, RNNs are able to take sensor data streams and predict surface degradation under different environmental and working conditions, thus optimizing maintenance schedules and increasing the life of critical components. For instance, time-series data can be used to classify the conditions of the road surface by using CNNs with LSTM layers, where temporal dynamics indicate a change in wear or in the impact of the environment. This is useful for understanding and avoiding the causes of surface failure [39].

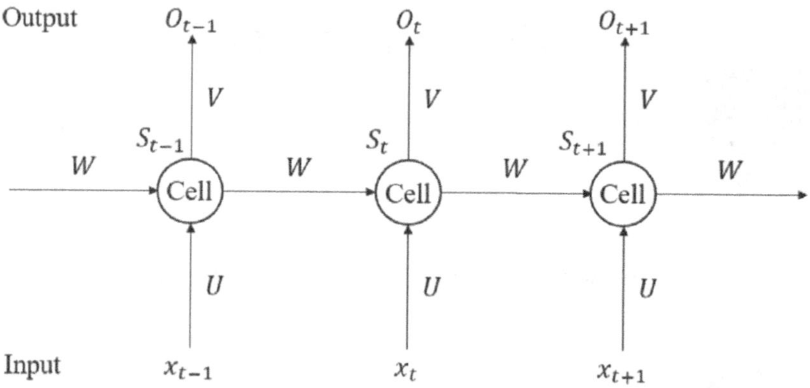

FIGURE 4.7 Structure of an RNN architecture [37]. (Credit statement: Creative Commons International Attribution 4.0.)

The fundamental benefit of DL, including CNNs and RNNs, is its feature learning and representation capabilities. In deep neural networks, an appropriate set of features is automatically learned directly from raw data, creating elaborate data representations without explicit programming or any significant amount of domain-specific manual intervention. This automated feature extraction is important in processing the high-dimensional, complex, and abstract data that often appear in surface characterization, both in microscopic images and sensor time-series data. DL, with its hierarchical structure, in which lower-level models learn simpler features (e.g., edges and textures) and higher-level models learn more abstract features (e.g., shapes and objects), can be used to give models state-of-the-art performance on a wide range of complex surface engineering tasks. Other approaches to unsupervised DL, such as deep belief networks and stacked autoencoders, also help to learn useful features and representations from large unlabeled datasets, which, in turn, can lead to a greater number of tasks at higher levels in other ML applications and enhance the perception and handling of material surfaces [32].

4.4.3 Feature Learning and Automated Representations

The advantage of modern AI, and in particular of DL, is that it can automatically extract patterns and form internal representations that are constructed directly from raw data. In classical ML, it took experts a long time to develop features such as surface roughness indices, grain size distributions, or texture statistics, and only after that could a model start learning. Although these handcrafted descriptors can generate some insights into material behavior, they are influenced by human bias and may fail to reveal the small or nonlinear trends that are likely to be lurking in complicated datasets.

The DL transformations are completely different. Rather than informing the model of what to look at, it is guided to learn the features involved by ranking them in a hierarchy. In CNNs, the simple visual features (e.g., edges or microscopy image local contrasts) may be initially identified in the initial layers. These cues are then aggregated into more significant structures by the middle layers, such as grain boundaries, pores, or surface defects. High-level patterns, such as those that typically directly relate to the mechanical or chemical performance of a material, are then captured in the deepest layers. Similarly, RNNs and LSTM models are particularly good at identifying time-dependent behavior, e.g., the time evolution of corrosion, fatigue, or surface treatment.

Automatic representation learning plays an important role in surface engineering. Data obtained using instrumentations such as SEM, atomic force microscopy, spectroscopy, or real-time sensors can be analyzed to give information showing how the microstructure correlates with material performance, a feature that cannot be determined using other engineering methods. An example is that CNNs are able to detect minute differences in microstructures of coatings, whereas RNNs are able to extrapolate how wear or degradation occurs under different environmental or operating conditions.

Besides making the model more accurate, the automated feature learning also improves the quality of the model and increases model flexibility. As the network is trained on a large and diverse set of noisy data, it can extrapolate over a wider range of materials, processes, and experimental conditions and does not require the design of features to be done by hand. Moreover, the representations acquired can expose the causes or significance of particular patterns and deliver facts that will instruct researchers to utilize optimal treatments, refine material compositions, and create next-generation surfaces [40,41].

In short, DL will make AI not just a predictive tool but a true research collaborator, which will not only reveal hidden knowledge but also learn more about materials and create a smarter and data-driven decision-making process in surface engineering.

4.5 DATA-CENTRIC AND MODEL-CENTRIC AI

The fundamental building blocks of AI are data and algorithms. The discipline has evolved under various paradigms, with the most significant being model-centric AI (MC-AI) and data-centric AI (DC-AI), both addressing different aspects of system development to enhance performance [42].

4.5.1 PRINCIPLES OF MC-AI

Much of AI research has been led by the model-centric approach. The benefit of this approach is that the algorithms are gradually refined under the assumption that the data remain largely fixed. Typical projects are re-architecturing network topologies, making them simpler, and hyperparameter tuning in order to perform better on current data. This has been the approach behind decades' worth of innovations, which has led to tremendous progress in ML, DL, and scalable neural architectures.

Among them are CNNs that have revolutionized image analysis through the ability to capture complex spatial features and RNNs along with advanced versions such as LSTM networks that have made it possible to model sequential or time-varying information with accuracy. The applications supported by these developments

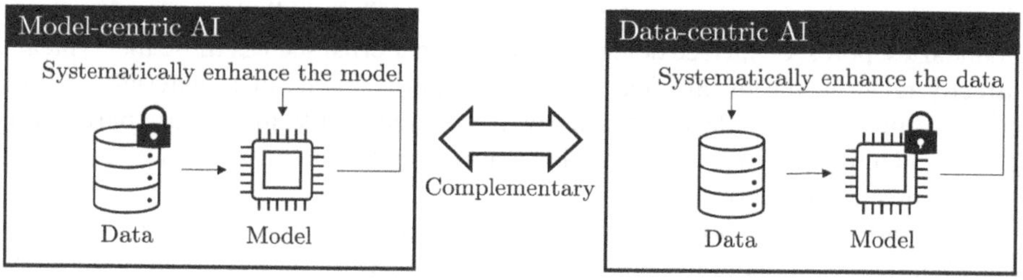

FIGURE 4.8 Data-centric and model-centric AI [42]. (Credit statement: Creative Commons International Attribution 4.0.)

include medical imaging, natural language processing, predictive maintenance, and autonomous systems.

Conventional MC-AI treated the datasets as fixed during model improvement and optimization (illustrated in Figure 4.8). Although these improvements have made the accuracy and generalization better, the performance of the model still primarily relies on the quality of data. Subpar, biased, or incomplete data may constrain the outcomes, regardless of the advanced model.

The awareness of these shortcomings has given rise to the emergence of DC-AI, which puts more emphasis on the advancement of data quality and representativeness, complementing model development and solving real-world problems more efficiently [43,44].

4.5.2 Principles of DC-AI

DC-AI is a significant change in the development of AI systems. It does not focus on refining the models themselves but rather on enhancing the quality, consistency, and usefulness of the data itself. Conventional, model-centric models would normally rely on the development of new architectures, hyperparameter optimization, and the optimization of algorithms with the assumption that the underlying data are known to be fixed and reliable. DC-AI defies that assumption, and it works on the principle that improved data will result in improved models. It is feasible to attain a significant performance improvement even with the utilization of ordinary architectures by improving the data quality, diversity, and representativeness. In this view, quality data are not a secondary consideration but among the most important determinants of model accuracy and reliability.

The basic practices of DC-AI are systematic data cleaning, error correction, label standardization, and conscientious data augmentation selection to ensure diversity and eliminate bias. It also focuses on analytical error and examination of datasets that replicate real-world conditions which helps in further generalization of the model. This type of emphasis is particularly needed when the data are, by nature, noisy, incomplete, or biased. An example would be that medical data may not generalize to some groups of patients, whereas in industrial sensors, the sensor data may be inaccurate or unstable. Having a focus on data quality, DC-AI allows AI models to generate more reliable and valid outcomes without adhering to sophisticated architectures. In contrast to technological developments, DC-AI promotes the application of better organizational and ethical practices, including effective systems of data governance, quality automated systems, and secure inter-institutional data-sharing procedures. These actions promote transparency, reproducibility, and cooperation in AI research and application. Finally, DC-AI also reinforces data as the major pillar of AI. It is also performance-enhancing, improving fairness, reliability, and applicability to the real world. Interestingly, data-centric and model-centric models are not conflicting philosophies but are complementary to each another. Advanced models are supported by quality data, and sometimes sophisticated algorithms help draw additional information out of quality data. This balance is becoming more accepted in modern AI practice as it is understood that the two can only achieve meaningful progress when they work together [42,45,46].

4.6 AI APPLICATIONS OF AI IN SURFACE ENGINEERING

The essential idea of AI is radically changing the field of surface engineering, allowing it to develop new opportunities in many key aspects of it, such as predicting material behavior, optimizing the manufacturing process, improving the quality control mechanism by detecting defects, and improving multiphysics modeling. All these developments are based on different AI methods, specifically ML and DL that are subdivisions of AI that learn using data to discover patterns and make predictions.

4.6.1 Predictive Modeling

Predictive modeling in surface engineering is a branch of surface engineering that employs ML, AI, or statistical methods to predict the behavior of surface-modified materials under different conditions. These models do not require much experimen tal testing; instead, they learn using existing experimental, simulation, or theoretical data to capture the relationships between processing parameters and material responses. Through this, they can make proper predictions of surface behavior, thereby saving time and cost, as well as being used to optimize coatings, surface treatments, and materials design to meet specific performance requirements. The different parameter predictions are discussed below.

4.6.1.1 Surface Roughness Prediction Using the AI Framework

Surface characterization is the analysis of the texture, roughness, microstructure, chemical composition, and mechanical properties in order to comprehend and improve the material performance. Surface roughness is one of them, which has a direct impact on functionality, efficiency, and durability. Industry 4.0 has shown the growing use of AI-based methods to predict and estimate surface roughness under data of machining parameters, sensor data, or surface images with reference measurements of profilometers (as shown in Figure 4.9). The AI-based system allows achieving surface quality evaluation on a high level but consumes less manual work and time [47].

The surface itself is frequently a diagnostic marker of the quality of the processes and the behavior of the materials, which can be viewed as the signature of its manufacturing history. Jiang et al. [48] proposed a feature spectrum to classify

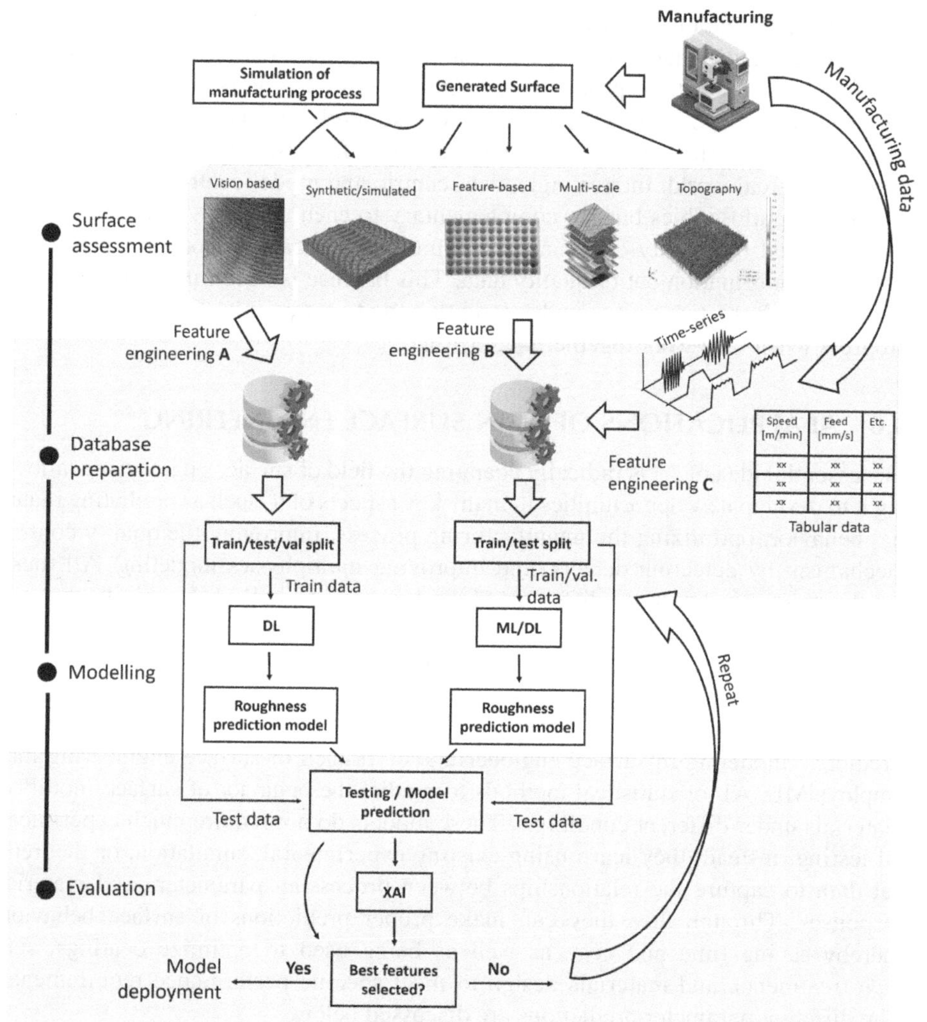

FIGURE 4.9 AI-based surface roughness assessment workflow [47]. (Credit statement: Creative Commons International Attribution 4.0.)

surfaces from stochastic to deterministic types, improving topography characterization beyond conventional International Organization for Standardization roughness parameters. Multiscale analysis also provides more insight by breaking down the surface features into roughness, waviness, and form by frequency bands and is useful in tracing changes in production. Methods such as Gaussian filtering and Fourier and wavelet transforms enable the study of topographical information at different scales in detail [49]. Modern vision techniques, for instance, machine vision with edge detection or laser speckle analysis, offer rapid and non-destructive approaches compared to the conventional stylus method. Such systems incorporate pre-processing measures like noise filtering, normalization, and data augmentation to guarantee reliable surface assessment [50].

4.6.1.2 Wear Prediction Using Support Vector Regression Method

Support vector regression (SVR) is an ML algorithm that is based on support vector machines, and it is aimed at the prediction of continuous values. It operates under the principle of fitting a ε-insensitive function, that is, any small errors do not matter, but any greater deviation is penalized. The balance enables the model to remain resistant to noise without compromising the accuracy. With the help of various kernel functions, i.e., linear, polynomial, or radial basis function (RBF), SVR is able to capture linear as well as nonlinear relationships. It is a well-known predictive method with good ability to generalize, and ε, C, and the type of kernel need to be carefully adjusted, but even with that, it is still very popular in finance, engineering, and bioinformatics due to its good predictive ability.

In this chapter, SVR was applied in predicting wear by taking input variables of load and sliding distance. The hyperparameters regularization factor (C), tolerance margin (ε), and the type of kernel (RBF or linear) were optimized using grid search with cross-validation to achieve optimal performance. After the optimum configuration was found, the model was trained on a normalized dataset and cross-validated with four-fold cross-validation. The consistency of the model was measured by the R^2 of each fold, and overall accuracy by the R^2, root mean square error (RMSE), and the mean absolute percentage error (MAPE).

The SVR model, which had been optimized, provided an impressive level of predictive accuracy with an R^2 of 0.9995 over the training set and 0.9999 over the testing set. The error values were considerably low; RMSE was 0.0616, and MAPE was only 0.10%, which shows that there was a lot of agreement between the predicted and experimental outcomes. The trained SVR model was also used in a new load and sliding distance condition set to further confirm its reliability. Table 4.2 indicates that the predicted values of wear were similar to the values of wear in the experiment, with slight differences in percentage. These findings firmly show the strength and usefulness of the SVR model in the accurate prediction and description of wear behavior in different circumstances [51,52].

4.6.1.3 Coefficient of Friction Prediction Using Gaussian Process Regression Method

Gaussian Process Regression (GPR), a probabilistic and nonparametric method, is best applicable to small- to medium-sized datasets because it not only gives

TABLE 4.2
Comparison of Experimental and SVR-Predicted Wear Values with Percentage Differences [52] (Credit statement: Creative Commons International Attribution 4.0)

Load (N)	Sliding Distance (m)	Experimental wear (μm)	Predicted wear (μm)	Difference (%)
20	3,000	47.83	47.7317	0.2055
30	14,000	61.4	61.3610	0.0636
20	14,000	57.64	57.6534	0.0232

predictions but also estimates of uncertainty [53]. It is also strong in its ability to capture complicated and nonlinear patterns with flexible kernels such as RBF, Matern, and rational quadratic. One of the strengths of GPR is that it tries to automatically tune hyperparameters by maximizing the marginal likelihood, which guarantees good model performance. In comparison, GPR is not highly susceptible to overfitting, and it is especially successful in approximating functions, optimization based on the Bayesian approach, and when noisy data are available [54]. Its primary weakness is that it is costly to compute, and its computational complexity is cubic to the size of the dataset ($O(n^3)$), hence impractical in large datasets. GPR is still widely used in scientific studies, materials science, and engineering simulations, where the degree of accuracy is very high and prediction error is provided in the form of a confidence interval [55].

The coefficient of friction (CoF) predictive model was also developed using GPR. The settings that were found to be the best are as follows:

$$\begin{aligned} \text{Kernel} &= 1.572 \times \text{Rational Quadratic}\left(\alpha = 3.25, \text{length\_scale} = 1.12\right) \\ &\quad + \text{White kernel}\left(\text{noise\_level} = 3.99 \times 10^{-5}\right) \end{aligned} \tag{4.1}$$

The regularization parameter was fixed to 1×10^{-7}, and the optimizer was restarted 20 times (n_restarts_optimizer=20) to prevent nonconvergence. To validate the model, 4-fold cross-validation was conducted, resulting in a total of 144 fits ($3 \times 4 \times 3 \times 4 = 144$) to systematically explore the hyperparameter space. The highest values of the corresponding R^2 of every fold show high levels of predictive performance. The trained model got an R^2 equal to 1.0 on the training set and equal to 0.9999 on the test set. Moreover, the errors in prediction were rather low; the RMSE is 0.0003, and the MAPE is 0.23%. These results are clear indications of the strength of the developed model. In order to compare the performance further, the optimized GPR model was applied to forecast the CoF in another set of load and sliding distance.

As presented in Table 4.3, the values of the predicted CoF were in great agreement with the experimental values, with only minor percentage deviations. This is a convincing analogy validating the precision and soundness of the GPR model on CoF forecasting.

TABLE 4.3
Comparison of Experimental and GPR-Predicted CoF Values with Percentage Differences [52] (Credit statement: Creative Commons International Attribution 4.0.)

Load (*N*)	Sliding Distance (m)	Experimental CoF	Predicted CoF	Difference (%)
20	3,000	0.1054	0.1056	0.19
30	14,000	0.1165	0.1168	0.26
20	14,000	0.1746	0.1749	0.17

4.6.1.4 Thickness Prediction Using ANN

In quality of surface engineering, correct determination and prediction of the coating layer are required since the efficiency and stability of the coating are often directly proportional to the thickness of the layer deposited. To achieve accurate predictions, researchers have turned to increasingly sophisticated means of calculation, and ANNs have become one of the most useful and effective tools in recent years. Numerous studies have reported that ANNs can be used with good results in nonlinear modeling of process parameters and coating depth.

As an example, Barletta et al. [56] modeled the working electrostatic fluidized-bed (EFB) coating process to predict the change in thickness as a function of time, applied voltage, and airflow. They used a MLP that was trained by a genetic algorithm and a backpropagation algorithm, which greatly enhanced the predictive performance and the ability to determine the best conditions of the EFB process. Similarly, Khalaj et al. [57] have applied a feed-forward backpropagation neural network to calculate the thickness of chromium carbonitride coating over pre-nitrided steels by considering 17 process parameters. Their model scored a very high correlation coefficient ($R^2 = 0.9948$), which showed that ANNs have great potential for reflecting nonlinear relationships among the variables.

Varol et al. [58] applied ANN modeling in another study to determine the effect of milling speed, milling time, and particle size on the thickness of Fe–Al intermetallic coatings. These findings demonstrated that the model is reliable as a predictor and optimizer of processes because longer milling times gave higher coating thickness and high predictive accuracy with an MAPE of 7.466%. Similarly, Shukla et al. [59] designed an ANN model to estimate the zinc coating thickness during hot-dip galvanizing, which uses six input variables and has a 6-(9-6-1)-1 network topology. Their model was very accurate, with the highest prediction of 10%.

4.6.1.5 Hardness Prediction Using ANNs

One of the main methods of increasing the wear resistance of a material is to make it hard, and one of the most effective methods of doing so is to coat it with an appropriate deposition technique. To analyze and optimize the performance of coating, it is important to accurately predict the resulting hardness. However, coating processes are complex and nonlinear, which require sophisticated computational tools to model the complex relationships between process parameters and material properties. ANNs are one of these techniques, and they have been especially effective when it comes to predicting coating hardness with very high accuracy.

Yazdi et al. [60] used an ANN model to examine the CVD/PVD coating of titanium thin films. They reached a high prediction accuracy of 98.3 using an MLP architecture. It was found that accurate prediction of hardness may be critical in achieving corrosion/erosion resistance, which directly impacts the working life of cutting tools. Similarly, Reyes-Mojena et al. [61] used the ANN model to predict the erosive wear rate of HVOF and flame spray/flex cords, which had a high-velocity hard coating. They discovered that microhardness and fracture toughness are the most critical factors to erosion rate, as well as porosity. The coatings that were tested included WC–CoCr cermet coatings, which showed good erosion resistance.

4.6.1.6 Microstructure Prediction Using ANNs

The microstructural properties of the coatings are important indicators that should be used to verify the efficiency of the deposition procedures. The most critical factors are phase composition, unmolten particle content, and porosity, which must be managed appropriately to achieve optimal coating performance. Therefore, their prediction is vital, and ANNs have been very effective in this regard. ANNs are capable of identifying complex nonlinear relationships among process parameters and material properties.

For example, Guessasma et al. [62] applied ANN modeling to examine the microstructural properties in atmospheric plasma-sprayed (APS) alumina–titania coatings. The process parameters that were taken into account included arc current, hydrogen flow rate, and total plasma rate and how these parameters influenced the phase composition, porosity, and unmolten particle fraction. Results indicated that changes in these parameters significantly impacted the content of the alumina and titania phase. Specifically, the alumina content increased with higher process parameters, whereas other microstructural features showed opposite trends. Guessasma et al. [63] used ANNs to predict the values of porosity and to determine the ideal APS conditions in relation to arc current, hydrogen ratio, plasma gas flow, carrier gas flow, and powder feed rate. Their findings showed that a rise in arc current and higher content of hydrogen minimized porosity, and higher powder feed rates tended to enhance the porosity. They also developed parabolic dependencies of carrier gas and plasma flows, and arc current and hydrogen ratio served as the control variables.

Likewise, Wang et al. [64] observed a nonlinear relationship between porosity and spraying parameters in plasma spraying using a backpropagation neural network. They were able to predict results consistent with the experimental values, and this demonstrated how the ANNs may be used to optimize the process and control quality. All these studies describe the importance of ANN-based modeling in the prediction and analysis of microstructures of coatings, and hence it is a useful tool that can improve the processes of surface engineering.

4.6.1.7 Grain Size Prediction Using ANN

The accurate prediction of grain size is very important in enhancing the surface gloss, mechanical strength, and general material behavior. ANNs are one of the modern computational tools that have demonstrated similarity in predicting the occurrence of process parameters in grain size across various coating processes.

Rashidi et al. [65] employed a feed-forward MLP ANN to examine how current density, the concentration of saccharin, and bath temperature determine the grain size of nanocrystalline nickel coating. The results also indicated that it was highly influenced by current density and the concentration of saccharin, whereas it was not as much influenced by bath temperature, and therefore, ANNs provide a powerful and accurate method of predicting the grain size of nanocrystalline coatings. Jarrah et al. [66] also used an ANN to predict the grain size of TiN coatings in a related work, where they used parameters such as nitrogen gas pressure, argon gas pressure, and turntable speed. Their network topology provided a high prediction accuracy of 96.02%, which highlights the ability of ANNs to optimize coating parameters and maximize the service life of the coated tools. In another study, Dobrzański et al. [67] used ANN modeling to both PVD and CVD coatings. They determined that variations in grain size had a significant influence on blade stability, with small grains improving durability and performance.

4.6.1.8 Oxidation Behavior Prediction Using ANN

Oxidation reactions accelerate corrosion and drastically lower the service life of components. To counter this, protective coatings are commonly employed to control high-temperature oxidation. Knowledge of their oxidation behavior is essential for assessing long-term performance. In recent years, ANNs have proven beneficial as tools for anticipating and examining oxidation-related phenomena in coatings.

Rashidi et al. [68] demonstrated the strong predictive capability of ANN models in the assessment of oxidation behavior for aluminized nanocrystalline coatings. Similarly, Godavarty et al. [69] employed ANN modeling in investigating mechanisms and kinetics of oxidation in laser surface-engineered composite boride coatings. Danaher et al. [70] also applied ANNs in examining high-power impulse magnetron sputtering and PVD coatings and thus effectively identifying systems with enhanced corrosion resistance.

Beyond oxidation research, ANNs have also been applied to predict a broad variety of coating properties, including thermal barrier behavior as well as optical properties of thin films, erosion resistance, texture coefficients, opacity of films, crack propagation in pharmaceutical coatings, and even magnetic properties, thermal conductivity, and diffusivity. These varied applications testify to the versatility of ANNs in modeling the complex interrelations that govern the coating performance and degradation behavior.

4.6.2 Parameter Optimization

The optimization of parameters is essential in achieving better performance in the process of surface treatment and coating. Thermal spraying, PVD, CVD, and laser surface processes are processes that rely on a set of parameters such as substrate temperature, rate of deposition, energy, and time of treatment. All these parameters affect the quality of coatings and surface properties. In the past, selection of these parameters was based on trial-and-error procedures, which were not efficient or even time-consuming.

With the introduction of AI and ML, this process has been transformed so that more precise and data-driven optimization is possible. These models can be programmed to identify complex, nonlinear relationships among process variables and surface responses, enabling researchers to determine the optimal conditions more precisely. For example, optimization of particle velocity and temperature for thermal spraying or laser power and scan speed for laser treatments can significantly enhance coating density, adhesion, and hardness. This model-based optimization is not only achieving high-quality and durable coatings but also minimizing the amount of extensive experiments, thus improving overall process efficiency.

4.6.2.1 Optimizing Coating Parameters for Tool Hardness via ANNs

Titanium-based and ceramic coatings are highly desired for their superior hardness and thermal stability, thereby being best suited for dies and cutting tools in metal-forming processes. Yet the quality of these coatings depends strongly on the accurate control of deposition parameters. To evaluate the hardness of thin titanium coatings, testing methods such as X-ray diffraction, ellipsometry, ultrasonic testing, and eddy current sensing are popular. While such tests give meaningful information on the coating structure and thickness, they tend to fail in measuring true maximum hardness due to the involved multiscale interactions between deposition parameters.

In order to solve these problems, a comprehensive experimental study was conducted on 192 high-speed steel tools with TiN, TiC, and TiC–N coatings under controlled conditions. Both CVD and PVD methods were used to apply the coatings with precisely controlled parameters such as ion bombardment time (20 minutes), substrate temperature (150°C), coating time (30 minutes), and voltage (65 V). Circular substrates were used for CVD, and rectangular ones, both polished and unpolished, were used to evaluate the influence of surface finish on coating performance. Microhardness testing revealed that polished substrates exhibited much greater hardness, with that of TiC–N coatings reaching as high as 900 Vickers, compared to 810 Vickers for unpolished surfaces.

Since coating processes are nonlinear in nature, ANNs, specifically the MLP, were proposed to model and predict coating hardness, as shown in Figure 4.10. The model utilized six input parameters, namely, ion bombardment time, substrate roughness, coating material, substrate temperature, working pressure, and chamber pressure. The model had two hidden layers consisting of 12 neurons and a single output neuron for hardness [60].

Training was conducted on 156 experimental data sets, and the remaining data were reserved for validation. The ANN had a correlation coefficient of 0.982 with an extremely low RMS error of 0.035, indicating superior predictability. This trained ANN model was further employed to find optimal coating parameters for maximum hardness. There were considerable improvements reported in different coating materials. For TiC coatings, the correlation accuracy improved by approximately 2%, and the RMS error decreased by 40% following optimization.

In order to test the ANN optimization, experiments were conducted on the conditions most likely to be predicted by the optimizer, and the results were quite close to the those of the model. Hardness improved by up to 2.3%, with that of TiN improving from 650 to 665 Vickers and of TiC–N from 900 to 915 Vickers. These improvements may seem insignificant, but in industrial uses, they are important, improving wear life, tool life, and reducing maintenance. The study demonstrates the sensitivity of the substrate roughness, ion bombardment duration, pressure in the chamber, and temperature on the hardness of the coatings. In all the experiments, polished substrates were found to be better when compared to unpolished ones, and so surface preparation/pretreatment is important. Overall, ANN modeling effectively captured the connections between process parameters and hardness, enabling accurate predictions and optimal parameter selection.

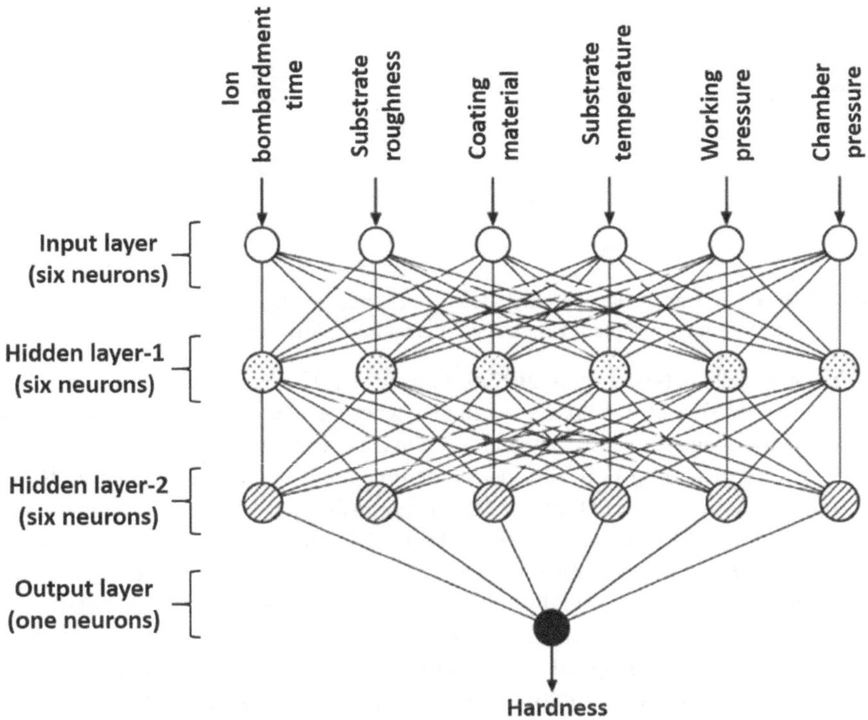

FIGURE 4.10 Suggested perceptron-based neural networks for the study [60]. (Credit statement: Reproduced with permission from Elsevier © 2011.)

4.6.3 Defect Detection

Defects in coatings and surface-treated parts, such as cracks, pores, delamination, and roughness variation, can significantly impact the performance, reliability, and lifespan of the components. This type of defect is typically caused by changes in the properties of materials, the processes of their applications, or process conditions that affect wear resistance, corrosion protection, and mechanical strength. It is necessary to detect failures at early stages to prevent failures, minimize maintenance costs, and preserve quality. Traditional inspection techniques such as the use of the human eye and nondestructive tests are time-consuming, labor-intensive, and subject to limitations in detecting tiny defects, especially on complex surfaces. To address these issues, image-based analysis using DL has been used. High-resolution images of the surfaces allow for autonomous detection and classification of defects with high accuracy, enabling real-time monitoring, quality control with consistency, and process optimization while reducing human intervention and improving efficiency.

4.6.3.1 Deep Learning-based Automatic Defect Detection

To meet the challenge of traditional inspection methods, DL is introduced employing the You Only Look Once version 8 (YOLOv8) algorithm to identify surface, nick, and edge defects on turbine and compressor blades, as outlined in Figure 4.11 [71].

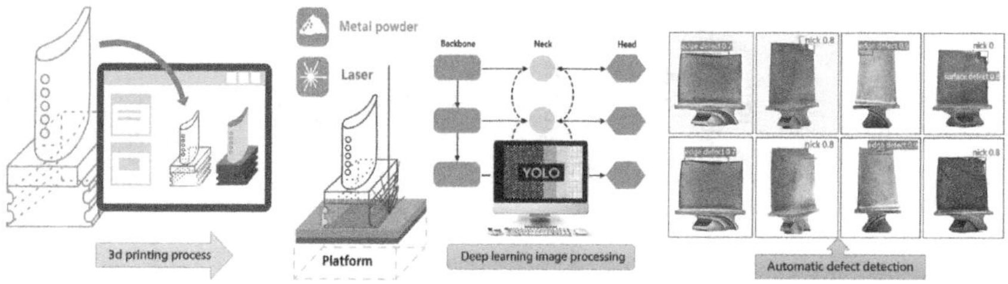

FIGURE 4.11 Surface, edge, and nick defects detected during turbine blade manufacturing [71]. (Credit statement: Creative Commons International Attribution 4.0.)

The approach is based on a custom dataset of 302 augmented images that were derived from 151 high-resolution originals, where every defect has been manually labeled using bounding boxes. The dataset consists of images of turbine and compressor blades, representing both defect-free samples and defective samples with three kinds of defects, as shown in Figure 4.12. Nicks, surface defects, and edge-related flaws in the dataset are illustrated in Fig. 4.12(a), Fig. 4.12(b), and Fig. 4.12(c), respectively.

The dataset was divided into training (67%), validation (18%), and testing (15%) sets. Images were preprocessed by standardized pixel values and resized to 640×640 pixels in RGB format for uniformity and enhanced model performance. YOLOv8 was used because of its speed and accuracy in detection. The model was trained with backpropagation over multiple epochs, repeatedly changing weights to reduce the loss function. The performance was calculated using matrices such as precision, recall, F1-score, mean average precision (mAP), and intersection over union (IoU). The model had 99.5% accuracy, 94% precision, 98% recall, and an F1-score of 96% on the validation set, with a detection rate of 150 frames per second.

Results obtained were visualized as confusion matrices, which ensured that the YOLOv8 model has accurately identified nicks, edge defects, and surface irregularities. It is also quicker, more dependable, and more economical than traditional methods. Although it is limited by the size of the datasets and computational requirements, these issues can be solved with the help of transfer learning and data expansion. The use of YOLOv8 in the inspection of turbine and compressor blades improves their precision, speed, and flexibility, which enhances the safety and reliability of aerospace [71].

4.7 CHALLENGES AND LIMITATIONS OF AI IN SURFACE ENGINEERING

AI has enormous opportunities in improving surface engineering, and its effective and efficient use requires some difficulties to be surmounted. The main practical constraints that are experienced when applying AI to detect defects, optimize processes, and characterize surfaces are described below.

a. **Availability and quality of data:** AI models use large and high-quality datasets. Labeled defect data, especially of rare or shallow surface defects, are hard to come by. Unregulated changes in imaging conditions and sensor properties are likely to bring noise and deteriorate the prediction quality.

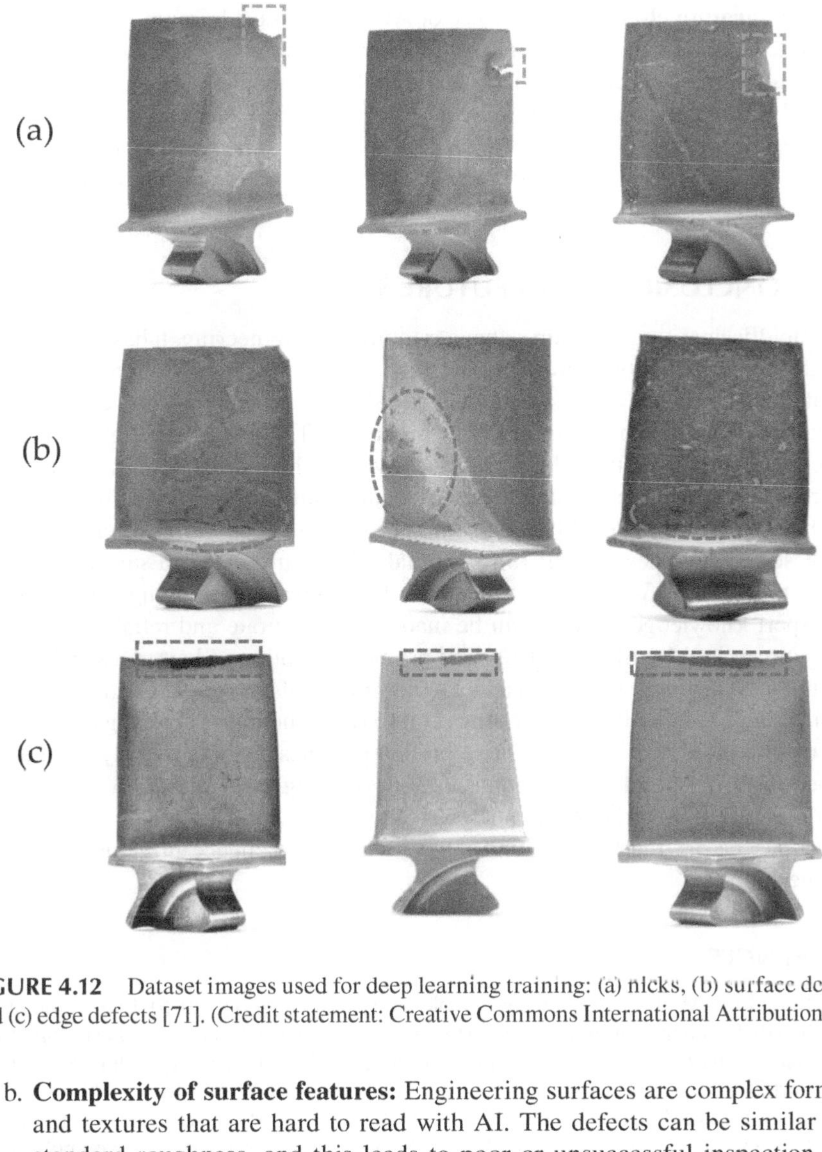

FIGURE 4.12 Dataset images used for deep learning training: (a) nicks, (b) surface defects, and (c) edge defects [71]. (Credit statement: Creative Commons International Attribution 4.0.).

b. **Complexity of surface features:** Engineering surfaces are complex forms and textures that are hard to read with AI. The defects can be similar to standard roughness, and this leads to poor or unsuccessful inspection of defects.
c. **Generalization across processes:** AI algorithms that have been trained on one type of material or coating can have great difficulty with new surfaces/processes without re-training, decreasing their flexibility/versatility.
d. **Computational demands:** DL models require a lot of computational resources, which makes it impossible to achieve real-time defect detection in an industrial setting.
e. **Interpretability and trust:** The majority of AI models can be considered black boxes. They make unfounded decisions; thus, human beings find it very hard to rely on them in safety-critical applications.

f. **Integration challenges:** The use of AI systems must be supported with the relevant amount of hardware and software, training of the team, and taking into account the current quality and regulation standards.
g. **Maintenance and adaptability:** AI models need to be changed after a certain period of time when new defects/issues emerge or the processes change. Unless they are retrained, their accuracy and reliability may decline with time.

4.8 CONCLUSIONS AND FUTURE SCOPE

AI is rapidly emerging as a game-changer in surface engineering. It has the capability to make defect detection more accurate, enhance coating processes, and improve the overall surface finish. AI-based models can shorten inspection time, aid predictive maintenance, and enable fine-tuning of processes. This results in higher efficiency and consistency of industrial processes. However, challenges such as limited data, difference in surface characteristics, computationally intensive requirements, and the "black box" aspect of AI models must be addressed carefully in order to leverage these technologies to their full potential. The future of AI in surface engineering is encouraging. With the combination of AI, advanced imaging, sensor fusion, and expert knowledge, models can be made more accurate and reliable. Increasing transparency in AI will instill trust and promote adoption in key industrial sectors. Continuous learning systems can adapt to new materials, processes, and defect types, keeping models up to date. Integrating AI into real-time manufacturing could enable smarter and more responsive/sensitive production lines, leading to fully autonomous surface inspection and quality control. With ongoing research and a close collaboration between material scientists, engineers, and AI professionals, AI is poised to become the driving force behind innovation and excellence in industrial surface engineering.

REFERENCES

1. Ramezani, M., Pasang, T., Chen, Z., Neitzert, T., and Au, D. (2015). Evaluation of carbon diffusion in heat treatment of H13 tool steel under different atmospheric conditions. *Journal of Materials Research and Technology*, *4*(2), 114–125. https://doi.org/10.1016/j.jmrt.2014.10.014
2. Giurlani, W., Zangari, G., Gambinossi, F., Passaponti, M., Salvietti, E., Di Benedetto, F., Caporali, S., and Innocenti, M. (2018). Electroplating for decorative applications: Recent trends in research and development. *Coatings*, *8*(8), 260. https://doi.org/10.3390/coatings8080260
3. Awang, M., Khalili, A. A., and Pedapati, S. R. (2019). A review: Thin protective coating for wear protection in high-temperature application. *Metals*, *10*(1), 42. https://doi.org/10.3390/met10010042
4. Riquelme, A., and Rodrigo, P. (2021). An introduction on the laser cladding coatings on magnesium alloys. *Metals*, *11*(12), 1993. https://doi.org/10.3390/met11121993
5. Fotovvati, B., Namdari, N., and Dehghanghadikolaei, A. (2019). On coating techniques for surface protection: A review. *Journal of Manufacturing and Materials Processing*, *3*(1), 28. https://doi.org/10.3390/jmmp3010028

6. Bonu, V., and Barshilia, H. C. (2022). High-temperature solid particle erosion of aerospace components: Its mitigation using advanced nanostructured coating technologies. *Coatings*, *12*(12), 1979. https://doi.org/10.3390/coatings12121979
7. Chami, A., Benabbou, R., Taleb, M., Rais, Z., and El haji, M. (2021). Analysis of survey data on corrosion in the automotive industry. *Materials Today: Proceedings*, *45*, 7636–7642. https://doi.org/10.1016/j.matpr.2021.03.113
8. Nikolova, M. P., and Apostolova, M. D. (2023). Advances in multifunctional bioactive coatings for metallic bone implants. *Materials*, *16*(1), 183. https://doi.org/10.3390/ma16010183
9. Wang, C., Jiu, H., Zhang, L., Xu, R., Che, S., Guo, Z., Han, Y., Ma, J., Li, H., and Guo, F. (2023). Carbon coating hollow NiS nanoparticles anchored on Ti_3C_2 nanosheet as high-performance anode for lithium-ion batteries. *Journal of Alloys and Compounds*, *956*, 170332. https://doi.org/10.1016/j.jallcom.2023.170332
10. Su, Y. L., Kao, W. H., and Chang, G. Y. (2023). Tribological, anti-corrosion, and electrical conductivity properties of CrC_x coatings deposited on stainless steel 316L and used as metal bipolar plates for fuel cells. *Journal of Materials Engineering and Performance*, *32*, 3739–3754. https://doi.org/10.1007/s11665-022-07333-y
11. Wang, J., Wang, L., Su, X., Xiao, R., and Cheng, H. (2022). Polarization maintaining fiber temperature and stress gradient sensitization sensor based on semiconductor-metal–polymer three-layer film coating. *ACS Applied Materials & Interfaces*, *14*(17), 20053–20061. https://doi.org/10.1021/acsami.2c03102
12. Sha, W., Guo, Y., Yuan, Q., Tang, S., Zhang, X., Lu, S., Guo, X., Cao, Y-C., and Cheng, S. (2020). Artificial intelligence to power the future of materials science and engineering. *Advanced Intelligent Systems*, *2*(4), 1900143. https://doi.org/10.1002/aisy.201900143
13. Tapeh, A. T. G., and Naser, M. Z. (2023). Artificial intelligence, machine learning, and deep learning in structural engineering: A scientometrics review of trends and best practices. *Archives of Computational Methods in Engineering*, *30*(1), 115–159. https://doi.org/10.1007/s11831-022-09793-w
14. Butt, M. A. (2022). Thin-film coating methods: A successful marriage of high-quality and cost-effectiveness—A brief exploration. *Coatings*, *12*(8), 1115. https://doi.org/10.3390/coatings12081115
15. Schalk, N., Tkadletz, M., and Mitterer, C. (2022). Hard coatings for cutting applications: Physical vs. chemical vapor deposition and future challenges for the coatings community. *Surface and Coatings Technology*, *429*, 127949. https://doi.org/10.1016/j.surfcoat.2021.127949
16. Ramezani, M., Mohd Ripin, Z., Pasang, T., and Jiang, C-P. (2023). Surface engineering of metals: Techniques, characterizations and applications. *Metals*, *13*(7), 1299. https://doi.org/10.3390/met13071299
17. Lee, S., Lee, J., Kim, W., and Hwang, N-M. (2020). Plasma etching behavior of YOF coating deposited by suspension plasma spraying in inductively coupled CHF3/Ar plasma. *Coatings*, *10*(11), 1023. https://doi.org/10.3390/coatings10111023
18. Wang, Q., Zhou, K., Zhao, S., Yang, W., Zhang, H., Yan, W., Huang, Y., and Yuan, G. (2021). Metal-assisted chemical etching for anisotropic deep trenching of GaN array. *Nanomaterials*, *11*(12), 3179. https://doi.org/10.3390/nano11123179
19. Aizawa, T., Rsadi, I., and Yunata, E. E. (2022). High density RF-DC plasma nitriding under optimized conditions by plasma-diagnosis. *Applied Sciences*, *12*(8), 3706. https://doi.org/10.3390/app12083706
20. Wieland, F., Bruch, R., Bergmann, M., Partel, S., Urban, G. A., and Dincer, C. (2020). Enhanced protein immobilization on polymers—A plasma surface activation study. *Polymers*, *12*(1), 104. https://doi.org/10.3390/polym12010104
21. Teranishi, N., Fuse, G., and Sugitani, M. (2018). A review of ion implantation technology for image sensors. *Sensors*, *18*(7), 2358. https://doi.org/10.3390/s18072358

22. Maawad, E., Brokmeier, H. G., and Wagner, L. (2010). Texture gradients in shot peened Ti-2.5 Cu. *Solid State Phenomena*, *160*, 141–146. https://doi.org/10.4028/www.scientific.net/SSP.160.141
23. Ryynänen, T., Toivanen, M., Salminen, T., Ylä-Outinen, L., Narkilahti, S., and Lekkala, J. (2018). Ion beam assisted E-beam deposited TiN microelectrodes—Applied to neuronal cell culture medium evaluation. *Frontiers in Neuroscience*, *12*, 882. https://doi.org/10.3389/fnins.2018.00882
24. Sun, L., Huang, L., Wu, P., Huang, R., Fang, N., Xu, F., and Xu, K. (2023). Progress on the effect and mechanism of ultrasonic impact treatment on additive manufactured metal fabrications. *Crystals*, *13*(7), 995. https://doi.org/10.3390/cryst13070995
25. Tandon, D., Li, H., Pan, Z., Yu, D., and Pang, W. (2023). A review on hardfacing, process variables, challenges, and future works. *Metals*, *13*(9), 1512. https://doi.org/10.3390/met13091512
26. Oliveira, E., Rodrigues, M., Pereira, J. P., Lopes, A. M., Mestric, I. I., and Bjelogrlic, S. (2024). Unlabeled learning algorithms and operations: Overview and future trends in defense sector. *Artificial Intelligence Review*, *57*(3), 66. https://doi.org/10.1007/s10462-023-10692-0
27. Ackoff, R. L. (1989). From data to wisdom. *Journal of Applied Systems Analysis*, *16*(1), 3–9.
28. Sarker, I. H., Kayes, A. S. M., Badsha, S., Alqahtani, H., Watters, P., and Ng, A. (2020). Cybersecurity data science: An overview from machine learning perspective. *Journal of Big Data*, *7*(1), 41. https://doi.org/10.1186/s40537-020-00318-5
29. Jhaveri, R. H., Ramani, S. V., Srivastava, G., Gadekallu, T. R., and Aggarwal, V. (2021). Fault-resilience for bandwidth management in industrial software-defined networks. *IEEE Transactions on Network Science and Engineering*, *8*(4), 3129–3139. https://doi.org/10.1109/TNSE.2021.3104499
30. Kaelbling, L. P., Littman, M. L., and Moore, A. W. (1996). Reinforcement learning: A survey. *Journal of Artificial Intelligence Research*, *4*, 237–285. https://doi.org/10.1613/jair.301
31. Yunita, A., Pratama, M. I., Almuzakki, M. Z., Ramadhan, H., Akhir, E. A. P., Mansur, A. B. F., and Basori, A. H. (2025). Performance analysis of neural network architectures for time series forecasting: A comparative study of RNN, LSTM, GRU, and hybrid models. *MethodsX*, *15*, 103462. https://doi.org/10.1016/j.mex.2025.103462
32. Taye, M. M. (2023). Understanding of machine learning with deep learning: Architectures, workflow, applications and future directions. *Computers*, *12*(5), 91. https://doi.org/10.3390/computers12050091
33. Fan, X., and Markram, H. (2019). A brief history of simulation neuroscience. *Frontiers in Neuroinformatics*, *13*, 32. https://doi.org/10.3389/fninf.2019.00032
34. Zadeh, S. S., Birgani, S. A., Khorshidi, M., and Kooban, F. (2023). Concrete surface crack detection with convolutional-based deep learning models. *International Journal of Novel Research in Civil Structural and Earth Science*, *10*(3), 25–35. https://doi.org/10.5281/zenodo.10061654
35. Suputra, P. H., Sensusiati, A. D., Artaria, M. D., Verkerke, G. J., Yuniarno, E. M., and Purnama, I. K. E. (2022). Automatic 3D cranial landmark positioning based on surface curvature feature using machine learning. *Knowledge Engineering and Data Science*, *5*(1), 27–40. https://doi.org/10.17977/um018v5i12022p27-40
36. Zhu, Z., Ning, W., Niu, X., Wang, Q., Shi, R., and Zhao, Y. (2023). Designing high elastic modulus magnesium-based composite materials via machine learning approach. *Materials Today Communications*, *37*, 107249. https://doi.org/10.1016/j.mtcomm.2023.107249
37. Huang, Y., Chen, C-H., and Huang, C-J. (2019). Motor fault detection and feature extraction using RNN-based variational autoencoder. *IEEE Access*, *7*, 139086–139096. https://doi.org/10.1109/access.2019.2940769

38. Islam, M. M., Fazle, A. B., and Prodhan, R. K. (2022). AI-based smart coating degradation detection for offshore structures. *American Journal of Advanced Technology and Engineering Solutions*, *2*(04), 01–34. https://doi.org/10.63125/1mn6bm51
39. Menegazzo, J., and Von Wangenheim, A. (2021). Road surface type classification based on inertial sensors and machine learning: A comparison between classical and deep machine learning approaches for multi-contextual real-world scenarios. *Computing*, *103*(10), 2143–2170. https://doi.org/10.1007/s00607-021-00914-0
40. Zhong, G., Wang, L-N., Ling, X., and Dong, J. (2016). An overview on data representation learning: From traditional feature learning to recent deep learning. *The Journal of Finance and Data Science*, *2*(4), 265–278. https://doi.org/10.1016/j.jfds.2017.05.001
41. Sowmya, A., and Trinder, J. (2000). Modelling and representation issues in automated feature extraction from aerial and satellite images. *ISPRS Journal of Photogrammetry and Remote Sensing*, *55*(1), 34–47. https://doi.org/10.1016/S0924-2716(99)00040-4
42. Jakubik, J., Vössing, M., Kühl, N., Walk, J., and Satzger, G. (2024). Data-centric artificial intelligence. *Business & Information Systems Engineering*, *66*(4), 507–515. https://doi.org/10.1007/s12599-024-00857-8
43. Jarrahi, M. H., Memariani, A., and Guha, S. (2023). The principles of data-centric AI. *Communications of the ACM*, *66*(8), 84–92. https://doi.org/10.1145/3571724
44. Whang, S. E., Roh, Y., Song, H., and Lee, J-G. (2023). Data collection and quality challenges in deep learning: A data-centric AI perspective. *The VLDB Journal*, *32*, 791–813. https://doi.org/10.1007/s00778-022-00775-9
45. Hamid, O. H. (2023). Data-centric and model-centric AI: Twin drivers of compact and robust industry 4.0 solutions. *Applied Sciences*, *13*(5), 2753. https://doi.org/10.3390/app13052753
46. Majeed, A., and Hwang, S. O. (2024). Towards unlocking the hidden potentials of the data-centric AI paradigm in the modern era. *Applied System Innovation*, *7*(4), 54. https://doi.org/10.3390/asi7040054
47. Ghosh, S., Knoblauch, R., El Mansori, M., and Corleto, C. (2024). Towards AI driven surface roughness evaluation in manufacturing: A prospective study. *Journal of Intelligent Manufacturing*, *36*, 4519–4548. https://doi.org/10.1007/s10845-024-02493-1
48. Jiang, X., Senin, N., Scott, P. J., and Blateyron, F. (2021). Feature-based characterisation of surface topography and its application. *CIRP Annals*, *70*(2), 681–702. https://doi.org/10.1016/j.cirp.2021.05.001
49. Jiang, X., and Blunt, L. (2004). Third generation wavelet for the extraction of morphological features from micro and nano scalar surfaces. *Wear*, *257*(12), 1235–1240. https://doi.org/10.1016/j.wear.2004.06.006
50. Kumar, B. M., and Ratnam, M. M. (2015). Machine vision method for non-contact measurement of surface roughness of a rotating workpiece. *Sensor Review*, *35*(1), 10–19. https://doi.org/10.1108/SR-01-2014-609
51. Shah, R., Pai, N., Thomas, G., Jha, S., Mittal, V., Shirvni, K., and Liang, H. (2025). Machine learning in wear prediction. *Journal of Tribology*, *147*(4), 040801. https://doi.org/10.1115/1.4066865
52. Shabana, S., Jatavallabhula, J. K., Kumar, R. N., Chinthalapudi, R., and Pappula, B. (2025). Towards intelligent tribology: Predicting wear and friction in WC–Co Coatings with machine learning. *Journal of Bio-and Tribo-Corrosion*, *11*(3), 88. https://doi.org/10.1007/s40735-025-01005-9
53. Huang, T., Dai, L., Li, S., Lin, M., and Xia, Y. (2025). Learning-based data-enable predictive control for trajectory tracking of unknown systems. *International Journal of Robust and Nonlinear Control*, *2025*, 7859. https://doi.org/10.1002/rnc.7859
54. Pan, G., Wu, X., and Wei, H. (2025). GrDBN-GPR: A next-gen road feature inference framework for traffic crashes frequency prediction. *IEEE Transactions on Intelligent Transportation Systems*, *99*, 1–11. https://doi.org/10.1109/TITS.2024.3510788

55. Patel, R. A., Kesharwani, S. S., and Ibrahim, F. (2025). Active learning and Gaussian processes for the development of dissolution models: An AI-based data-efficient approach. *Journal of Controlled Release*, *379*, 316–326. https://doi.org/10.1016/j.jconrel.2025.01.003
56. Barletta, M. (2006). Electrostatic fluidized bed (EFB) coating of heat sensitive and electrical insulating substrates with low-curing thermoset epoxy-polyester (EP) powders. *Progress in Organic Coatings*, *56*(2–3), 185–198. https://doi.org/10.1016/j.porgcoat.2006.04.002
57. Khalaj, G. (2013). Artificial neural network to predict the effects of coating parameters on layer thickness of chromium carbonitride coating on pre-nitrided steels. *Neural Computing and Applications*, *23*(3), 779–786. https://doi.org/10.1007/s00521-012-0994-2
58. Varol, T., Canakci, A., Ozsahin, S., Erdemir, F., and Ozkaya, S. (2017). Artificial neural network-based prediction technique for coating thickness in Fe-Al coatings fabricated by mechanical milling. *Particulate Science and Technology*, *36*(6), 742–750. https://doi.org/10.1080/02726351.2017.1301607
59. Shukla, S. K., Sadhukhan, A. K., and Gupta, P. (2017). Development of ANN model for prediction of coating thickness in hot dip galvanizing process. *International Journal of Material Science and Engineering*, *5*(2), 60–68. https://doi.org/10.17706/ijmse.2017.5.2.60-68
60. Yazdi, M. R. S., Khorasani, A. M., and Faraji, M. (2011). Optimization of coating variables for hardness of industrial tools by using artificial neural networks. *Expert Systems with Applications*, *38*(10), 12116–12127. https://doi.org/10.1016/j.eswa.2011.03.002
61. Reyes-Mojena, M. Á., Sánchez-Orozco, M., Carvajal-Fals, H., Sagaró-Zamora, R., and Camello-Lima, C. R. (2017). A comparative study on slurry erosion behavior of HVOF sprayed coatings. *DYNA*, *84*(202), 239–246. https://doi.org/10.15446/dyna.v84n202.56542
62. Guessasma, S., and Coddet, C. (2004). Microstructure of APS alumina–titania coatings analysed using artificial neural network. *Acta Materialia*, *52*(17), 5157–5164. https://doi.org/10.1016/j.actamat.2004.07.022
63. Guessasma, S., and Coddet, C. (2005). Neural computation applied to APS spray process: Porosity analysis. *Surface and Coatings Technology*, *197*(1), 85–92. https://doi.org/10.1016/j.surfcoat.2005.01.124
64. Wang, L., Fang, J. C., Zhao, Z. Y., and Zeng, H. P. (2007). Application of backward propagation network for forecasting hardness and porosity of coatings by plasma spraying. *Surface and Coatings Technology*, *201*(9–11), 5085–5089. https://doi.org/10.1016/j.surfcoat.2006.07.088
65. Rashidi, A. M., Eivani, A. R., and Amadeh, A. (2009). Application of artificial neural networks to predict the grain size of nano-crystalline nickel coatings. *Computational Materials Science*, *45*(2), 499–504. https://doi.org/10.1016/j.commatsci.2008.11.016
66. Jarrah, M. I., Jaya, A. S. M., Azam, M. A., Alsharif, M. H., and Muhamad, M. R. (2016). Intelligence integration of particle swarm optimization and physical vapour deposition for tin grain size coating process parameters. *Journal of Theoretical and Applied Information Technology*, *84*(3), 355–369.
67. Dobrzański, L. A., Staszuk, M., and Honysz, R. (2010). Application of artificial neural networks in properties modelling of PVD and CVD coatings. *Archives of Computational Materials Science and Surface Engineering*, *2*(3), 141–148.
68. Rashidi, A. M., Hayati, M., and Rezaei, A. (2012). Application of artificial neural network for prediction of the oxidation behavior of aluminized nano-crystalline nickel. *Materials & Design*, *42*, 308–316. https://doi.org/10.1016/j.matdes.2012.06.011

69. Godavarty, A., Agarwal, A., and Dahotre, N. B. (2000). Neural networks in studies on oxidation behavior of laser surface engineered composite boride coatings. *Applied Surface Science*, *161*(1–2), 131–138. https://doi.org/10.1016/S0169-4332(00)00279-8
70. Danaher, S., Dudziak, T., Datta, P. K., Hasan, R., and Leung, P. S. (2013). Long-term oxidation of newly developed HIPIMS and PVD coatings with neural network prediction modelling. *Corrosion Science*, *69*, 322–337. https://doi.org/10.1016/j.corsci.2012.12.016
71. Zubayer, M. H., Zhang, C., Liu, W., Wang, Y., and Imdadul, H. M. (2024). Automatic defect detection of jet engine turbine and compressor blade surface coatings using a deep learning-based algorithm. *Coatings*, *14*(4), 501. https://doi.org/10.3390/coatings14040501

5 Optimization Techniques for Lubrication Design

Parminder Singh, Adityabir Singh, Sunirmit Verma, Pushpendra Gupta, and Manpreet Singh

5.1 INTRODUCTION

Lubrication plays a very crucial role in the performance and longevity of manufacturing machinery, both large and small scale, including industrial and microelectromechanical devices [1]. Its main function is to decrease friction and wear between moving parts, thereby safeguarding materials, enhancing efficiency, and lowering energy consumption. This is achieved by forming a separating layer, in the form of a liquid, solid, or gas, between two surfaces and hence avoiding direct contact between the moving parts. This is necessary because a lack of adequate lubrication may lead to abrasive, adhesive, and fatigue wear, which may result in premature failure and costly operations [2]. Friction management is also done with the addition of lubricants that are used to dissipate heat, clean off debris, and corrosion protection. Indicatively, lubricants are applied in engines to absorb the heat in the hot engine components, in combating particulate matter, and in corrosion prevention [3].

Lubrication design has to be optimized so as to attain a high level of efficiency, wear reduction, and energy conservation. Poor lubrication design could result in excessive friction, premature wear of parts, and high power usage, which could impact the reliability and cost of operation of the system [4]. For example, the effectiveness of turbochargers is very sensitive to the proper bearing lubrication, and optimization of lubrication design might be applied in order to minimize energy dissipation [5]. Similarly, axial pivoted pumps require specific lubrication plans that can cope with the combined effects of viscosity, temperature, and pressure, along with the progressive wear that occurs at the swash plate/slipper interface [6].

The ultimate aim of the optimal lubrication design is to find the best balance between friction minimization to enhance the energy efficiency and film thickness that is adequate to prevent component wear under different load, speed, and temperature [4]. This is a highly sensitive balance particularly in severe conditions like extreme temperatures or heavy loads, where conventional lubricants will tend to become ineffective. Metal-based lubricating coating has been developed to address these issues, with characteristics of thermal stability, oxidation resistance, increased wear resistance, and fine microstructures that aid in lowering friction during extreme conditions [7]. Other than its technical benefits, the optimal lubrication design has

DOI: 10.1201/9781003624165-5

enormous economic and environmental benefits. For example, lower friction enables more efficient operation of components in terms of energy with reduced carbon emissions, while increased wear control extends the life of components, reduces the replacements, and saves resources [7].

The earlier designs of lubrication were mostly based on correlations and experimental data of simplified bearings or surface contact prototypes [8]. These techniques were instructive but not flexible enough to deal with problems with complicated geometries, high operating rates, and multiphysics interactions, involving temperature, deformation, and surface roughness effects [4].

Nowadays, engineers can investigate the behavior of lubrication in realistic operating conditions due to the development of computational modeling and numerical simulation. It has been possible to predict, through the use of response surface methodology (RSM), computational fluid dynamics (CFD), and optimization algorithms, the distribution of pressure, film formation, and temperature increase in lubricated interfaces. Some modern optimization algorithms may be used to assist the researcher in systematically searching through design variables and performance objectives, for example, genetic algorithms (GA), swarm intelligence, and machine learning (ML). This paradigm shift from empirical and computational methodologies can be considered a significant step toward data-assisted and predictive design in lubrication [9,10].

This chapter focuses on the latest optimization methods which can be used to improve the design and effectiveness of lubricating systems. This chapter begins with an explanation of the principles of lubrication and then explains how optimization problems are formulated. Conventional and modern optimization techniques are discussed, including evolutionary algorithms, artificial intelligence (AI)-based techniques, and hybrid computational techniques.

This chapter concludes by summarizing the current issues and emerging types of lubricants, including nanoparticle-based lubricants, sustainable lubricants, as well as intelligent lubricants. Through this chapter, the reader will get a clear insight on how optimization methodologies can turn the conventional lubrication design into very efficient and environmentally friendly engineering systems.

5.2 FUNDAMENTALS OF LUBRICATION DESIGN

The basic principles involved in designing a lubrication system are of paramount importance in enhancing the performance, life cycle, and efficiency of mechanical systems. Lubrication is meant to separate the relative motion of interacting surfaces and thus lessen friction and wear. It does this by operating in a variety of lubrication regimes, and each one of them exhibits different mechanisms and physical processes.

5.2.1 Lubrication Regimes and Stribeck Curve

Centuries of observation and experiment have resulted in the study of friction and lubrication. The initial works of Leonardo da Vinci and Charles-Augustin de Coulomb paved the way for the knowledge of frictional forces [11]. This was systematized in

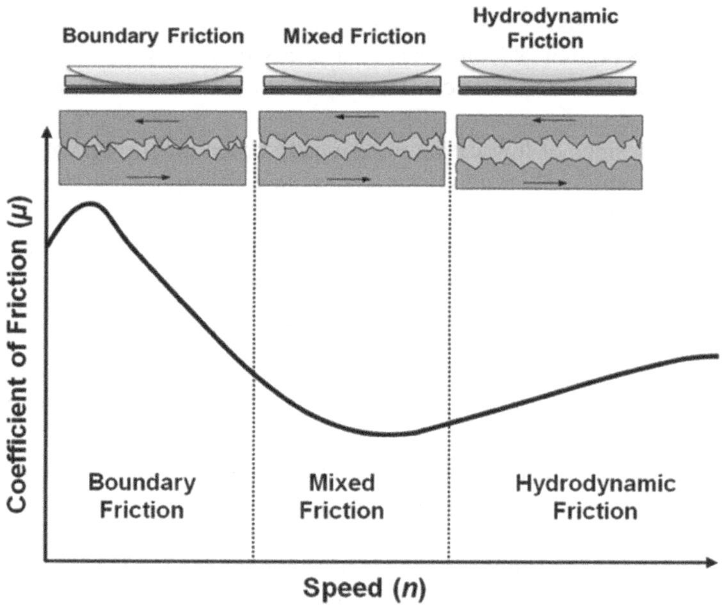

FIGURE 5.1 Simplified illustration of the Stribeck curve [15]. (Credit statement: Creative Commons International Attribution 4.0.)

the early 20th century, when Richard Stribeck studied journal bearings and determined the dependence of the coefficient of friction (μ) on viscosity, speed, and load, which led to the development of the Stribeck curve [12]. This curve illustrates the correlation between the coefficient μ with the Hersey number, which is a dimensionless ratio of the relationship between lubricant film thickness and surface roughness [13,14]. In a more general way, the Stribeck curve can be drawn between coefficient μ and sliding speed, as shown in Figure 5.1.

The above figure shows that the lubrication regime can be categorized into boundary, mixed, and hydrodynamic as a function of sliding speed. The boundary regime exists at lower speeds, and the lubrication film is very thin and insufficient to keep the interacting surfaces apart. In the mixed regime, the lubrication thickness is just enough to partially separate the surfaces, whereas in the hydrodynamic regime, the lubrication film is sufficiently thick to keep the interacting parts apart from each other. The knowledge of these regimes is the key to predicting and maximizing the tribological performance of lubrication systems.

5.2.2 Lubricant Properties

The performance of any lubrication system is fundamentally governed by intrinsic properties, such as viscosity, viscosity index, additive composition, thermal and oxidative stability, density, and lubricity. Each of these contributes to the lubricant's ability to perform effectively under a wide range of temperatures, loads, and working conditions.

a. **Viscosity:** It plays a major role in defining the ability of a lubricant to create and maintain a separating film between the interacting surfaces. An increase in the viscosity of the lubricant tends to produce thicker films that can support heavier loads; however, this advantage often comes with the disadvantage of increased viscous drag, which results in higher energy consumption and thermal inefficiency. For most liquid lubricants, viscosity decreases with increasing temperature and has a direct correlation with pressure [16]. The addition of nanoparticles in the lubricant has been found to increase both the lubricant viscosity and tribological performance of the lubricant [17].
b. **Viscosity index:** The viscosity index is the measure of the dependency of the viscosity of the lubricant on temperature. It is desirable that the lubricant's viscosity does not significantly alter across the operating temperature range. A significant improvement in the viscosity index is observed (up to 8%), when 5% by mass lithium bis(trifluoromethylsulfonyl)amine was added to an ester-based lube oil [18].
c. **Additive composition:** Modern lubricants are not made out of pure base oils; they also include a range of chemical additives intended to enhance the performance, stability, and protection. Nanoparticles such as Al_2O_3 and ZnO, when added in polyolester-based nanolubricants, can positively impact viscosity and thermal conductivity [19]. Other lubricity-enhancing additives can be polymer additives [20], organic peroxides [21], and ionic liquids [22].
d. **Thermal stability:** Good thermal stability of a lubricant means its ability to resist decomposition when exposed to high temperatures. Synthetic lubricants, such as polyolester-based lubricants, often show superior thermal stability under extreme temperatures compared to petroleum-based oils [23]. Epoxidation of mustard oil has been shown to improve its thermal stability, thus making it a more suitable bio-lubricant [24].
e. **Oxidative stability:** Oxidative stability refers to the ability of a lubricant to resist chemical degradation on reaction with oxygen. Biolubricants derived from iso-stearic acid exhibit oxidative stability when used in an oxidative environment [25].

5.2.3 Influence of Design and Operating Parameters

In addition to lubricant properties, the effectiveness of the lubrication design also depends on the design features and operating conditions including geometry, surface finish, applied load, speed, and temperature. A brief discussion on these factors is given below.

a. **Geometry:** It plays a fundamental role in defining the contact area and the pathways for lubricant flow. For example, in RV reducers (rotary vector gearbox), the optimization of the turning arm bearing structure plays a very crucial role in avoiding lubrication failures [26]. Additionally, the geometric parameters of textures on interacting surfaces, such as chevron textures, significantly influence the pressure distribution and flow velocity vectors, thereby affecting lubrication performance [27].

b. **Surface finish:** Recent studies on high-speed helical gears used in aviation engines indicate that roughness wavelength and amplitude have a significant influence on the lubrication properties [28]. Surface texturing, such as the creation of micro-dimples/grooves, is a common strategy used to modify the surface topography of the contacting surfaces to enhance lubrication, particularly under starved, boundary, and mixed lubrication conditions [29].
c. **Load:** It is another critical parameter that profoundly influences lubrication. Figure 5.2 shows the impact of load on the performance of different types of solid lubricants, namely, *G*-type (80% copper, 10% silica, and 10% graphite), *B*-type (80% copper, 10% silica, and 10% h-boron nitride), and *M*-type (80% copper, 10% silica, and 10% MoS_2).
d. **Speed:** Speed has a great influence on the shear rate within the lubricant, which impacts its tribological performance. From Figure 5.2, it can be seen that the wear loss increases linearly with speed for *B*- and *G*-type solid lubricants, whereas for *M*-type, the wear loss decreases when the speed exceeds 6 m/s.

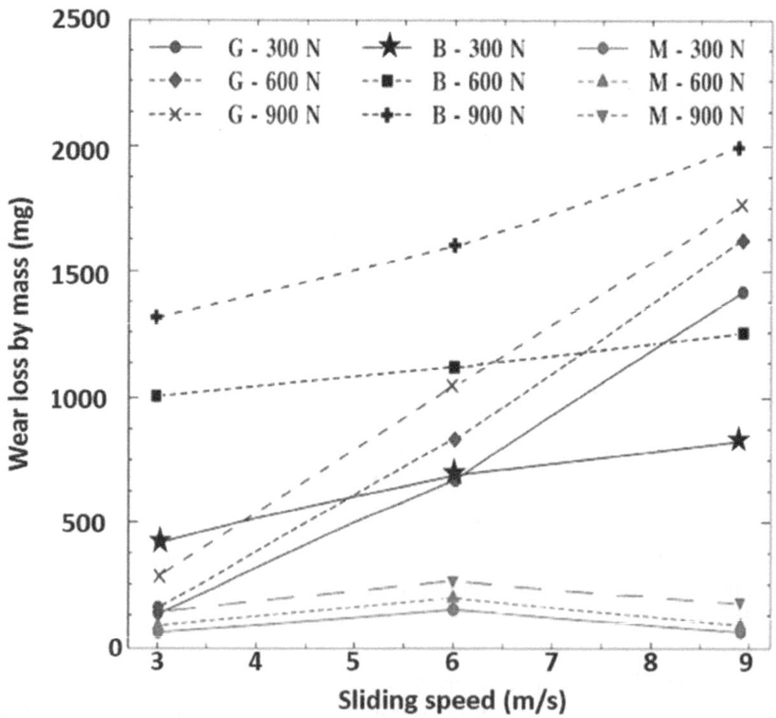

FIGURE 5.2 Impact of load/speed on wear loss [30]. (Credit statement: Reproduced with permission from Elsevier © 2015.)

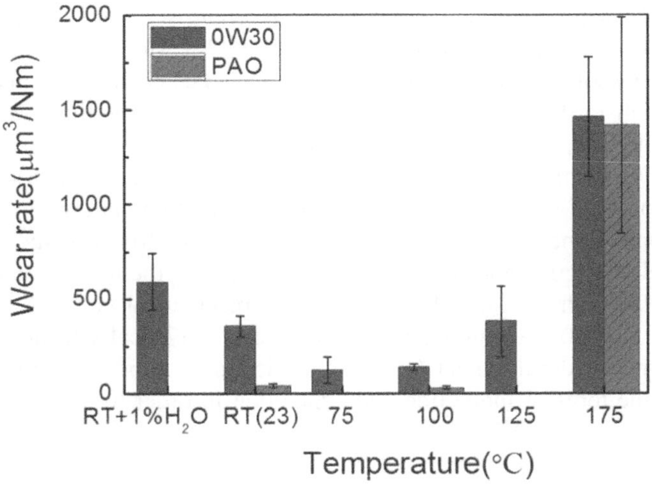

FIGURE 5.3 Impact of temperature on wear loss [31]. (Credit statement: Reproduced with permission from Elsevier © 2015.)

e. **Temperature:** High temperatures in lubrication often lead to a significant increase in both coefficient of friction and wear rates. Figure 5.3 shows that the wear increases as the operating temperature increases (100°C–175°C) due to a decrease in lubricant viscosity.

Clearly, several parameters have an influence on the lubrication system, hence indicating a potential scope for optimization. The following sections describe the methods by which this can be achieved.

5.3 OPTIMIZATION IN LUBRICATION: CONCEPTS AND NEEDS

The intricacy of contemporary tribological systems, in which load, temperature, speed, and material interactions are dynamically affected by one another, creates the need for a scientific and methodical design of a lubrication system instead of relying on experience alone. Optimization is a way to offer this kind of framework, that is, a systematic approach to finding the most effective combination of parameters that can be used to achieve high performance, reliability, and sustainability [32]. The following points further support this claim.

5.3.1 Why Optimization is Necessary in Lubrication Design

Lubrication design is to be optimized to meet complicated operating conditions and material interactions. For example, the frictional response of two environments (Venusian CO_2 atmosphere and atmospheric air of the Earth) to different operating temperatures is discovered to vary significantly in aerospace engineering [33].

Another technology where optimization is important is surface texture technology to accomplish optimal performance and durability [34]. Modern combustion engines have extremely sophisticated lubrication systems, which need thorough optimization in order to deliver the perfect combination among long engine life, low friction, and effective heat dissipation [35]. Minimum quantity lubrication (MQL) and nanofluid MQL (NMQL) technologies are also optimizable in machine tool lubrication to minimize friction, tool wear, and heating of the machine tool cutting [36]. Lubrication design optimization, in short, is a complex undertaking driven by the need to manage the level of friction and wear, save energy, enhance the life of components, and ensure reliable operation of a broad variety of mechanical systems operating in a multitude of and often changing conditions. It entails critical selection of lubricant, complex material development (e.g., self-lubricating composites and textured surfaces), and the tools of modelling and simulation in order to predict and optimize tribology.

5.3.2 Key Performance Indicators

The effectiveness of lubrication optimization is evaluated using certain key performance indicators that measure the magnitude of the performance of the system in accordance with its intended purpose. The indicators are used to establish the objective functions in the optimization process. Some key indicators are given as follows:

a. **Friction reduction or reduction in energy consumption:** Friction has a direct effect on energy loss, temperature increase, and wear of the components. Optimizing the lubricant viscosity, film thickness, and the choice of additives to reduce the coefficient of friction contributes to an increase in mechanical efficiency and a reduction in energy use. For example, in internal combustion engines, the objective of lubrication system optimization is to reduce the friction losses and, consequently, the power required to drive the oil pump [37]. Similarly, sustainable machining processes, such as MQL-assisted multi-roller burnishing (MQLAMRB), are optimized to reduce total energy consumption [38].
b. **Reduced wear or extended equipment life span:** Wear rate is a measurement that describes the loss of material at the contact interface over time. Lubrication optimization seeks to reduce wear through the selection of a suitable viscosity grade, proper film formation, and the inclusion of anti-wear additives that promote the formation of a protective tribofilm. An optimal design should prevent direct asperity interactions and still maintain acceptable fluid drag [39].
c. **Low lubricant consumption:** This is more of an economic performance indicator and plays a critical role in aero-engines, where accurately predicting and analyzing lubricant consumption rates are essential for maintaining proper lubricant levels while in operation [40].
d. **Improved reliability of operation:** Reliability refers to the capacity of the lubrication system to produce consistent performance over long durations without failure. It relies on aspects such as the rate of lubricant degradation,

resistance to contamination, and thermal–oxidative stability. An optimized lubrication system maintains consistent viscosity, prevents film breakdown, and provides constant protection even when operating loads and environmental conditions vary [41].

These key performance indicators, in combination, characterize the multiobjective nature of lubrication optimization as the enhancement of one factor (e.g., wear reduction) can be in conflict with another (e.g., energy efficiency). Therefore, optimization is associated with identifying the optimal compromise or Pareto-optimal solution among conflicting performance objectives.

5.3.3 Challenges in Conventional Design Approaches

The traditional approaches to designing lubrication are mostly empirical, having been constructed on the foundations of experiment, analytical approximations, and working rules. However, these old methodologies face serious constraints when dealing with current engineering systems that have high accuracy, reduced size, and the capacity to perform in extreme conditions.

a. **Empirical limitations:** By its nature, empirical correlations tend to be valid only within limited operating regimes, and they are not concerned with nonlinear and dynamic interactions between lubricant properties, surface properties, and environmental factors. Thus, designs that are built on these methods can be associated with low predictive accuracy in real-world applications.
b. **Parameter interdependence:** The number of interdependent variables in lubrication systems is large. Simultaneously, a change in one parameter, e.g., temperature, affects viscosity, film thickness, and friction. The conventional one-factor-at-a-time experimental approaches do not take such interdependencies into account, which results in partial or incomplete, and sometimes deceptive, design information.
c. **Time and cost constraints:** Experimental testing of lubricants at varying speeds, loads, and temperatures is time-consuming and costly. More sophisticated measurement techniques, such as high and *in situ* film thickness and temperature friction testing, are expensive and time-consuming. This renders the trial-and-error approach impractical where design turnaround is required in a short time.
d. **Inadequate adaptability for advanced lubricants:** The growing demand for non-Newtonian lubricants (e.g., nano-lubricants, ionic liquids, and bio-based oils) has made the situation more complicated. These sophisticated lubricants exhibit complex rheological and tribochemical behaviors that standard models do not predict accurately.
e. **Environmental and sustainability constraints:** The emphasis of modern lubrication systems is on the difficulties presented by rigorous biodegradability, toxicity, and energy expenditure standards. Advanced, holistic optimization is hugely required as a way of optimizing both technical performance and environmental impact concurrently.

Contemporary optimization models contain CFD and finite-element approaches as computer-aided tools to model the behavior of lubrication in real-life settings [42,43]. These simulations give rather precise approximations of the pressure distribution, thickness, and temperature increase, therefore decreasing the use of intensive physical testing. The computation outcomes are then checked in experiments in order to perfect models and promote reliability.

5.4 SINGLE AND MULTIOBJECTIVE OPTIMIZATION IN LUBRICATION

The optimization process includes various techniques such as gradient-based methods, GA, or simulated annealing [44]. Out of these techniques, gradient-based methods use information about the slope of the objective function to iteratively move toward an optimal solution. On the other hand, GAs, inspired by natural selection, explore a set of candidate solutions simultaneously, making them more effective in a complex, multidimensional search space.

5.4.1 Single-Objective Optimization of the Lubrication Designs

Single-objective optimization in lubrication design focuses on optimizing a single performance parameter, which is of critical importance in improving the efficiency as well as longevity of a mechanical system. For example, Simunovic et al. [45] studied the face-milling process of structural steel, with the objective of reducing lubrication fluid usage, thereby leading to cleaner and more sustainable manufacturing. The optimal machining parameters were obtained by using minimum surface roughness as the objective function. To achieve this, the RSM is applied to establish a correlation between the roughness (R_a) and the key operating parameters, namely, A (depth of cut, mm), B (feed rate, mm/min), C (Number of revolutions, rev/min), and D (cooling/lubrication). The relationships are established below Eq. (5.1),

$$\begin{aligned} R_a = 1.57 + 0.05A + 0.2B - 0.14C - 0.22D[1] - 0.2D[2] - 0.06BD[1] \\ - 0.1BD[2] - 0.18CD[1] + 0.07CD[2] - 0.3A^2 - 0.3B^2 \end{aligned} \tag{5.1}$$

For the single-objective optimization of minimizing the surface roughness, the RSM plot is shown in Figure 5.4. It can be seen that for dry cutting, when the depth of cut is 0.5 mm, the optimal feel rate was found to be 100 mm/min and the number of revolutions 800 rev/min, corresponding to a minimum possible surface roughness of 0.27 μm.

In another example, Singh et al. [46] studied the lubrication characteristics of a blend of *Phyllanthus emblica* seed oil and SAE 20*W*40 oil and conducted an RSM analysis to determine the minimum possible values of specific wear rate and coefficient of friction. The operating parameters include the blend ratio (BR), load (L), and the sliding velocity (SV) of the pin-on-disk tester. From the Box–Behnken design, the influence of the operating parameters on the wear rate and friction coefficient is expressed in the correlations mentioned below (Eqs. 5.2 and 5.3).

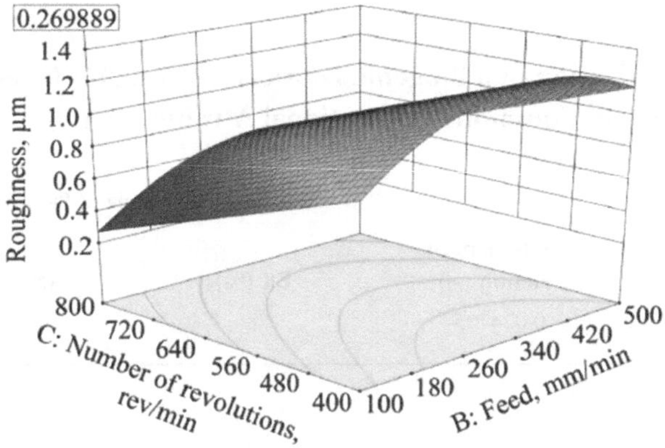

FIGURE 5.4 Response surface methodology plot for minimal (optimal) surface roughness [45]. (Credit statement: Reproduced with permission from Elsevier © 2015.)

$$\begin{aligned}\text{Wear rate}: 1.37E-005-4.96E-(007\times \text{BR})\\ +6.83E-(008\times L)-3.39E-(006\times \text{SV})\\ -2.45\text{E}-(011\times \text{BR}\times L)+1.18E\\ -(008\times \text{BR}\times \text{SV})-8.2E-(010\times L\times \text{SV})\\ +2.9E-\left(008\times \text{BR}^2\right)-1.06E\\ -\left(101\times L^2\right)-1.34E-\left(007\times \text{SV}^2\right)\end{aligned} \quad (5.2)$$

$$\begin{aligned}\text{Coefficient of friction}: 0.039-4.77E-(004\times \text{BR})\\ +4.48E-(005\times L)-3.94E\\ -(003\times \text{SV})-5.0E-(008\times \text{BR}\times L)\\ +1.2E-(005\times \text{BR}\times \text{SV})+(0.0\times L\times \text{SV})\\ +2.79E-\left(005\times \text{BR}^2\right)-5.0E-\left(009\times L^2\right)-2.4E-\left(005\times \text{SV}^2\right)\end{aligned} \quad (5.3)$$

The results of the optimization (minimization) of the wear rate and the coefficient of friction are presented in Table 5.1. The optimal operating factors were 8% BR (8% *Phyllanthus emblica* seed oil in SAE 20W40 oil), an SV of 3.8 m/s, and a load of 50 N.

TABLE 5.1
Response Optimization of a Biogenic Lubricant Blend [46] (Credit statement: Creative Commons International Attribution 4.0)

Minimum Values of Objective Function Parameters		Optimized Parameter Set		
Wear Ratio (mm^3/Nm)	Coefficient of Friction (–)	BR (%)	*L* (N)	SV (m/s)
1.84×10^{-7}	0.024725	8	50	3.8

5.4.2 Multiobjective Optimization of the Lubrication Designs

In contrast to single-objective optimization, multiobjective optimization problems are more realistic and involve several conflicting objective functions. Therefore, multiobjective optimization seeks to find a set of optimal solutions, known as the Pareto front, where no single objective can be improved without degrading at least one other objective [47]. For example, Najiha et al. [48] performed a multiobjective optimization on the end milling of aluminum alloy. Three objectives are considered (see Eq. 5.4), which are optimized simultaneously using GAs. In Eq. (5.4), the symbols R_a, MRR, FW, CS, FR, DoC, and MQL depict surface roughness (μm), metal removal rate (mm^3/min), flank wear (μm), cutting speed (rpm), feed rate (mm/min), depth of cut (mm), and MQL flow rate (mL/min), respectively. Figure 5.5 shows a Pareto diagram of the input operating parameters, represented using a bubble chart.

$$\text{Multi}-\text{objective function} = \begin{cases} \text{Minimize } R_a = f\left(\text{CS, FR, DoC, MQL}\right) \\ \text{Maximize MRR} = f\left(\text{CS, FR, DoC, MQL}\right) \\ \text{Minimize FW} = f\left(\text{CS, FR, DoC, MQL}\right) \end{cases} \tag{5.4}$$

In the above Figure 5.5, the MQL flow rate (shown on the x-axis) was plotted against the three objectives mentioned in Eq. (5.4), where y-axis represents surface roughness (first objective), the bubble diameter represents the material removal rate (second objective), and the bubble colour represents the flank wear (third objective).

In another study, Chinchanikar et al. [49] performed turning experiments using chemical vapor deposition-coated TiCN-Al_2O_3 carbide inserts. The lubricant used was a blend of coconut oil and hexagonal boron nitride (hBN) nanoparticles. The operating parameters included were CS (m/min), FR (mm/rev), DoC (mm), and nanoparticle concentration (NC) in the lubricant (%). Regression equations were formulated for R_a and resultant force (RF), as shown below (Eqs. 5.5 and 5.6).

$$\text{Surface roughness } (R_a): 13.29(\text{CS})^{-0.0248}(\text{FR})^{1.4085}(\text{DoC})^{-0.0805}(\text{NC})^{-0.006} \tag{5.5}$$

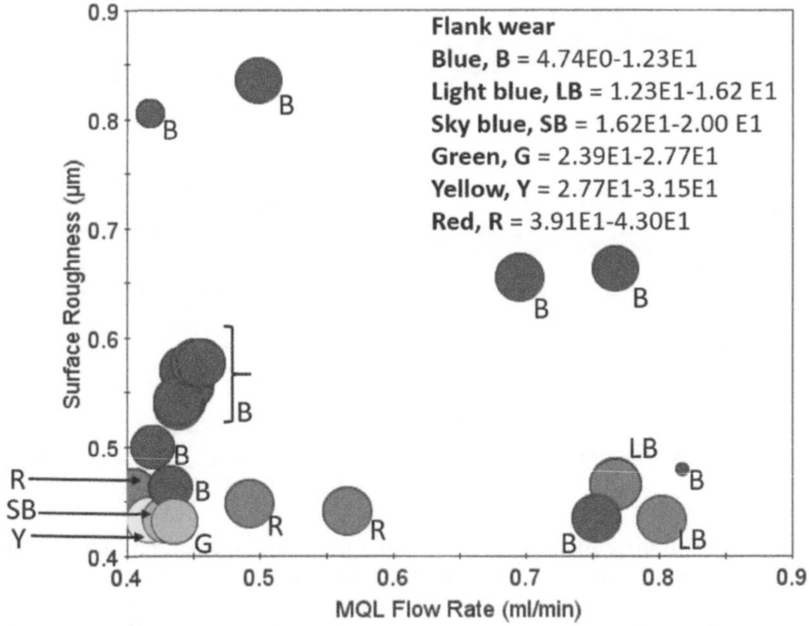

FIGURE 5.5 Bubble chart representing the Pareto diagram drawn with respect to lubrication requirements [48]. (Credit statement: Creative Commons International Attribution 4.0.)

$$\text{Resultant force (RF)}: 642.73(\text{CS})^{-0.0612}(\text{FR})^{0.054}(\text{DoC})^{0.1429}(\text{NC})^{0.0402} \tag{5.6}$$

The above study involved a multiobjective function consisting of three distinct, contradictory objectives, as shown in Eq. (5.7).

$$\text{Multi}-\text{objective function} = \begin{cases} \text{Minimize } R_a = f(\text{CS, FR, DoC, NC}) \\ \text{Minimize RF} = f(\text{CS, FR, DoC, NC}) \\ \text{Maximize MRR} = f(\text{CS, FR, DoC, NC}) \end{cases} \tag{5.7}$$

The optimization was performed using the nondominated sorting genetic algorithm-II (NSGA-II), with the following parameters: population size = 1,000, generation = 100, crossover probability = 0.8, and crossover constant 0.5, mutation probability = 0.5, and mutation constant = 20. Figure 5.6 shows the two-dimensional Pareto plots between the objective functions given in Eq. (5.7).

From the plots, it can be seen that the value of R_a tends to decrease with an increase in the value of RF, which implies that higher cutting forces lead to smoother surfaces. Second, the value of MRR increases initially with an increase in RF, reaches a maximum around the mid-range of RF, and then declines. This implies that moderate cutting forces provide the best balance between efficient material removal and process

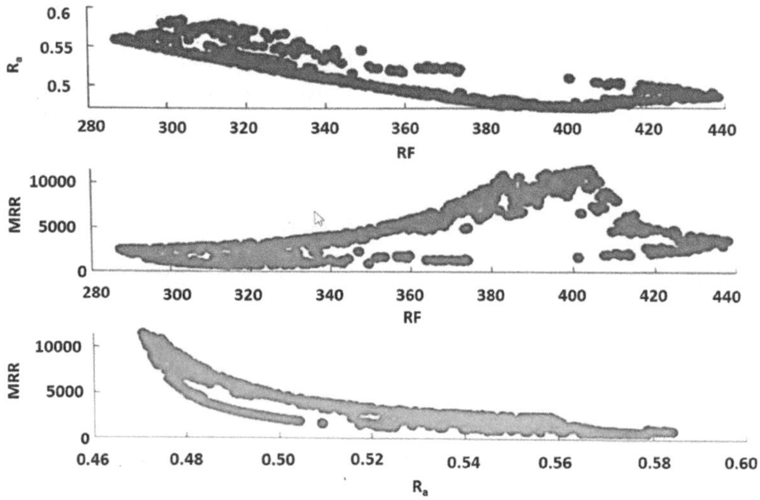

FIGURE 5.6 Two-dimensional Pareto plots of a multiobjective optimization [49]. (Credit statement: Creative Commons International Attribution 4.0.)

stability. Third, an inverse relationship can be plotted between MRR and R_a, meaning that a higher metal removal rate comes at the expense of surface quality.

Next, the study aimed to find the optimum single set of operating parameters (CS, FR, DoC, and NC) such that all the three objectives of Eq. (5.7) were satisfied. One of the methods adopted in the study was the desirability function approach, in which individual desirability values, namely, desirability for R_a (D_{R_a}), desirability for RF (D_{RF}), and desirability for MRR (D_{RF}), were calculated. These values were then combined to determine the overall desirability function (D_M). The mathematical expression for these parameters is shown below.

$$D_{R_a} = \begin{cases} 0,\ R_a \geq 1.58 \\ \left(R_{a,\max} - R_{a,i}\right) / \left(R_{a,\max} - R_{a,\min}\right) \\ 1,\ R_a \leq 0.4 \end{cases}$$

$$D_{\mathrm{RF}} = \begin{cases} 0,\ \mathrm{RF} \geq 444 \\ (\mathrm{RF}_{\max} - \mathrm{RF}_i) / (\mathrm{RF}_{\max} - \mathrm{RF}_{\min}) \\ 1,\ \mathrm{RF} \leq 301 \end{cases} \tag{5.8}$$

$$D_{\mathrm{MRR}} = \begin{cases} 0,\ \mathrm{MRR} \leq 350 \\ (\mathrm{MRR}_i - \mathrm{MRR}_{\min}) / (\mathrm{MRR}_{\max} - \mathrm{MRR}_{\min}) \\ 1,\ \mathrm{MRR} \geq 15{,}000 \end{cases}$$

$$D_M = \left(D_{R_a} \times D_{\mathrm{RF}} \times D_{\mathrm{MRR}}\right)^{1/3}$$

TABLE 5.2
Best Optimal Set of Operating Parameters Obtained from the Analysis Using the DFA [49] (Credit statement: Creative Commons International Attribution 4.0)

	Optimal Operating Parameters				Desired Objective Functions		
S. No.	CS (m/min)	FR (mm/rev)	DoC (mm)	NC (%)	R_a (μm)	RF (*N*)	MRR (mm^3/min)
1	125	0.1	0.75	0.325	0.47	388	9,375
2	125	0.1	0.75	0.25	0.47	383	9,375
3	120.5	0.1	0.75	0.25	0.48	384	9,037.5
4	125	0.1	0.75	0.2875	0.47	386	9,375

Table 5.2 shows four such optimal operating sets, which are listed in decreasing order of the desirability function value. It is observed that the cutting speed can be set in the range 120.5–125 m/min, the feed rate at 0.1 mm/rev, the depth of cut at 0.75 mm, and the nanoparticle concentration in the lubricant between 0.25% and 0.325%. It can also be seen that the value of R_a and RF lie toward the lower side of their respective range (see Eq. 5.8), whereas the MRR lies toward the higher end of its range (see Eq. 5.8).

In another study, a more specific multiobjective optimization problem was considered, involving the volume of the lubricant required in a limited slip differential (LSD) under thermal influence [50]. The Lattice Boltzmann method was used to model the heat-carrying capacity of lubricant flow under the influence of gear speed (ν, rev/s) and oil depth (H, m). Results from the simulation showed that both large and small immersion depths affected the heat dissipation capacity of the lubricant (see Figure 5.7, redrawn by authors as a three-dimensional graph). Therefore, this study focused on obtaining the optimal values of oil depth at different gear speeds. The three objectives of the study were the time (t) required to reach equilibrium temperature, the average temperature of all monitoring points (T), and the volume of the lubrication oil in the LSD. The multiobjective function is shown below (Eq. 5.9).

$$\text{Multi}-\text{objective function} = \begin{cases} \text{Minimize } t = f(\nu, H) \\ \text{Maximize } T = f(\nu, H) \\ \text{Minimize } V = f(\nu, H) \end{cases} \tag{5.9}$$

Using the RSM, the optimal oil depth was predicted to be −0.02 m, corresponding to an oil volume of 3.554 L. The predicted oil depth clearly shows good heat dissipation capacity at all gear speeds, as can be seen in Figure 5.7.

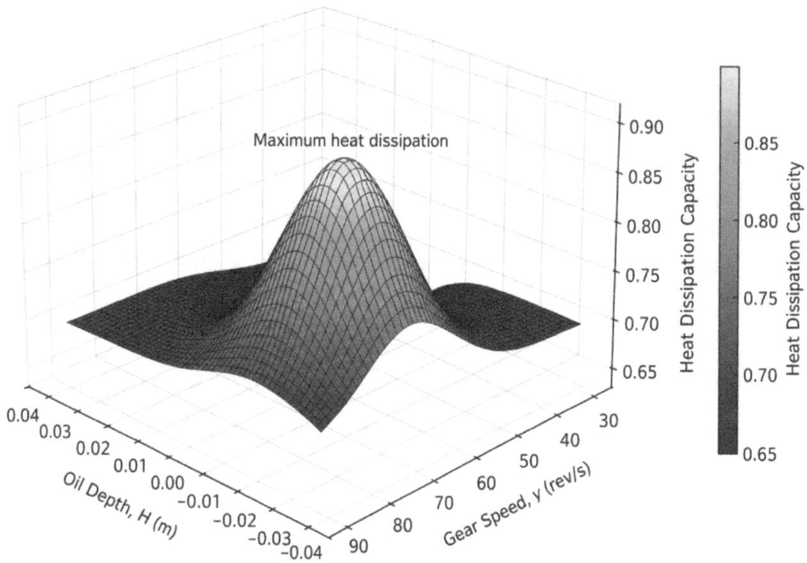

FIGURE 5.7 Three-dimensional diagram showing the impact of ν (rev/s) and H (m) on the lubricant's heat dissipation capacity. (Credit statement: Created by the authors.)

5.5 AI AND DATA-DRIVEN OPTIMIZATION IN LUBRICATION

AI and data-driven optimization has transformed modern lubrication system design by improving performance, enabling predictive maintenance, and reducing the operational costs across various applications [51,52]. Out of the three primary applications of AI, predictive maintenance appears to be the most important as it aims to predict equipment failures before they actually occur in the real system [53]. For example, Rigolli et al. [54] applied predictive maintenance to industrial gearboxes by assessing lubricant quality through analyses of fluid properties, contamination, and wear. The condition of the lubricant was tested using Fourier-transform spectroscopy (FT-IR), optical emission spectroscopy, particulate counting, and measurements of fluid properties such as lubricant density and viscosity. Figure 5.8 shows the steps followed in the diagnostic model, where the input data were sent to the AI model to generate diagnostic interpretations.

The performance of the above model (Figure 5.8) was also evaluated by comparing the model output with the expert review provided by the machine operator. The use of AI has led to a decrease in diagnostic time by almost 60%, thereby increasing the performance and efficiency of the gear system.

ML algorithms also play a central role in enhancing the performance of the lubrication systems by analyzing real-time data obtained from some common sensors such as temperature sensors, dielectric sensors, and pH sensors. For example, Zhu et al. [55] used an array of electrochemical sensors to train artificial neural network architectures to determine four lubricant properties: water content, total acid number,

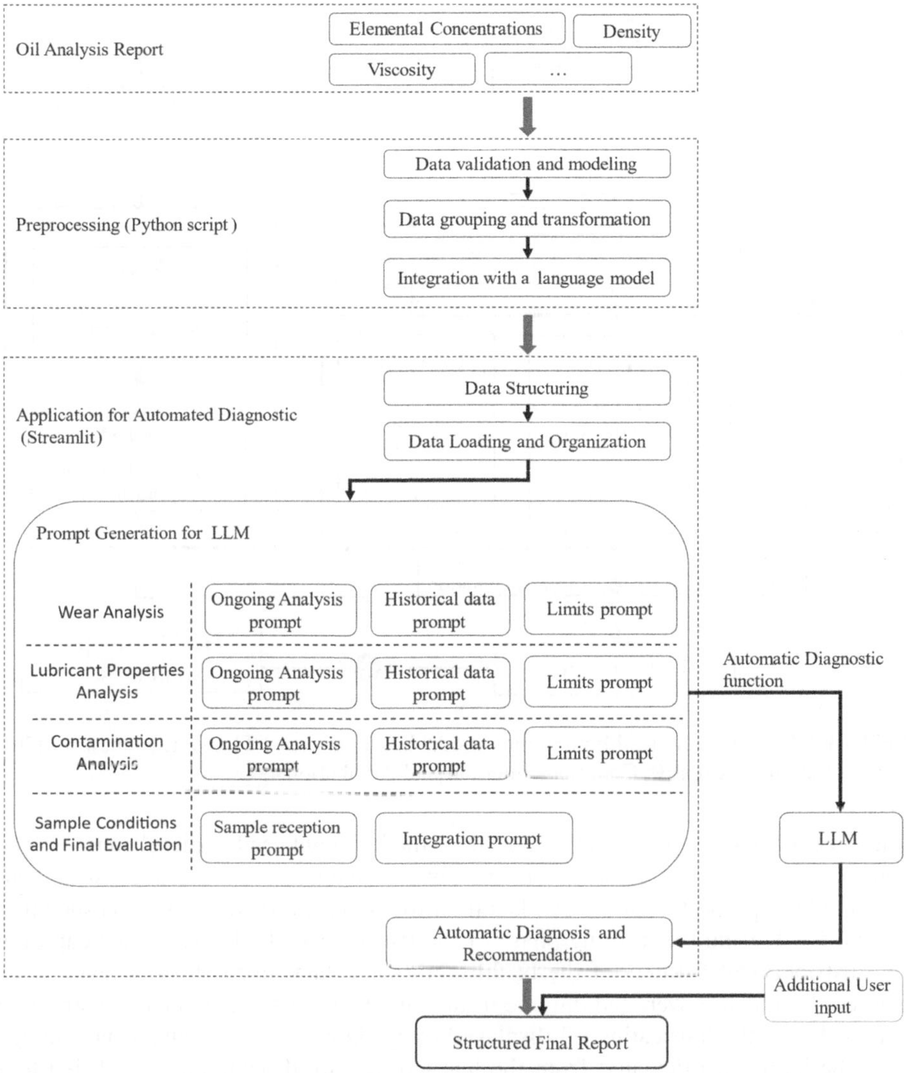

FIGURE 5.8 Workflow for diagnostic analysis of lubricating oil [54]. (Credit statement: Creative Commons International Attribution 4.0.)

soot, and sulfur content. The trained model could be very useful in online monitoring of lubricating oil.

Deep learning techniques such as long short-term memory (LSTM) networks are particularly helpful in capturing sequential/temporal patterns in the lubrication system design. For example, Chen et al. [56] developed an SFTI-LVAE framework which consists of seasonal-trend decomposition using loess (S), a factor attention network (F), a temporal convolutional network (T), an informer (I), and an LSTM variational

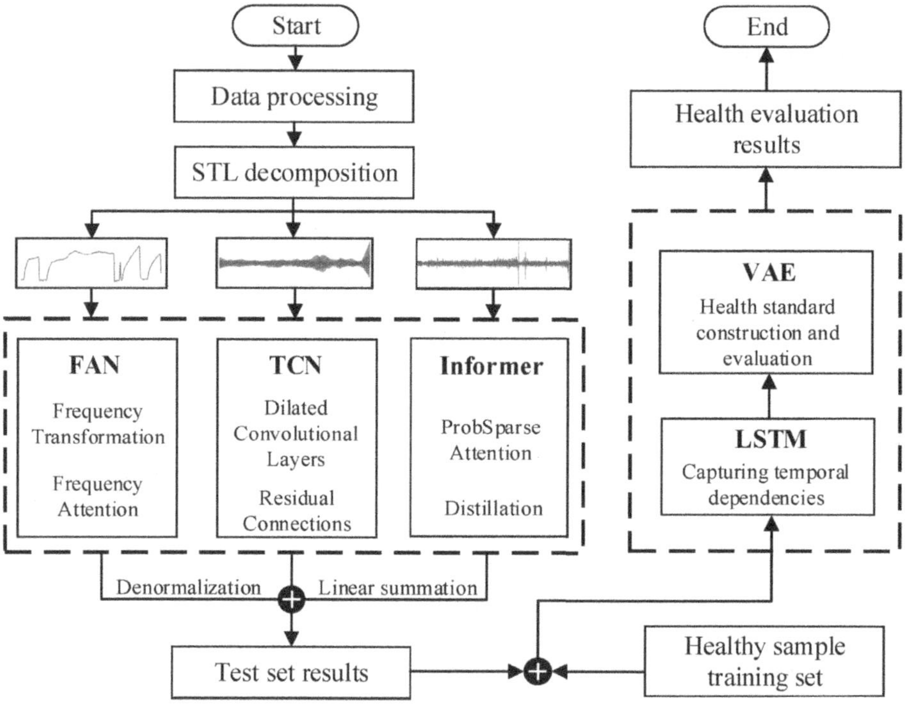

FIGURE 5.9 Lubricant health prediction model based on the SFTI-LVAE framework [56]. (Credit statement: Creative Commons International Attribution 4.0.)

autoencoder (LVAE) to predict the health of the lubricant used in diesel engines. The sensors used in the study were particle contaminator sensor, micro-moisture sensor, and fluid property sensor. The layout of the SFTI-LVAE framework is shown in Figure 5.9. During the training phase of the framework, the LVAE model captured the distributional features of the healthy lubricant data. In the testing phase, new observed sequences were fed into the trained model so as to compute reconstruction errors. When the lubricating oil displayed early signs of wear or abnormal impurities, the input data deviated from the learned healthy distribution, which led to a notable increase in the reconstruction error. The study claimed that the proposed SFTI-LVAE framework has 96.67% defect prediction accuracy with zero false alarm, and it generates a warning almost 6.47 hours before the actual failure of the lubrication system. Thus, the SFTI-LVAE framework provided an innovative solution in the era of digital tribology.

5.6 EXISTING CHALLENGES AND FUTURE TRENDS

One prominent challenge faced during the application of AI to lubrication system design is the inherent complexity of the system itself. These systems involve complex interactions among contacting surfaces, lubricants, and environmental factors under diverse operating conditions, making it very difficult to develop a comprehensive and accurate AI model. Another critical challenge was faced in acquiring quality data for training the AI models. However, it is costly, time-consuming, and sometimes even impractical to obtain such high-quality data, particularly in harsh environments such as space applications where there is a high vacuum and vastly different temperatures. The requirement to come up with biodegradable and environmentally friendly lubricants is another problem of AI-driven optimization. The AI models should take into account the aspects of the controlling factors of lubricant biodegradability, toxicity, and overall life cycle impact, which bring one more complexity to the AI-based optimization process.

To address such challenges, current research is looking into emerging opportunities to improve the role of AI in the design of lubrication systems. Merging of AI with state-of-the-art computational modeling and simulation tools such as CFD and multi-body dynamics is one important future trend in the design of lubrication. This strategy can result in healthier prediction of lubricant behavior, which will allow improvements in the design through interactive design. Another potential opportunity is the adoption of GAs and other optimization methods and integrating them with AI so that they can establish the best operating parameters of lubrication.

Moreover, modeling and optimization are only some of the opportunities for future research. These include studies on nanoparticle-based lubricants, attainment of superlubricity, decentralized lubrication systems, and lubrication of electrified vehicles. Collectively, they have the power to transform the science of lubrication and build the next generation of intelligent and sustainable tribological systems.

5.7 CONCLUSIONS

Lubrication design optimization has assumed much more complex, informative, and multiobjective designs as opposed to the conventional trial-and-error methods. The principles of lubrication regimes, the properties of lubricants, and the working parameters will be a stepping stone to higher levels of tribological performance and system efficiency. The increasing demands for enhanced reliability, energy efficiency, and environmental sustainability have caused optimization to become an important component of modern lubrication system design. New computational algorithms and AI now positively influence this industry and provide an opportunity to conduct predictive modeling, automation of design, and optimization of the process. However, some problems still exist, including the difficulty in modeling lubrication systems, the fact that training data are not always of the highest quality, and the challenge of creating environmentally friendly lubricants. The combination of AI and computational modeling tools (such as CFD and multibody dynamics) and the implementation of advanced optimization algorithms (genetic and neural networks) should be resorted to in order to solve these issues. The future of lubrication design is the convergence of AI, nanotechnology, and sustainable materials. New research on nanoparticle-based lubricants, superlubricity, decentralized lubrication systems, and lubrication in electrified automobiles will probably change the boundaries of tribological science. With these smart optimization structures, engineers and researchers can come up with more efficient, more reliable, and eco-friendly lubricant systems, and this will be the first step toward the next generation of smart and sustainable mechanical systems.

REFERENCES

1. Zhang, W. (2022). Tribology of SiC ceramics under lubrication: Features, developments, and perspectives. *Current Opinion in Solid State and Materials Science*, 26(4), 101000. https://doi.org/10.1016/j.cossms.2022.101000
2. Peng, H., Zhang, H., Shangguan, L., and Fan, Y. (2022). Review of tribological failure analysis and lubrication technology research of wind power bearings. *Polymers*, 14(15), 3041. https://doi.org/10.3390/polym14153041
3. Menezes, P. L., Reeves, C. J., and Lovell, M. R. (2013). Fundamentals of lubrication. In *Tribology for Scientists and Engineers: From Basics to Advanced Concepts*. Menezes, P. L., Nosonovsky, M., Ingole, S. P., Kailas, S. V., Lovell, M. R. (Eds.), Springer, New York. https://doi.org/10.1007/978-1-4614-1945-7_10
4. Hu, Q., Zeng, K., Han, S., Xu, J., Hu, W., and Li, J. (2024). Synthesis and performance evaluation of Metallocene Polyalphaolefins (mPAO) base oil with anti-friction and anti-wear properties. *Polymers*, 16(19), 2828. https://doi.org/10.3390/polym16192828
5. Novotný, P., Vacula, J., and Hrabovský, J. (2021). Solution strategy for increasing the efficiency of turbochargers by reducing energy losses in the lubrication system. *Energy*, 236, 121402. https://doi.org/10.1016/j.energy.2021.121402
6. Long, J., Lu, Y., Zhang, H., Qian, X., and Zhang, J. (2024). Life cycle assessment of a slipper/swash plate friction pair based on thermal-fluid-structure lubrication state dynamic recognition. *Tribology International*, 192, 109256. https://doi.org/10.1016/j.triboint.2024.109256
7. Wu, S., Tang, Y., Gu, J., Li, R., Liang, Y., Liu, P., Wang, H., Cuihua, A., Deng, Q., Zhao, L., and Hu, N. (2024). Controllable preparation of metal-based lubrication coatings in extreme environmental applications. *Materials & Design*, 241, 112922. https://doi.org/10.1016/j.matdes.2024.112922
8. Peng, H., Chen, Y., Shangguan, L., Cheng, R., Li, Y., and Yang, C. (2023). Multi-objective-based intelligent lubrication system performance evaluation technology for construction machinery. *Applied Sciences*, 13(21), 11768. https://doi.org/10.3390/app132111768
9. Zhang, K., Wang, J., Ban, Y., Sun, C., Gao, P., and Jin, D. (2020). Multi-field coupling simulation of gear: A review. *Journal of Failure Analysis and Prevention*, 20(4), 1323–1332. https://doi.org/10.1007/s11668-020-00938-2
10. Pouzar, J., Kostal, D., Westerberg, L-G., Nyberg, E., and Krupka, I. (2025). Labyrinth seal design for space applications. *Vacuum*, 232, 113882. https://doi.org/10.1016/j.vacuum.2024.113882
11. Besson, U. (2013). Historical scientific models and theories as resources for learning and teaching: The case of friction. *Science & Education*, 22(5), 1001–1042. https://doi.org/10.1007/s11191-012-9502-4
12. Jacobson, B. (2003). The Stribeck memorial lecture. *Tribology International*, 36(11), 781–789. https://doi.org/10.1016/S0301-679X(03)00094-X
13. Hersey, M. D. (1914). The laws of lubrication of horizontal journal bearings. *Journal of the Washington Academy of Sciences*, 4(19), 542–552. https://www.jstor.org/stable/24520857
14. He, T., Zhu, D., Wang, J., and Jane Wang, Q. (2017). Experimental and numerical investigations of the Stribeck curves for lubricated counterformal contacts. *Journal of Tribology*, 139(2), 4051. https://doi.org/10.1115/1.4034051

15. Pondicherry, K. S., Rummel, F., and Laeuger, J. (2018). Extended stribeck curves for food samples. *Biosurface and Biotribology*, 4(1), 34–37. https://doi.org/10.1049/bsbt.2018.0003
16. Bair, S. (2016). The temperature and pressure dependence of viscosity and volume for two reference liquids. *Lubrication Science*, 28(2), 81–95. https://doi.org/10.1002/ls.1302
17. Sharif, M. Z., Azmi, W. H., Ghazali, M. F., Zawawi, N. N. M., and Ali, H. M. (2023). Viscosity and friction reduction of double-end-capped Polyalkylene Glycol nano-lubricants for eco-friendly refrigerant. *Lubricants*, 11(3), 129. https://doi.org/10.3390/lubricants11030129
18. Yeh, F.-M., Volli, V., Laiwang, B., Tung, P.-H., and Shu, C.-M. (2019). Oxidative stability and thermal performance of ester based lube oil with lithium salt additives. *Applied Thermal Engineering*, 150, 1328–1336. https://doi.org/10.1016/j.applthermaleng.2019.01.061
19. Kedzierski, M. A., Brignoli, R., Quine, K. T., and Brown, J. S. (2017). Viscosity, density, and thermal conductivity of aluminum oxide and zinc oxide nanolubricants. *International Journal of Refrigeration*, 74, 3–11. https://doi.org/10.1016/j.ijrefrig.2016.10.003
20. Dickmann, J. S., Avery, K., Devlin, M. T., and Kiran, E. (2020). High-pressure density, viscosity, and modeling of mixtures of a poly(α-olefin) base oil lubricant with polymeric additives. *Industrial & Engineering Chemistry Research*, 59(16), 7926–7942. https://doi.org/10.1021/acs.iecr.9b05757
21. Chen, R.-Y., Tung, P.-H., Laiwang, B., Liaw, H.-J., and Shu, C.-M. (2022). Oxidation and thermal stability analysis of hexadecyl mercaptan added to engine oil. *Journal of Thermal Analysis and Calorimetry*, 147(7), 4685–4696. https://doi.org/10.1007/s10973-021-10792-5
22. Lazarenko, D., and Khabaz, F. (2022). A molecular dynamics study on the tribological performance of Imidazolium–based ionic liquids mixed with oil in comparison to pure liquids. *Fluids*, 7(12), 384. https://doi.org/10.3390/fluids7120384
23. Law, D. A., Lohuis, J. R., Breau, J. Y., Harlow, A. J., and Rochette, M. (1984). Development and performance advantages of industrial, automotive and aviation synthetic lubricants. *Journal of Synthetic Lubrication*, 1(1), 1–33. https://doi.org/10.1002/jsl.3000010102
24. Sikdar, S., and Menezes, P. L. (2024). The role of epoxidation process on improving the oxidative, thermal stability, and tribological performance of mustard oil nano lubricants. *Journal of Renewable and Sustainable Energy*, 16(2), 46. https://doi.org/10.1063/5.0190046
25. Sarker, M. I., Mainali, K., Sharma, B. K., Yadav, M. P., Ngo, H., and Ashby, R. D. (2023). Synthesized biolubricants from naturally derived oleic acid: Oxidative stability and cold flow performance. *Industrial Crops and Products*, 204, 117315. https://doi.org/10.1016/j.indcrop.2023.117315
26. Zhang, Z., Wang, J., Zhou, G., and Pei, X. (2018). Analysis of mixed lubrication of RV reducer turning arm roller bearing. *Industrial Lubrication and Tribology*, 70(1), 161–171. https://doi.org/10.1108/ILT-12-2016-0311
27. Shen, Z., Wang, F., Chen, Z., Ruan, X., Zeng, H., Wang, J., An, Y., and Fan, X. (2021). Numerical simulation of lubrication performance on chevron textured surface under hydrodynamic lubrication. *Tribology International*, 154, 106704. https://doi.org/10.1016/j.triboint.2020.106704
28. Hou, Y., Liu, Z., Hu, D., Guo, H., and Liu, Z. (2025). Influence of roughness on lubrication characteristics of high-speed helical gears in aviation engines. *Journal of Northwestern Polytechnical University*, 43(1), 1–12. https://doi.org/10.1051/jnwpu/20254310001
29. Tian, X., Zhang, C., Li, D., Li, L., Jin, L., and Chen, Y. (2026). The Influence of surface micro-texture on the performance of magnetic fluid sealing. *Tribology International*, 214, 111330. https://doi.org/10.1016/j.triboint.2025.111330

30. Prabhu, T. R. (2015). Effects of solid lubricants, load, and sliding speed on the tribological behavior of silica reinforced composites using design of experiments. *Materials & Design*, 77, 149–160. https://doi.org/10.1016/j.matdes.2015.03.059
31. Cai, Z., Zhou, Y., and Qu, J. (2015). Effect of oil temperature on tribological behavior of a lubricated steel–steel contact. *Wear*, 332–333, 1158–1163. https://doi.org/10.1016/j.wear.2015.01.064
32. Shamini, R., Jadhav, P., Deshpande, S., and Chavan, S. (2023). Design optimization methods for forced lubrication system used in automotive transmissions. *SAE International Journal of Commercial Vehicles*, 16(4), 2–16. https://doi.org/10.4271/02-16-04-0025
33. Tsigkis, V., Lan, P., and Polycarpou, A. A. (2021). Dynamic tribo-pair of PS400 and DLC for planet Venus application. *Tribology International*, 164, 107237. https://doi.org/10.1016/j.triboint.2021.107237
34. Qiang, L., Qinglei, L., Yujing, D., Bin, L., Yujun, W., and Weiwei, X. (2021). Advances in optimization design and application of textured surfaces. *China Surface Engineering*, 34(6), 59–73.
35. Haas, A., Esch, T., Fahl, E., Kreuter, P., and Pischinger, F. (1991). Optimized design of the lubrication system of modern combustion engines. *SAE Technical Paper*, 1991, 912407. https://doi.org/10.4271/912407
36. Li, Y., Chen, H., Zhang, J., Xiao, G., Yi, M., Chen, Z., and Xu, C. (2024). Synergetic lubrication between self-lubricating ceramic tools and nanofluids: Advancing the application of self-lubricating ceramic cutting tools. *International Journal of Refractory Metals and Hard Materials*, 120, 106622. https://doi.org/10.1016/j.ijrmhm.2024.106622
37. Zhang, J., Qiu, Y., Li, M., and Xu, M. (2017). Sequential multi-objective optimization for lubrication system of gasoline engines with bilevel optimization structure. *Journal of Mechanical Design*, 139(2), 493. https://doi.org/10.1115/1.4035493
38. Nguyen, T.-T., Nguyen, T.-A., Trinh, Q.-H., and Le, X.-B. (2022). Multi-performance optimization of multi-roller burnishing process in sustainable lubrication condition. *Materials and Manufacturing Processes*, 37(4), 407–427. https://doi.org/10.1080/10426914.2021.1962533
39. Dmitrichenko, M., Milanenko, O., Savchuk, A., Turytsia, Y., Pavlovskiy, M., Kushch, O., and Bobro, A. (2024). The conceptual model for increasing wear resistance and lubrication efficiency for non-conformal and conformal friction units from the standpoint of micro-EHD theory. In *TRANSBALTICA XIV: Transportation Science and Technology*. Prentkovskis, O., Yatskiv (Jackiva), I., Skačkauskas, P., Karpenko, M., Stosiak, M. (Eds.), Springer, Cham. https://doi.org/10.1007/978-3-031-52652-7_16
40. Zhou, Q., Guo, Y., Xu, K., Chai, B., Li, G., Wang, K., and Dong, Y. (2024). Research on the prediction algorithm of aero engine lubricating oil consumption based on multi-feature information fusion. *Applied Intelligence*, 54(22), 11845–11875. https://doi.org/10.1007/s10489-024-05759-6
41. Xiao, H., Qi, L., Shi, J., Li, S., Tang, R., Zuo, D., and Da, B. (2025). Reliability assessment of ship lubricating oil systems through improved dynamic bayesian networks and multi-source data fusion. *Applied Sciences*, 15(10), 5310. https://doi.org/10.3390/app15105310
42. Maccioni, L., and Concli, F. (2020). Computational fluid dynamics applied to lubricated mechanical components: Review of the approaches to simulate gears, bearings, and pumps. *Applied Sciences*, 10(24), 8810. https://doi.org/10.3390/app10248810
43. Singh, K., Sadeghi, F., Peterson, W., Lorenz, S. J., Villarreal, J., and Jinmon, T. (2022). A CFD-FEM based partitioned fluid structure interaction model to investigate surface cracks in elastohydrodynamic lubricated line contacts. *Tribology International*, 171, 107532. https://doi.org/10.1016/j.triboint.2022.107532
44. Mohammadi, A., and Sheikholeslam, F. (2023). Intelligent optimization: Literature review and state-of-the-art algorithms (1965–2022). *Engineering Applications of Artificial Intelligence*, 126, 106959. https://doi.org/10.1016/j.engappai.2023.106959

45. Simunovic, K., Simunovic, G., and Saric, T. (2015). Single and multiple goal optimization of structural steel face milling process considering different methods of cooling/lubricating. *Journal of Cleaner Production*, 94, 321–329. https://doi.org/10.1016/j.jclepro.2015.02.015
46. Singh, Y., Singh, P., Sharma, A., Choudhary, P., Singla, A., and Singh, N. K. (2018). Optimization of wear and friction characteristics of Phyllanthus Emblica seed oil based lubricant using response surface methodology. *Egyptian Journal of Petroleum*, 27(4), 1145–1155. https://doi.org/10.1016/j.ejpe.2018.04.001
47. Cai, X., Zou, T., and Gao, L. (2023). Surrogate-assisted operator-repeated evolutionary algorithm for computationally expensive multi-objective problems. *Applied Soft Computing*, 147, 110785. https://doi.org/10.1016/j.asoc.2023.110785
48. Najiha, M. S., Rahman, M. M., Kadirgama, K., Noor, M. M., and Ramasamy, D. (2015). Multi-objective optimization of minimum quantity lubrication in end milling of aluminum alloy AA6061T6. *International Journal of Automotive and Mechanical Engineering*, 12, 3003–3017. https://doi.org/10.15282/ijame.12.2015.15.0250
49. Chinchanikar, S., Katiyar, J. K., and Manav, O. (2022). Multi-objective optimizati of turning titanium alloy under minimum quantity lubrication. *Journal of Optimization in Industrial Engineering*, 15, 243–260. https://doi.org/10.22094/JOIE.2021.1937743.1886
50. Zhang, Z., Jia, X., Yang, T., Gu, Y., Wang, W., and Chen, L. (2021). Multi-objective optimization of lubricant volume in an ELSD considering thermal effects. *International Journal of Thermal Sciences*, 164, 106884. https://doi.org/10.1016/j.ijthermalsci.2021.106884
51. Rahman, M. H., Shahriar, S., and Menezes, P. L. (2023). Recent progress of machine learning algorithms for the oil and lubricant industry. *Lubricants*, 11(7), 289. https://doi.org/10.3390/lubricants11070289
52. Han, K., Shao, Y., Han, J., Song, W., and Sun, Z. (2025). Digital design and optimization of intelligent parameter-driven lubrication systems. *Advances in Engineering Technology Research*, 13(1), 1607. https://doi.org/10.56028/aetr.13.1.1607.2025
53. Bala, A., Rashid, R. Z. J. A., Ismail, I., Oliva, D., Muhammad, N., Sait, S. M., Al-Utaibi, K. A., Amosa, T. I., and Memon, K. A. (2024). Artificial intelligence and edge computing for machine maintenance-review. *Artificial Intelligence Review*, 57(5), 119. https://doi.org/10.1007/s10462-024-10748-9
54. Rigolli, D., Pompei, L., Manfredini, M., Vignoli, M., la Battaglia, V., and Giorgetti, A. (2025). The development of a predictive maintenance system for gearboxes through a statistical diagnostic analysis of lubricating oil and artificial intelligence. *Machines*, 13(8), 693. https://doi.org/10.3390/machines13080693
55. Zhu, X., Du, L., Liu, B., and Zhe, J. (2016). A microsensor array for quantification of lubricant contaminants using a back propagation artificial neural network. *Journal of Micromechanics and Microengineering*, 26(6), 065005. https://doi.org/10.1088/0960-1317/26/6/065005
56. Chen, Y., Shen, J., Zhou, F., Li, H., Yang, K., and Wang, L. (2025). Multi-sensor fusion and deep learning for predictive lubricant health assessment. *Lubricants*, 13(8), 364. https://doi.org/10.3390/lubricants13080364

6 Tribological and Wear Analysis Using Neural Networks

Jashanpreet Singh

6.1 INTRODUCTION

Wear is a progressive deformation of a solid surface that occurs as a result of mechanical interaction which may involve frictional contact between two bodies or, in exceptional cases, between a body and a rolled surface. It is a multiscale, complicated phenomenon that is dependent on material characteristics, surface roughness, pressure of contact, temperature, lubrication, and environmental factors [1]. Adhesive, abrasive, corrosive, and fatigue wear are the primary wear processes that are regulated by a variety of physical and chemical activities [2–4]. The knowledge of these mechanisms is most crucial in predicting material degradation, designing better components, and increasing service life in tribological systems. Quantitative evaluation of wear behavior is, therefore, a key factor in mechanical engineering and materials science studies.

Wear study is an important feature in several industrial uses, including automotive engines, aerospace parts, machining tools, biomedical implants, and renewable energy systems [5]. Loss of material to components by wear can lead to reduced system performance, inefficiency, and unforeseen downtime. Predictive wear modeling is employed in order to create materials and surfaces that have the capability of sustaining high loads, temperatures, and other environmental demands. Also, as industries shift to smart manufacturing and Industry 4.0 paradigms, the data-informed wear prediction is necessary for predictive maintenance and a digital twin that simulates the behavior of components under dynamic conditions [6].

Conventional wear prediction methods, including equation-based wear prediction introduced by Archard or empirical regression methods, heavily depend on simplified assumptions that do not necessarily reflect the process of wear behavior (nonlinear and stochastic) in reality [7]. These procedures are frequently not able to scale out to other materials, contact conditions, and environmental variables. Moreover, they need a lengthy period of experimental calibration, which is both time-consuming and expensive. Their inability to deal with complex interdependencies among tribological parameters is an indication that classical models require more adaptation and intelligent adaptation methods which can gain unobservable relationships directly through the information.

DOI: 10.1201/9781003624165-6

Neural networks (NNs) have become effective computational devices to approximate nonlinear systems, thus becoming the best tool in wear and tribological analysis [8]. The artificial neural networks (ANN) computer models are based on the networks of biological neurons which compose human brains. The input vector is the dendrite, as illustrated in Figure 6.1, which enables the cell body (or soma) to get the signal of a considerable number of neighboring neurons. The neurons send signals to the other cells via their axons. The impulses that are received by the dendrites are processed in the cell body, where the electrical state of the membrane is modified as a result of sodium–potassium exchange with the surrounding environment. When this electric potential reaches a certain threshold, the neuron causes an action potential that moves along the axon to the synapses and the adjacent neurons [9]. Using either experimental or time-simulation data training, the NNs can identify multifaceted core linkages involving input parameters, such as load, sliding speed, hardness, temperature, and wear result [10]. The NNs are not designed in a functional way as the empirical models; they may be optimized as more information becomes available and can, therefore, be better in prediction. They are further integrated with optimization algorithms and hybrid intelligent systems, which enhance the robustness of the model considerably and deliver a step toward reliable data-driven wear prediction in complex engineering environments.

A broad range of puzzling physical processes (TS) characterize the coating process in thermal spraying. During thermal spraying methods, there are intricate nonlinear interdependencies between the structure of the coating, in-flight particle properties, and process parameters. Computer-aided methodologies are used to quantify these intricate linkages and enhance the repeatability of the operation. Conventional modeling tools are routinely used to understand these relationships. It is a requirement that has prompted researchers to come up with a reliable method that could use modern digital tools, including artificial intelligence (AI), to counteract the issues with

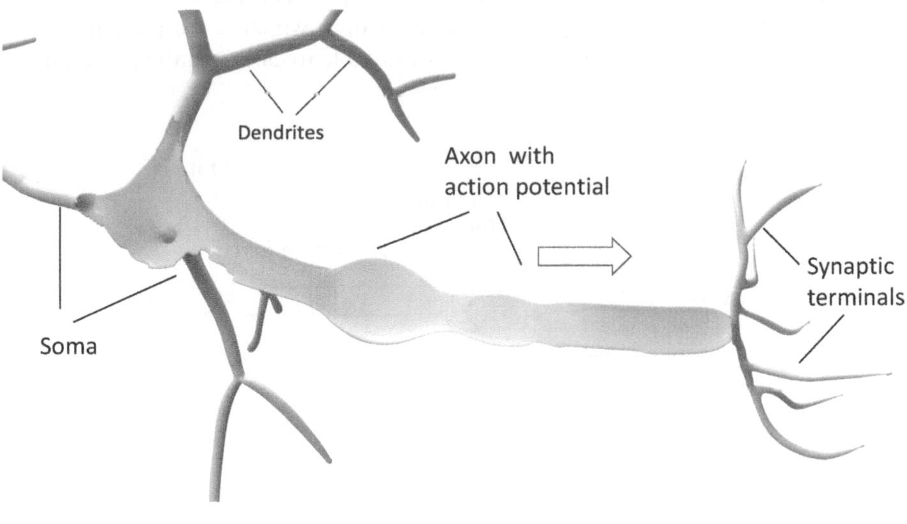

FIGURE 6.1 Illustration of the signal flow between input at dendrites and output at axon terminals [11]. (Creative Commons Attribution 4.0.)

TS-deposited coatings in different working conditions. The goal is to participate in a thorough conversation about AI and the many models connected with AI, including ANN, convolutional neural network (CNN), and hyperspectral imaging, which may help ease the digital transformation of the modern surface engineering sector and aid in the creation of a lasting TS coating system. This chapter has offered a succinct summary of the features of CNNs and ANNs in order to encourage the utilization of these reliable machine learning (ML) techniques for the optimization of TS parameters and process modeling.

6.2 FUNDAMENTALS OF NEURAL NETWORKS

6.2.1 Architecture and Working Principles

ANNs are computer models that are based on the design and functionality of the human brain. An ANN works in a similar fashion to a biological neuron, in that it converts an input space to an output space through a grouping of operational components, referred to as neurons, and connections (weights), which compares to a biological brain in terms of having an analogy of synapses. Neurons transmit messages among themselves by transmitting action potentials via their axons, which is determined by a transfer function that reflects the firing rate of the action potential. Depending on characteristics such as whether the neuron should be engaged, the inputs to the neuron might be more or less important, resulting in a bigger or smaller impact. The weight is adjusted to achieve this. The neuron can, therefore, be likened to a small computer that takes the incoming information, processes it, and produces an output. As shown in Figure 6.2, the simple neuron has an output and inputs designated R (p1, p2, p3, etc.). The bias and weighted input are combined to create the transfer function input, and each input is allocated the appropriate weight (*w*). In order to establish a correlation between the input and output, learning algorithms are necessary. The feedforward back-propagation algorithm, which is the most frequently used among the various learning algorithms available to carry out the processing, is

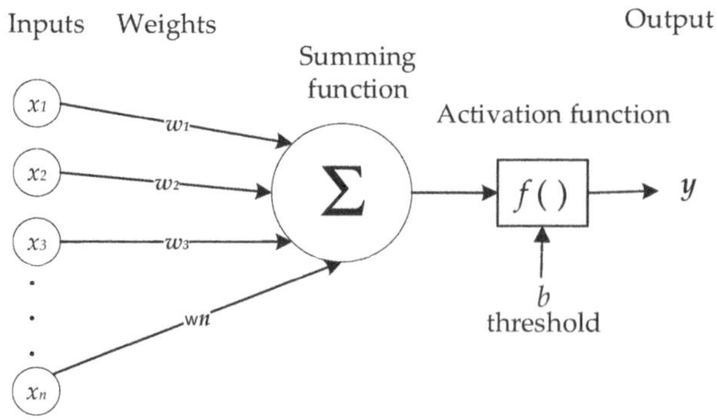

FIGURE 6.2 Illustration of the construction of a simple artificial neuron with R inputs [12]. (Creative Commons Attribution 4.0.)

the most common. The nonlinear transfer function is used to define the connection between the input and output in the back-propagation learning algorithm.

They consist of neurons that are interconnected and interwoven into layers. Overall, such layers are an input, one or more hidden, and an output layer [13]. By processing input signals via weighted connections and a transfer function, each neuron produces an output that serves as input to subsequent layers. The neuron j output in the lth layer can be mathematically defined as follows:

$$y_k^i = f\left(\sum_{i=1}^{n}\left(\omega_{i,j}^i x_i^{l-1} - b_j^l\right)\right) \tag{6.1}$$

where x_i^{l-1} is the input from neuron i in the previous layer, $\omega_{i,j}^i$ is the weight connecting neuron i to neuron j in layer l, b_j^l is the bias term, and f is the activation function. These linked neurons' aggregate activity determines the network's overall output. Learning is to adjust the connection weights so as to reduce the prediction errors according to a loss function. This is used in tribological modeling to model nonlinear relationships among wear parameters, which involve sliding velocity, applied load, and material hardness.

6.2.2 Activation Functions and Learning Algorithms

NNs with activation functions are nonlinear and therefore able to capture more complex relationships between two variables compared to linear mappings. The most common example of activation functions is the sigmoid function:

$$F(x) = \frac{1}{1+e^{-x}} \tag{6.2}$$

the hyperbolic tangent (tanh) function:

$$F(x) = \tan(x) = \frac{e^x - e^{-x}}{e^x + e^{-x}} \tag{6.3}$$

$$F(x) = \frac{1-e^{-2x}}{1+e^{-2x}} \tag{6.4}$$

and the rectified linear unit (ReLU):

$$F(x) = \max(0, x) \tag{6.5}$$

Gradient descent learning algorithms and their variants (e.g., Adam, RMSProp, and stochastic gradient descent) do this by repeatedly updating the connection weights to minimize the loss function. The gradient descent basic weight update equation is as follows:

$$\omega_{i,j}^{(t+1)} = \omega_{i,j}^{(t)} = -\eta \frac{\partial \text{MSE}}{\partial \omega_{i,j}} \tag{6.6}$$

where η is the learning rate and $\partial MSE/\partial\omega_{i,j}$ is the gradient of the loss function with respect to the weight. Efficient learning algorithms are crucial to achieving rapid convergence and high prediction accuracy when dealing with large experimental or simulation-derived datasets [14].

Parameters that are utilized in training the model are the number of neurons in the hidden layers, the rate of learning, the number of hidden layers, and the momentum factor. The performance evaluation is done due to the analysis of the root mean square error (RMSE) between the predicted output and the result. The ANN determines the error by comparing the predicted values with the target values after each iteration. When the error is more than the allowable error, the weights are again altered so as to minimize the error and the network re-run. To counter the problem of overfitting, the data are separated into test, training, and validation datasets.

6.2.3 Supervised vs. Unsupervised Learning for Wear Prediction

During supervised learning, networks are trained with labeled data, in which all input parameters and output values, e.g., wear rate or coefficient of friction, are known [15–17]. The paradigm finds extensive application in wear modeling [10]. On the other hand, unsupervised learning is concerned with determining unseen trends or groupings in unspecified data, including classifying the material conduct in different operational circumstances. As examples, in self-organizing maps, the clustering is based on the distance between input vectors and neuron weights:

$$D_{i,j} = \left\| x_i - \omega_j \right\| \tag{6.7}$$

where x_i is the input vector and ω_j is the weight vector of neuron j. Such methods can reveal intrinsic relationships among tribological parameters, aiding in feature extraction and dimensionality reduction prior to model training.

6.2.4 Performance Metrics and Model Evaluation

The performance of neural net models is important to evaluate predictive reliability. Common statistical measures are as follows:

- RMSE:

$$\text{RMSE} = \sqrt{\frac{1}{N}\sum_{k=1}^{N}\left(y_k^{\text{pred}} - y_k^{\text{true}}\right)^2} \tag{6.8}$$

- Mean absolute error (MAE):

$$\text{MAE} = \frac{1}{N}\sum_{k=1}^{N}\left|y_k^{\text{pred}} - y_k^{\text{true}}\right| \tag{6.9}$$

- Coefficient of determination (R^2):

$$R^2 = 1 - \frac{\sum_{k=1}^{N} \left(y_k^{\text{pred}} - y_k^{\text{true}}\right)^2}{\sum_{k=1}^{N} \left(y_k^{\text{pred}} - \bar{y}\right)^2} \tag{6.10}$$

- Mean squared error (MSE) defined as follows:

$$\text{MSE} = \frac{1}{N} \sum_{k=1}^{N} \left(y_k^{\text{pred}} - y_k^{\text{true}}\right)^2 \tag{6.11}$$

A high value of R^2 that approaches 1 implies that the model strongly correlates with the predicted and observed values, whereas a lower value of MSE or MAE depicts better precision. Cross-validation methods are commonly used to provide model generalization, like *k*-fold validation. In the evaluation of categorical wear states, a variety of methods, including confusion matrices and receiver operating characteristic (ROC) curves, are useful. The correctness and generalizability of NN models in real-world tribological applications are contingent upon their robust performance analysis.

6.3 DATA ACQUISITION AND PRE-PROCESSING

6.3.1 Experimental and Simulation-Based Wear Datasets

Any wear prediction algorithm using NNs relies on accurate data collection. Controlled laboratory experiments, *in situ* monitoring systems, or numerical simulations (e.g., finite element (FE) and computational fluid dynamics (CFD) models) can be used to gather wear data. The experimental data are frequently associated with the measurements of such parameters as applied load (L), sliding distance (s), hardness (H), surface roughness (R_a), and temperature (T). This is commonly done using tribometers, pin on disk testers, or reciprocating sliding devices. In addition to physical testing, simulation-based datasets can be used to provide high-resolution information on contact stress, temperature fields, and wear depth evolution and synthesize large datasets at different operating conditions [18]. The integration of the two sources of data increases the diversity and representativeness of training samples, allowing NNs to effectively generalize across a variety of wear mechanisms.

$$w = k\frac{Ls}{H} \tag{6.12}$$

Here, k is the wear coefficient, L is the applied load, s is the distance of the slide, and H is the hardness of the softer material. Integrating experimental and simulation data increases the variety and representativeness of training samples, allowing NNs to generalize well across the multiple wear mechanisms.

6.3.2 Feature Selection and Dimensionality Reduction

In wear modeling, feature selection is essential because the redundancy or irrelevance of input parameters may reduce the performance of the NN as well as the complexity of calculation. The approaches that can be employed to determine the most effective factors to be used to affect friction and wear rate include principal component analysis (PCA), correlation analysis, and mutual information [19]. Indicatively, in PCA, correlated variables are converted into a few principal components that are not correlated without losing most of the variance of the data. The process not only reduces dimensions but also reduces overfitting, allowing faster, more efficient training. The choice of features like hardness, load, and temperature due to their statistical relevance will make sure that the network considers physically significant inputs to process accurate wear prediction.

In wear modeling, feature selection is important because redundant or irrelevant input parameters may affect NN performance negatively and add complexity to its computation. The methods that may be applied to establish the strongest parameters contributing to wear rate and friction are correlation analysis, PCA, and mutual information. In PCA, the transformation of the original correlated variable X into uncorrelated principal component Z may be written as follows:

$$Z = XW \tag{6.13}$$

Principal Axis Variance (PAV) covariance matrix of X W is the matrix of eigenvectors of the covariance matrix of X. PCA retains the majority of the variance in the data, but dimensionality is also reduced. The use of features like hardness (H), load (L), and temperature (T), which are chosen on the basis of statistical relevance, will guarantee that the network will be interested in physically significant features to make a proper prediction of wear.

6.3.3 Data Normalization, Encoding, and Noise Reduction

Preprocessing is necessary to guarantee numerical stability and enhance the convergence of learning algorithms. Normalization techniques like min–max scaling and z-score standardization are used to bring the input features onto a shared scale to ensure that no one variable prevails in the learning process [20]. When there are categorical variables (e.g., material type and lubrication condition), during the encoding process, such categories are transformed into numerical representations (i.e., one-hot encoding). Another crucial preprocessing method is noise reduction, typically performed by signal filtering or smoothing, or outlier detection. Eradicating experimental errors and noise on the sensor improves the soundness of the data and the model it generates.

Preprocessing is necessary to ensure numerical stability and enhance the convergence of learning algorithms. Normalization puts features on a shared level. For min–max scaling:

$$x_i' = \frac{x_i - x_{\min}}{x_{\max} - x_{\min}} \tag{6.14}$$

or for z-score standardization:

$$x_i' = \frac{x_i - \mu}{\sigma} \tag{6.15}$$

where μ and σ are the mean and standard deviation of the feature, respectively. In the case of categorical variables (e.g., material type and lubrication condition) one-hot encoding is used to encode a category C into a vector:

$$C = [0,0,1,0,\ldots,0] \tag{6.16}$$

The reduced noise can be assessed through filters or smoothing algorithms. When considering an example of a time-series wear data ($w(t)$) that is moving on average (average), a simple moving average filter is as follows:

$$w(t) = \frac{1}{n}\sum_{i=0}^{n-1} \omega(t-i) \tag{6.17}$$

This minimizes measurement errors and sensor noise and enhances the strength of the dataset.

6.3.4 Handling Imbalanced and Missing Data in Wear Datasets

Wear datasets are commonly disproportionate, especially in case some operating conditions are overrepresented, which results in prejudiced learning outcomes. This can be solved by using oversampling methods like the Synthetic Minority Oversampling Technique (SMOTE) and undersampling of the majority classes [21]. Missing or incomplete data are typical of tribological experiments because sensors malfunction or data are corrupted in some manner, so interpolation, imputation with regression, or deep learning implicative methods, such as autoencoders, can be used to manage them. The correct management of skewed and incomplete information will make NNs biased and predictable in different working conditions and, therefore, enhance the predictability rates. This strict data preparation step is essential in developing credible and high-performing models in tribological systems.

Wear datasets are often skewed, especially when there is overrepresenting of some operating conditions, which results in biased learning results. Methods of oversampling like SMOTE generate artificial minority samples:

$$x_{\text{new}} = x_i + \delta.\left(x_j - x_i\right) \tag{6.18}$$

where x_i is the minority sample, x_j is the k-nearest neighbor of the sample, and $\delta \in [0,1]$ is the random number. Undersampling will decrease the number of samples in the overrepresented classes. Interpolation can be used to deal with missing or incomplete data or regression-based imputation:

$$x_{\text{miss}} = \frac{x_{i-1} + x_{i+1}}{2} \tag{6.19}$$

Autoencoders in deep learning can reconstruct missing features as $x_{\text{reconstruct}}$ or regression-imputation. The deep learning autoencoders can be used to replicate missing features as $x_{\text{reconstruct}}$

$$x_{\text{reconstruct}} = f_{\text{decoder}}\left(f_{\text{encoder}}\left(x_{\text{input}}\right)\right) \tag{6.20}$$

The correct management of skewed and missing data enables the NNs to be nonbiased and be able to work with a wide range of operating conditions, thus enhancing the reliability of prediction. Strict data preparation is essential in the creation of high-performance, reliable models within tribological systems.

6.4 PREDICTION OF WEAR USING NEURAL NETWORK MODELS

The prediction of wear is an important part of tribology and materials engineering as proper prediction of the degradation of materials can greatly contribute to the improvement of the performance, the reliability, and the maintenance planning of engineering components. Traditional wear prediction models are usually based on empirical or mechanistic models, which might not have the ability of capturing interactions with complex nonlinearities among the operational parameters, material properties, and environmental factors. Over the past few years, NN models have become a potent instrument in making predictions of wear behavior because they can learn complex, nonlinear relationships between experimental and operational data. This chapter discusses some of the NN structures that have been broadly used in predicting wear, with an emphasis on their concepts, mathematical equations, benefits, weaknesses, and uses.

6.4.1 Feedforward Neural Networks (FNN)

The FNNs consist of an input layer, hidden layer(s), and output layer. Neurons of one layer are fully connected with neurons of the next one. The reaction of each neuron is defined as the weighted sum of the synaptic inputs, which is subsequently converted into an activation function, as in Eq. (6.1). In the context of wear prediction, the x_i in Eq. (6.1) may include operating characteristics such as temperature, load, sliding speed, and lubricating qualities. The FNN is configured to optimize a loss function, which, in many cases, is the MSE. FNNs are effective in situations where the association between the inputs and outputs is moderately nonlinear but can be problematic with extremely complicated or high-dimensional information.

6.4.2 Backpropagation Neural Networks (BPNN)

The backpropagation neural networks (BPNNs) are a further development of FNNs, which apply the backpropagation algorithm to adjust the weights in the course of training. The goal of the training process is to reduce the error of the network by computing the gradient of the loss function of the network backward [22]. Equation (6.6) dictates the weight update ($\omega_{i,j}^{(t+1)}$). In wear prediction, BPNNs are employed to create multifaceted nonlinear correlations amid input characteristics like material hardness, sliding distance, and lubrication situations and the resulting wear rates. The network is adjusted iteratively to reach the best possible prediction results by adjusting the weights. BPNNs are quite useful in multivariate and nonlinear data, but the network parameters must be tuned to avoid overfitting.

6.4.3 Radial Basis Function Neural Networks (RBFNN)

The activation function of the hidden layer in neural networks (RBFNNs) is often a Gaussian function, which is described as the radial basis function (radial basis).

$$\varphi_j(x) = \exp\left(-\frac{\|x_i - c_j\|^2}{2\sigma_j^2}\right) \tag{6.21}$$

Here, x represents the input vector, c_j represents the center of the radial activity of neuron j, and σ_j represents the spread parameter. The RBFNN is estimated by computing the output as a weighted sum of the radial basis functions:

$$y(x) = \sum_{j=1}^{M} w_j \varphi_j(x) \tag{6.22}$$

where M is the number of hidden neurons and w_j is the output weights. RBFNNs are especially useful in wear prediction when the dataset has localized trends, e.g., set of load–temperature combinations that result in particular wear behavior. The radial basis functions are localized, which enables the network to make a nice interpolation between the known data points, and it therefore becomes very useful with clustered datasets.

6.4.4 Convolutional Neural Networks (CNN) for Surface Image Analysis

ANNs are constructed based on a layered architecture that usually has multiple processing layers at the input and one processing layer in the output. It is a concept that has evolved beyond the traditional multilayer perceptron (MLP) [23] models and advanced deep CNNs [24]. A CNN has the ability to establish meaningful feature relationships when given visual stimuli, and these resemble the human nervous system interconnection. Activation functions – like the sigmoid – transform the mean signal of one layer of neurons into a stimulus of the next in a nonlinear and adaptive way to allow nonlinear mapping between the inputs of two neurons and the outputs [25]. CNNs can effectively learn spatial and temporal correlations of the data by use of learnable filters. Convolution is done through sliding a chosen filter across an image and performing element-wise multiplications of filter values with the values in a particular region of the image to create feature maps [26]. The filters may be used directly to modify images. Photos may be examined and classified by a CNN [27].

CNNs are specialized networks that are specifically built to process grid-like data, such as pictures of worn surfaces that are collected by microscopy or profilometry. The core operation in CNNs is the convolution, defined as follows:

$$S(i,j) = (I * K)(i,j) = \sum_m \sum_n I(m,n) K(i-m, j-n) \tag{6.23}$$

where I is the input image, K is the convolutional kernel (or filter), and $S(i,j)$ is the feature map at location (i,j). Pooling layers are used to minimize the dimensionality of feature maps, while fully linked layers are employed to map the retrieved features to wear metrics. CNNs can automatically extract complex surface features such as roughness, scratches, or pits and correlate these with quantitative wear measures. The ability of CNNs to handle large-scale, high-dimensional datasets and learn hierarchical features makes them particularly effective for image-based wear prediction, though they require substantial labeled image data and significant computational resources.

6.4.5 Convolutional Neural Network (CNN) in Thermal Spray

Zhu et al. [28] were eager to eliminate the constraints of ANN models; hence, they used a basic CNN model to predict control parameters during the process of plasma spraying, as shown in Figure 6.3. The filter size of the pooling layer is 4×4, whereas that of the convolutional layer is 6×6. The results depict the ability of CNN models to generalize, which is useful in the process of setting control parameters, to achieve desired target coatings. Traditional methods, such as numerical simulation, are also much slower than CNN models in forecasting the outcome. The 3D dispersion of the plasma jet was formed in the simulation program within a time span of about 10 hours where control parameters had been selected. The CNN model also took very short time (0.01 seconds) to predict the needed control parameters after having been trained on a sequence of in-flight types of particle characteristic distributions. This, in turn, allows saving computer resources and time by applying an NN model in computing the control parameters in the process of plasma spraying. To train and update the parameters of the model such as weight coefficients and biases, the BP technique will be utilized [29]. The training of various parameters of the CNN model is used to minimize the error or loss.

In order to find the best answer with the least amount of loss, the Adam method [30], which is also known as the stochastic gradient descent technique, was widely used during model training. One popular optimization technique, gradient descent, finds the parameters' partial derivatives. With an increase in the gradient, the slope becomes steeper. Instead of using the whole dataset, stochastic gradient descent uses a randomly chosen subset to calculate the gradient. This speeds up iterations while decreasing the computing cost of large optimization tasks. Furthermore, some study has shown that CNNs may provide results with adequate accuracy when used for microstructure measurement in TS coatings. In Chen et al. [31], to create sub-images of different sizes with their centers placed at the pixel, the pixels of the top coat layer were randomly chosen. This allowed them to train a CNN with different topologies.

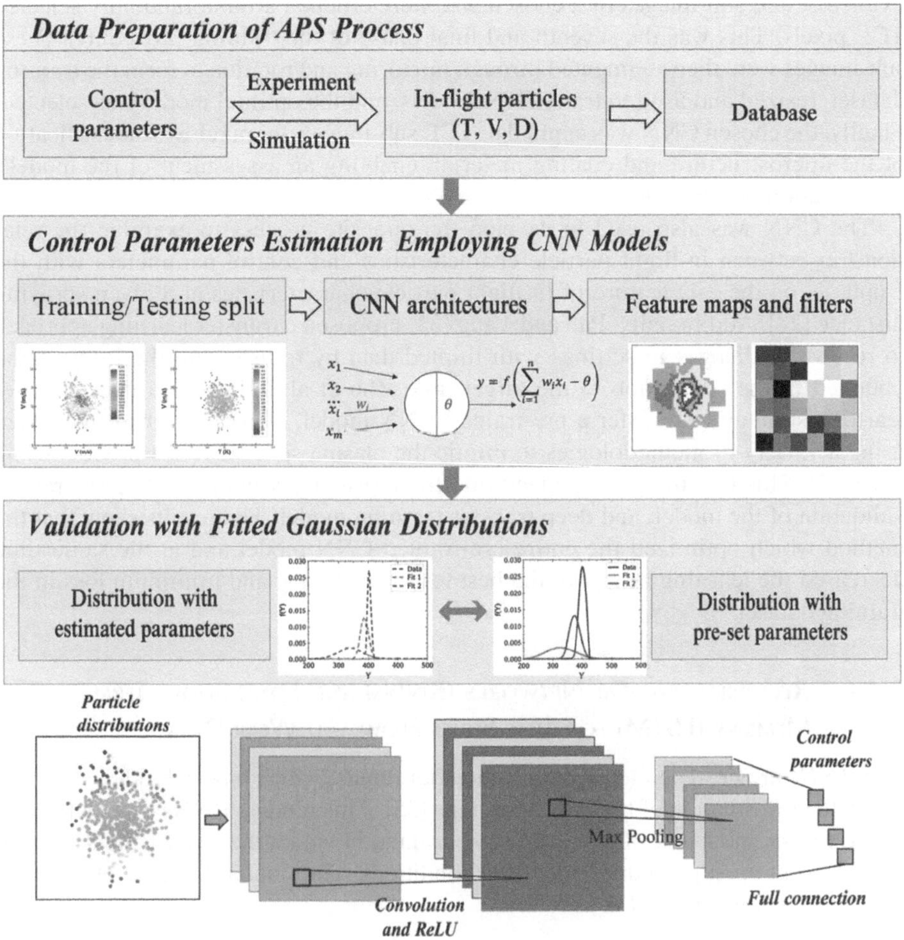

FIGURE 6.3 Illustration of a fundamental CNN model [28]. (Reproduced from © Elsevier.)

They used 150 photos of size 100 by 100, randomly chosen from a batch of 30 high-resolution photographs of thermal barrier coating as their dataset to evaluate their approach. Through the results, CNN-based models demonstrated a higher average of classification accuracy of 100% with the confidence level at 90% and reduced average relative error of 0.113. They were based on a VGG16 model.

Building on the work of Chen et al. [31], Lu et al. [32] created a better CNN-based method by adding new processes such as data augmentation and transfer learning. Their dataset comprised 159 raw images of coatings produced using three powder types – Type A: Metco 601NS (50 images), Type B: Metco 995C (49 images), and Type C: Metco 204BNS (60 images). A ground truth mask was created by manually classifying each raw image pixel into four categories: amount material, top coat layer (TCL) microstructure, TCL coating material, and substrate. Then, the TCL was extracted, and sub-images of a chosen size were cropped around randomly selected TCL pixels. This was the seventh and final phase of the training procedure. These sub-images were then augmented through mirroring and rotation to form the training dataset, resized and used to train CNN models, and the optimal model was selected. Finally, the chosen CNN was applied to TCL sub-images for pixel-wise identification of the microstructure and coating material, enabling an assessment of the model's classification performance.

The CNN was also used to develop the opposite process to examine the relationship between in-flight particle characteristics and control parameters with the emphasis on the 2-D feature of in-flight particle characteristics at a given spraying distance [28]. Additionally, Pan and Yang [33] proposed a transfer learning approach to resolve challenges in settings with limited data by transferring the information obtained in the source job to the target task. Zhu et al. [34] used a deep transfer learning strategy to transfer a pre-trained CNN model, which was then fine-tuned using a variety of methodologies to mimic the plasma spray reversal process using NiCrAlY. This research is organized into three primary sections: data preparation, validation of the model, and deep transfer learning model. The results show that the method which optimized the entire pre-trained CNN model and at the same time decreased the learning rate gave the best testing accuracy and minimum loss in the training dataset.

6.4.6 Recurrent Neural Networks (RNNs) and Long Short-Term Memory (LSTM) for Time-Series Flow and Wear Data

The RNNs are designed to process sequential inputs, which have a hidden state that stores the information of the past time steps [35]. This renders RNNs especially well suited for the modeling of time-series wear data, in which the development of wear is contingent upon preceding operating conditions. The standard update of a hidden state when using an RNN is as follows:

$$h_t = f\left(W_{xh}x_t + W_{hh}h_{t-1} + b_h\right) \tag{6.24}$$

where h_t is the hidden state at time step t; x_t is the input vector at time t; W_{xh} and W_{hh} are input-to-hidden and hidden-to-hidden weight matrices, respectively; b_h is the bias term; and f is an activation function such as tanh or ReLU. Subsequently, the network output y_t is determined as follows:

$$y_t = g\left(W_{hy}h_t + b_y\right) \tag{6.25}$$

where W_{hy} is the hidden-to-output weight matrix, b_y is the output bias, and g is the output activation function, often linear for regression tasks like wear prediction.

As much as the RNNs can be used to capture the temporal dependencies, they are prone to disappearance and inflation of gradients during the training process. This constraint is resolved by the implementation of gating mechanisms in LSTM networks, which govern the flow of information. Key equations for an LSTM cell include the following:

Forget gate:

$$f_t = \sigma\left(W_f x_t + U_f h_{t-1} + b_f\right) \tag{6.26}$$

Input gate:

$$i_t = \sigma\left(W_i x_t + U_i h_{t-1} + b_i\right) \tag{6.27}$$

Cell candidate:

$$\tilde{c}_t = \tanh\left(W_c x_t + U_c h_{t-1} + b_c\right) \tag{6.28}$$

Cell state update:

$$c_t = f_t \Theta c_{t-1} + i_t \Theta \tilde{c}_t \tag{6.29}$$

Output gate:

$$o_t = \sigma\left(W_o x_t + U_o h_{t-1} + b_o\right) \tag{6.30}$$

Hidden state output:

$$h_t = o_t \Theta \tanh\left(c_t\right) \tag{6.31}$$

In this context, σ is the sigmoid activation function, Θ signifies element-wise multiplication, and W, U, and b are weight matrices and biases for their respective gates.

In wear prediction, LSTM networks are capable of modeling the time-dependent sequence of operational parameters including load, speed, temperature, and lubrication conditions to make predictions of wear. Their long-term dependence retention capacity renders them very effective in predicting wear in systems that are exposed to variable or cyclic loading conditions. Madhavan et al. [36] developed an LSTM time-series model to forecast the response of force on valves during hydrogen-driving vehicles and infrastructure. The experiment aimed at testing the valve performance and durability at conditions of extreme temperatures (−40°C, 25°C, and 85°C) on the basis of the European Commission Regulation (EU) No. 406/2010 on vehicle type approval of the hydrogen vehicle. The LSTM network was trained and tested using experimental endurance data, which are equal to 75,000 operational cycles in 20 years of service life. The predicted accuracy of the proposed model, which includes three input parameters and one output, was high, with R^2 values of more than 0.96 at all the different temperatures tested, and was optimally at 25°C after 40,000 cycles of training with an $R^2 = 0.969$. It was discovered that the final layer in LSTM architecture consisting of a dense layer interestingly equipped with a ReLU activation function was necessary to transform sequential outputs into a desired predictive format, as shown in Figure 6.4. Madhavan et al. [36] concluded that the LSTM

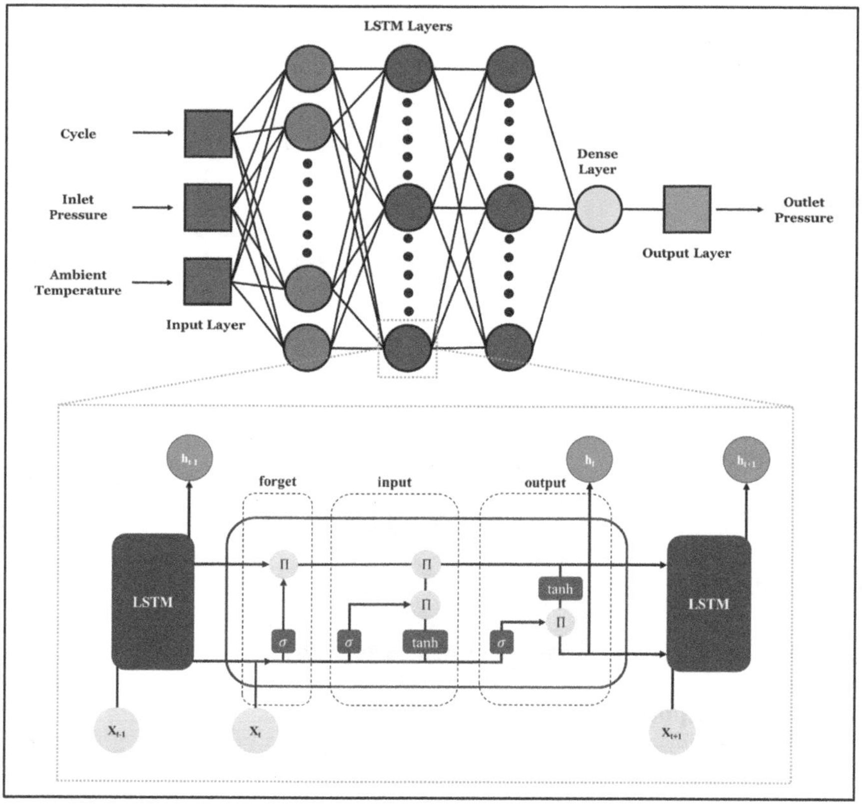

FIGURE 6.4 A long short-term memory (LSTM) model with three inputs and one output, developed in this study to predict valve behavior (The detailed model architecture is highlighted within the dotted box) [36]. (Creative Commons Attribution 4.0.)

framework provides an effective, stable, and real-time diagnostics to fast-track valve tests and development in hydrogen systems to promote sustainable hydrogen mobility technologies.

6.5 CASE STUDIES AND APPLICATIONS

6.5.1 Artificial Neural Network for TS Coatings

ANNs have become a potent method of forecasting the behavior and characteristics of thermal sprays. The coatings have been popularized to augment the surface properties of wear resistance, corrosion protection, and thermal insulation. The equation of state between process parameters, including spray distance (*ds*), feed rate (*F*), plasma power (*P*), particle velocity (v_p), and substrate temperature (*Ts*), and coating properties, such as hardness (*H*) and porosity (ϕ), are too complicated and nonlinear to be adequately represented by simple models, which make ANNs a particularly useful application to such systems.

For thermal spray coating, a feedforward ANN with input (process parameters), hidden (nonlinear interactions), and output (coating characteristics) layers is one possible implementation. To determine the output of the jth neuron in the hidden or output layer, the calculation in Eq. (6.1) is done. The variables x_i, w_{ij}, b_j, and f in Eq. (6.1) represent the input features, connection weights, bias term, and activation function, respectively, such as sigmoid, tanh, or ReLU. The network is optimized to reduce the error in prediction, which is usually defined as MSE, as shown in Eq. (6.11).

For the purpose of forecasting thermal spray coatings, two typical types of NNs are used: feedforward and BP. They are able to measure multiples of process variable interactions to forecast both important coating properties, optimize spray parameters, and reduce experimental trial reasons.

Feedforward and BP neural networks are often employed for thermal spray coating predictive modeling. They can record intricate relationships between process variables, allowing for the prediction of crucial coating characteristics, optimization of spray settings, and a decrease in experimental trial-and-error. Advanced architectures like RBFNN or CNN can also be employed when image-based characterization, such as porosity or surface morphology, is available. The combination of ANNs with experimental and simulation data allows for robust, rapid, and cost-effective prediction of coating performance, enabling optimization of thermal spray processes with minimal trial-and-error experimentation.

The process responses comprise the characteristics of in-flight particles, and NC is a smart system that connects them to the process parameters. One of such systems is based on an ANN, a set of interconnected neurons that serve as processing units. A number of technical fields have made use of ANNs, a new kind of technology with strong learning capabilities and flexible topologies [37]. There are two main types of ANN: multilayer networks and single-layer NNs. The most commonly used ANN model in the application of TS is the multilayered feed-forward network due to its outstanding accuracy. A multilayered network was used in a limited number of studies published on the uses of surface coatings. It typically consists of three layers, namely, input, hidden, and output layers.

As shown in Figure 6.3, the process parameters' settings are represented by the input pattern. One or more neurons define each parameter in this manner, dependent upon the kind of parameter. A single neuron is enough to represent a single real number, such the argon gas flow rate. In contrast, different types of injectors and other parameter categories can only be explained by x neurons. The size of the input pattern might change depending on the settings that are specified. The temperature, diameter, and particle velocity are shown in the output pattern. Each neuron describes one of these parameters. The number of hidden layers gives a vivid description of the relations between the processing parameters and the in-flight particle properties. The exact number of neurons of buried layers cannot be identified by the help of gross principles. A network that has many neurons in the hidden layer gives the leeway to optimize a wider variety of parameters. The network, however, becomes under-characterized at some point beyond which the hidden layer becomes large. The outcome of an insufficient number of neurons in the buried layers is under-fitting. The single hidden layer is used in order to determine the best number of hidden layers in the simulation. The buried layer comprises a range of four to 20 neurons.

The intensity of a two-neuron connection is represented by a numerical figure named as weight. Using a transfer function, each neuron in the ANN structure adjusts the flux that it accepts by the other neurons, which is a weighted number. This change guarantees that it is nonlinear at the neuronal level. The optimization of the ANN structure is achieved by including experimental sets into the learning process. This approach enables the development of a predetermined number of neurons and a weight population. An assumed population of the weights and a predetermined number of neurons p are used to compute and compare the system's response to the experimental results. The structural weights will be adjusted in such a way that the residual error of every layer is minimized.

Over the last several years, the ANN has gradually increased its engagement in TS coatings. ANNs are commonly employed in TS coatings to carry out a broad range of activities, which comprise calculating micro-hardness, porosity, coating thickness, coating microstructure, tribological characteristics, and in-flight behavior of particles. The ANN is also a crucial component of the process modeling of surface coatings because it compensates the altered variables and eliminates unwanted noise. These were capable of more accurately optimizing such complex nonlinear processes and predicting more favorable conditions for coating deposition on the material's surface. The further choice of TS process parameters, in particular, when it comes to the creation of ceramic coating with predetermined functional properties, depends on whether AI methods will be implemented. Most recently, ANNs have been applied in replicating features on tracks while additive manufacturing via cold spray. In future studies, the data-efficient ANN model will be incorporated into the tool-path planning algorithm to improve the geometric control and create more complex product designs with the help of cold spray additive manufacturing (CSAM) [38].

6.5.2 Neural Network Prediction of Wear Rate

The tendency to learn complex, nonlinear correlations between the working conditions and the behavior of materials has made ANNs a popular prediction tool in tribological systems in a number of ways. The input parameters usually consist of load (L), sliding speed (v), temperature (T), hardness (H), lubrication properties (μ), and surface roughness (R_a), and the network output is the wear rate (W). In any case, a wear prediction model based on NNs can be simply summarized as follows:

$$W_{\text{pred}} = F\left(L, v, T, H, \mu, R_a; W, b\right) \tag{6.32}$$

Here, F is the mapping that is learned of the input parameters to wear rate, and W and b are all the network weights and biases, respectively. The approach will facilitate the selection of materials, design optimization, and maintenance scheduling in tribological systems, as well as enable the rapid and trustworthy prediction of wear.

To ensure protective coatings on different materials is the primary objective of atmospheric plasma spraying. In this method of deposition, the particles are injected into a large heated plasma plume where they fuse and then fall onto a substrate. The Mo coatings have been of interest in areas where a high temperature is involved [39]. The aerospace, pulp, automotive, and paper industries are the primary applications with PS Mo-based coatings due to their superior wear resistance characteristics. In Cetinel et al. [40], a wear-friction device was used as a general wear testing apparatus, where the Mo coating was created through the use of plasma spraying. The wear behavior of PS Mo coatings (ANN) was simulated by using an NN approach. This study has an NN consisting of the input layer, two hidden layers, and the output layer. The input factors were the average load, time, and dry and acidic atmosphere. The two output variables were microhardness and wear loss. Consequently, ANNs can be applied in the assessment of the wear of Mo coating, and the degree of wear loss may be determined without calculation, which is time-consuming. Singh and Singh [41] have constructed an ANN that was used to predict the behavior of slurry erosion of 18Cr–8Ni, 16Cr–10Ni–2Mo, super duplex 24Cr–6Ni–3Mo-N, and gray cast iron heavy-duty pump impeller and casing material. A pot tester was used in experiments where sand was the erodent, and training (70%), validation (15%), and testing (15%) sets were used. The ANN gave an accurate result with a high R^2 of 0.99 variation in the data. Figure 6.5 indicates that the ANN model showed a high correlation between the predicted and experimental values of erosion, and the data points were closely compressed along the diagonal line, which supports the high predictive reliability of the ANN model. Figure 6.6 shows that the ANN produced the least residues among other regression models, meaning that there was a negligible gap between the predicted outcomes and the actual ones. The model was much better than PCR and PLSR methods as it has generated least errors in prediction, and it has shown great agreement ($R^2 = 0.99$ and Lin's concordance coefficient$=$0.99). In the laboratory, the authors found that 18Cr-8Ni steel exhibited the greatest erosion resistance, which confirms the strength of the designed ANN structure.

Satpathy et al. [42] also report that ANNs have been successfully used to explore the wear behavior of plasma spray coatings. The authors proposed the ANN is the best strategy to use with regard to time and resource conservation. Pati et al. [43] adopt an ANN and experimental design to examine the erosion wear behavior of a plasma-sprayed coating. The authors concluded that experimental and ANN-predicted data were in very good agreement, and the error does not exceed 7%.

6.5.3 Wear Modeling in Cutting Tools and Bearings

In industry, wear prediction of cutting tools and bearings is essential in optimization of performance, increase of life of service, and minimization of maintenance expenses. The cutting tools undergo progressive wear as a result of abrasive, adhesive, and thermal influences during machining processes, and bearings are exposed to rolling and sliding contact, which results in fatigue and surface deterioration. ANNs offer an effective system of modeling wear in such systems due to their ability to model nonlinear dependence between operating parameters and wear results.

In cutting tools, the main input factors are cutting speed (v_c), feed rate (f), depth of cut (d), tool material hardness (H_t), workpiece hardness (H_w), and lubrication conditions (μ). An ANN mapping tool can be used to predict the wear of the tool (W_t):

The primary input parameters for cutting tools are as follows: an ANN mapping function may be used to estimate the tool's wear:

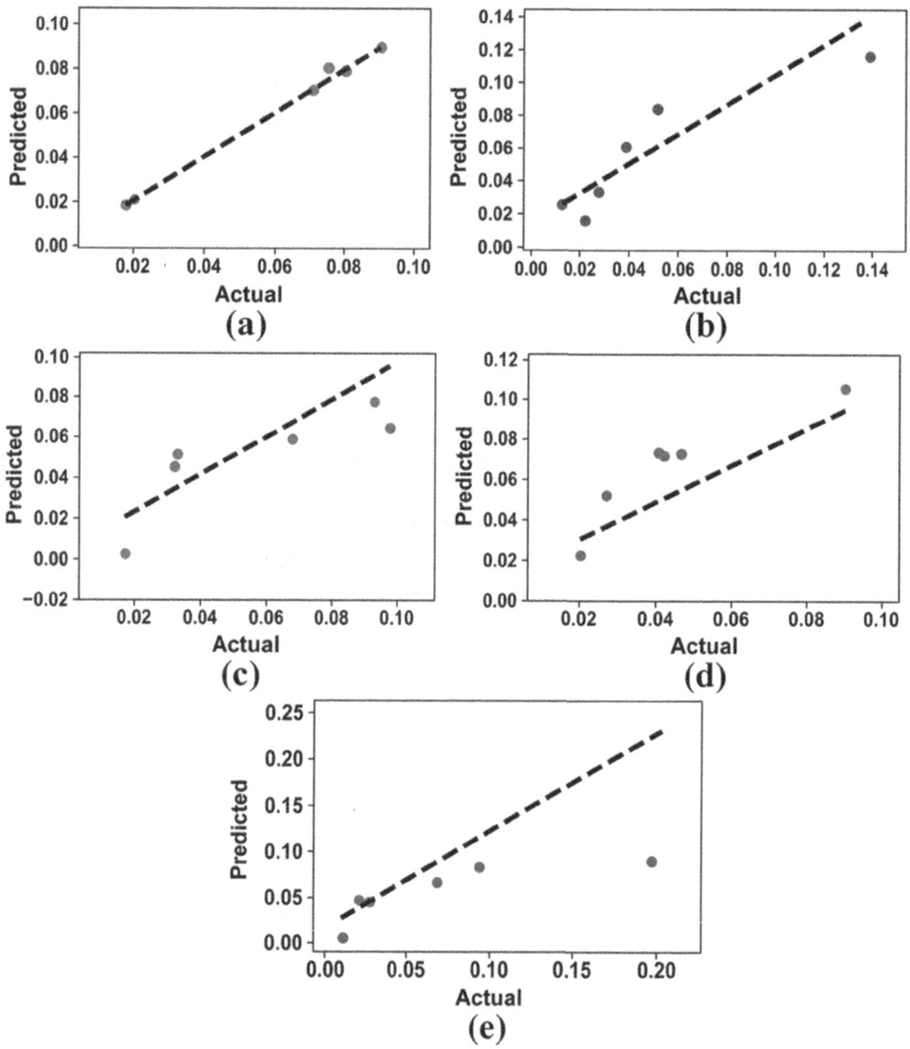

FIGURE 6.5 Comparison between the experimentally measured and model-predicted values using different regression approaches: (a) ANN, (b) GPR, (c) MLR, (d) PLSR, and (e) PCR [41]. (Reproduced from © Elsevier.)

$$W_{\text{pred}} = F\left(v_c, f, d, H_t, H_w, \mu; W, b\right) \tag{6.33}$$

During training, reported MSE can be used to quantify the predicted rate of tool wear (with Eq. 6.11). These predictive models are useful in optimization of machining, minimization of downtime, and enhancement of surface finish of the manufactured parts.

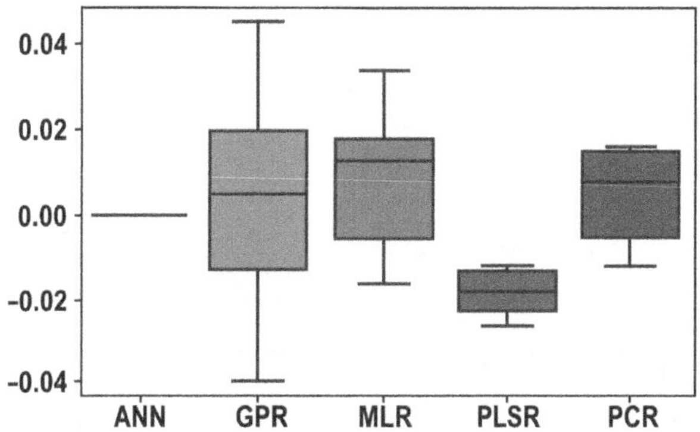

FIGURE 6.6 Comparison of residual errors for the ANN model against other regression models, highlighting the superior accuracy of ANN predictions [41]. (Reproduced from © Elsevier.)

The wear of bearings is determined by the following factors: load (L), rotational speed (ω), lubricant viscosity (η), and operating temperature (T). Bearing wear (W_b) is predicted by the ANN output as follows:

$$W_b = f_{\text{ANN}}\left(L, \omega, \eta, T; W, b\right) \tag{6.34}$$

The wear is a common result of Archard law of sliding and rolling contacts as given in Eq. (6.12).

Rong et al. [44] developed an effective tool wear prediction framework that combines the use of Brownian motion stochastic process, Bayesian inference, and support vector regression (SVR) by which the framework becomes capable of making accurate and uncertainty-sensitive predictions. The method measures uncertainties during machining processes by probabilistically biased predictions about the tool wear and future trends of wear prediction. The strategy comprises offline and online stages, as shown in Figure 6.7. During the offline stage projection, the Brownian motion stochastic process is used to model degradation features, and then the SVR model is trained to predict wear progression by these features. Internally, during the online phase, Bayesian inference is used to continually refresh the model parameters using real-time data, and random simulations are used in order to predict the future degradation trends. New features are then entered into the trained SVR model so as to predict the tool wear distribution. They reported that such a hybrid Bayesian–SVR model is very useful in terms of predicting stochastic degradation behavior of tools with accuracy and to a large extent uncertainty measurement; thus, it is of great use in adaptive machining, predictive maintenance, and adaptive control.

ANNs improve this classical model by adding more parameters of operations and materials that are used in practice that influence wear (i.e., the quality of lubrication, contamination, and vibrations). Examples of application-based benefits of ANN-based wear modeling include the following:

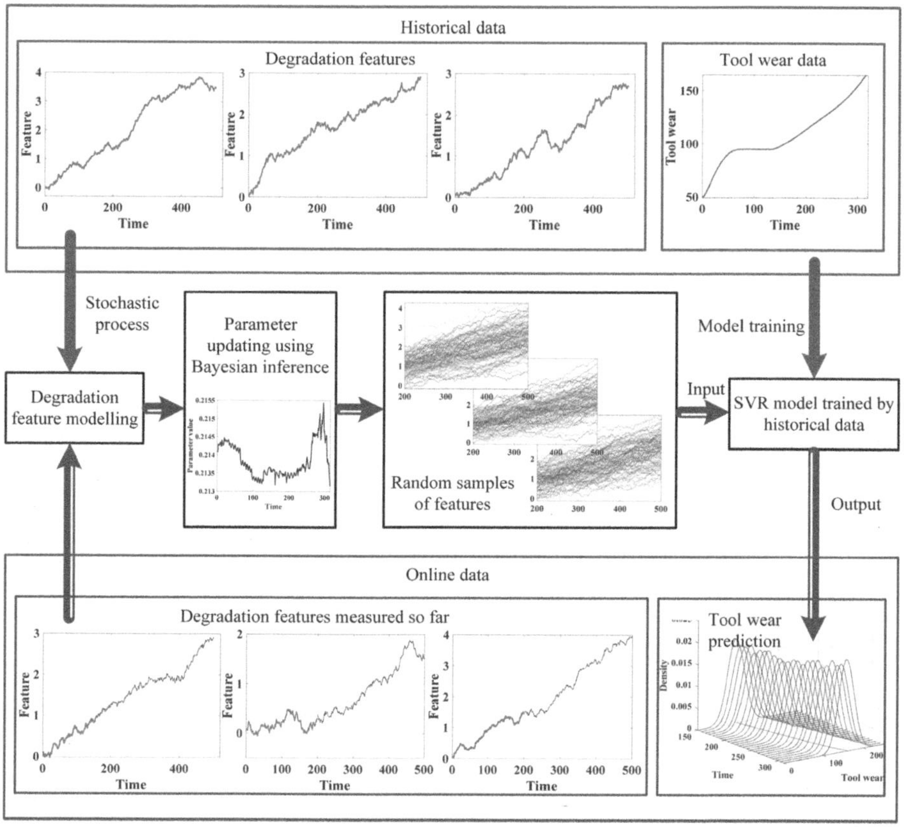

FIGURE 6.7 Procedure of the Bayesian Inference and Support Vector Regression method [44]. (Creative Commons Attribution 4.0.)

- **Cutting tools:** Predicting flank and crater wear for preventative maintenance; optimizing feed, speed, and tool material to preserve dimensional accuracy and maximum tool life.
- **Bearings:** The anticipation of maintenance details, reduction of unforeseen failure, and the choice of suitable lubricants and materials in particular operating loads and speeds.
- **Industry integration:** ANN-based wear models may be integrated into digital twins to facilitate predictive maintenance in aircraft components, automotive systems, and manufacturing lines.

The prediction of wear in cutting tools and bearings is rendered more precise, cost-effective, and adaptive to intricate industrial settings by the integration of experimental observations, simulation data, and ANN-based models. This predictive capacity enhances the design of more durable tribological systems, extends the life of components, and lowers downtime.

6.5.4 Validation against Experimental and Numerical Results

Validation of NN models is a critical step to ensure predictive reliability in wear analysis. The only way to achieve accuracy in the prediction of the wear is to ensure that the model not only fits the training data, but it also extrapolates effectively to the unobserved operational conditions. The validation is generally compared to experimental measurements and numerical simulation outcomes, e.g., the finite-element analysis (FEA) or the CFD types of studies.

6.5.4.1 Numerical Validation

The data from the simulation may also be used to verify the predictions of the neural network. FE models of contact surfaces or components used in bearings can mimic wear behavior for a variety of loads, speeds, and temperatures. Forecasted wear depths may be used to verify the ANN.

$$\text{Error}_k = W_k^{\text{Pr}} - W_k^{\text{FE}} \tag{6.35}$$

Cross-validation provides the model with local and global trends in wear. Such a two-way assurance (experimental and numerical) contributes to detecting anomalies caused by noise in measurements and sensor effects and simplified simulation models.

6.5.4.2 Sensitivity and Parametric Analysis

Sensitivity analysis establishes the effect of varyingly input parameters (e.g., load L, sliding speed v, temperature T, and material hardness H) on the offered wear predictions. Regarding every input xi, the sensitivity Si can be determined as follows:

$$S_i = \frac{\partial W_{\text{prd}}}{\partial x_i} \tag{6.36}$$

Parameter analysis builds upon this principle by the systematic variation of several parameters to discuss their interaction to wear behavior. This allows practitioners to prioritize critical variables, streamline operation circumstances, and inform the choice of material or design.

6.5.4.3 Model Robustness and Uncertainty

The quantification of uncertainty guarantees that NN prediction considers variability of data, measurement error, as well as model approximations. The comparison of the experimental and numerical wear to the ANN output allows establishing confidence intervals or error bands:

$$W_{\text{prd}} \pm \sigma \tag{6.37}$$

where σ is the standard deviation of predictive errors. This makes sure that the predictions of ANN-based wear are accurate as well as reliable in the use of industrial decisions. In certain cases, Bayesian NNs or Monte Carlo dropout can also be used to get probabilistic distributions of wear rates that were predicted and have confidence limits that can be used in risk-conscious decision-making in industrial applications.

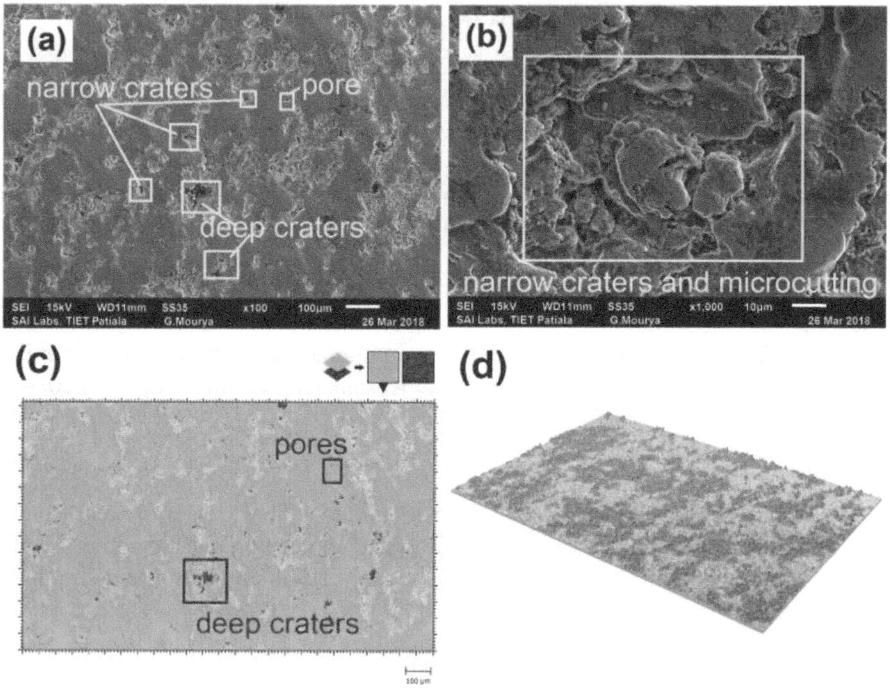

FIGURE 6.8 (a and b) Scanning electron microscope (SEM) micrographs, (c) 2D topology, and (d) 3D surface topography analyses illustrating the erosion behavior of the Colmonoy 88 coating produced by the HVOF process [46]. (Reproduced from © Elsevier.)

6.5.5 Image Processing in Characterization of Thermal Spray Coatings

When it comes to characterizing erodent, IP approaches are quite successful. The IP instruments offer a complete range of characterizations, including three-dimensional topography, porosity measurement, two-dimensional profilometry, surface roughness, fracture analysis, and other similar measures. Experiments were conducted by Singh et al. [45] to analyze the topography, porosity, and 2D profilometry of the WC-10Co4Cr coating. When it comes to testing for surface erosion on coatings, imaging may be very successful. Imaging technologies make it easy to test wear signs. To investigate the processes of surface wear on the Colmonoy 88 HVOF coating, Singh et al. [46] conducted an IP experiment. Figure 6.8 shows the fundamental processes that have been identified to explain the phenomena of wear, which include craters, microcutting, and pores.

6.6 CHALLENGES AND FUTURE PROSPECTS

NN-based wear prediction has shown significant potential in tribological systems, but several challenges must be addressed to ensure robust and reliable deployment. However, the novel computational methodologies and their incorporation into sophisticated digital structures offer promising perspectives for future studies.

6.6.1 Data Quality and Generalization Issues

The quality, diversity, and representativeness of the training datasets are one of the main issues of the NN-based prediction of wear. Sensor inaccuracies, variability in operation, and incomplete measurements make experimental wear data limited, noisy, or imbalanced. Similarly, the simulation-based datasets, though large, can be based on simplifications that restrict their real-world application. The quality of the data directly influences network generalization and results in overfitting or poor performance in terms of predicting wear in novel unknown operating conditions. These problems can be addressed with the help of data augmentation, transfer learning, and hybrid experimental-simulation datasets, which enhance the robustness of models in a variety of material systems and operating conditions.

6.6.2 Interpretability and Explainability in Neural Network Models

Although NNs have a predictive aspect, they are commonly referred to as black boxes, which means that it is very hard to understand the effects of certain inputs on wear predictions. This nontransparency restricts trust in industrial use especially where safety and reliability are of paramount importance. SHapley Additive explanations (SHAP), Local Interpretable Model-agnostic explanations (LIME), and sensitivity analysis have partial interpretability and quantify the impact of inputs. The additional research of explainable AI techniques that are specifically applied to tribological requirements will contribute to the increased practical implementation of NNs in predictive maintenance, tool design, and optimization of bearings.

6.6.3 Integration with Digital Twins and AI-Based Tribological Systems

The combination of NN wear models with digital twins and tribological platforms with AI is a radical solution to predictive maintenance and system optimization. It is possible to simulate real-time wear progression with the help of digital twins, which operate using sensor data of the working equipment, and continuously revise predictions with the help of NNs, which enables proactive interventions. This type of integration needs smooth data pipelines, a strong ability for model generalizability, and efficient real-time computation. The application of NN-based wear models in such systems represents the opportunities of smart decision-making, adaptive control of processes, and optimized use of material in manufacturing, automotive solutions, and aerospace.

6.6.4 Opportunities for Reinforcement Learning and Self-Adaptive Systems

Reinforcement learning (RL) and self-adaptive neural systems could form the basis of tribological wear prediction in the future. In the RL-based structure, the model can be optimized for operation strategies (e.g., load, speed, and lubrication) through iterative optimization to reduce wear and at the same time along with the simulated or real-world interactions. Self-adaptable networks have the ability to dynamically change their architecture and learning parameters when there are variations in materials, environmental conditions, or operational loads. These measures can be used to enable decentralized optimization of tribological systems, provide predictive information in real time, and improve the resilience of systems in the presence of relevant industrial conditions.

6.7 CONCLUSIONS

Wear modeling with NNs is a revolutionary method in tribology, which has strong predictive power compared to conventional empirical and analysis models. NNs are capable of predicting the wear of cutting tools, bearings, and surface finishes accurately by capturing complex nonlinear relationships between the operational parameters, material properties, and environmental factors. Experimental and simulation data are carefully combined and preprocessed, and features are selected to guarantee that the model is reliable and generalized. Predictive performance can also be assessed by statistical measures, sensitivity analysis, and quantification of uncertainty, which can then be used to inform decision-making in the industrial context. The issues of data quality, interpretability, and integration with digital twins will not be left behind, but the application of RL and self-adaptive systems has great potential in further research. NN-driven wear analysis provides an effective, flexible, and scalable framework for optimizing component life, maintenance schedules, and operational efficiency, advancing the field of predictive tribology.

REFERENCES

1. Grützmacher PG, Profito FJ, Rosenkranz A. Multi-scale surface texturing in tribology—Current knowledge and future perspectives. *Lubricants*. 2019;7:95.
2. Singh J, Kumar S, Mohapatra SK. An erosion and corrosion study on thermally sprayed WC-Co-Cr powder synergized with $Mo_2C/Y2O_3/ZrO_2$ feedstock powders. *Wear*. 2019;438–439:102751. Available from: https://doi.org/10.1016/j.wear.2019.01.082.
3. Singh J, Kumar S, Mohapatra SK. Erosion wear performance of Ni-Cr-O and NiCrBSiFe-WC(Co) composite coatings deposited by HVOF technique. *Ind Lubr Tribol*. 2019;71:610–619.
4. Bolelli G, Berger LM, Bonetti M, et al. Comparative study of the dry sliding wear behaviour of HVOF-sprayed WC-(W,Cr)2C-Ni and WC-CoCr hardmetal coatings. *Wear*. 2014;309:96–111. Available from: https://doi.org/10.1016/j.wear.2013.11.001.
5. Hussein MA, Mohammed AS, Al-Aqeeli N. Wear characteristics of metallic biomaterials: A review. *Materials*. 2015;8:2749–2768.
6. Imandi R, Chethana B, Prasad BMP, et al. Enhancing data management in industry 5.0: the role of digital twins in optimizing industrial operations. In: Sarkar I, Hazra A, Maurya P (Eds) *Industry 50 Key Technol Drivers*. Cham: Springer; 2025:211–236.
7. Tian Y, Khan M, Yuan H, et al. Wear modeling and friction-induced noise: A review. *Friction*. 2025;272:10–16.
8. Villarrubia G, De Paz JF, Chamoso P, et al. Artificial neural networks used in optimization problems. *Neurocomputing*. 2018;272:10–16.
9. Cohen LB. Changes in neuron structure during action potential propagation and synaptic transmission. *Physiol Rev*. 1973;53:373–418.

10. Marian M, Tremmel S. Current trends and applications of machine learning in tribology—A review. *Lubricants*. 2021;9:86.
11. Kotthaus JP. An acoustoelectric approach to neuron function. *Acoustics*. 2023;5:601–618.
12. Engin M, Duymazlar O, Engin D. An artificial neural network-based data-driven embedded controller design for a pneumatic artificial muscle-actuated pressing unit. *Appl Sci*. 2024;14:4797.
13. Shafi I, Ahmad J, Shah SI, et al. Impact of varying neurons and hidden layers in neural network architecture for a time frequency application. *2006 IEEE International Multitopic Conference*, Islamabad, Pakistan. 2006:188–193.
14. Kim KS. Machine learning for accelerating energy materials discovery: Bridging quantum accuracy with computational efficiency. *Adv Energy Mater.* 2025;16:e03356.
15. Gahane S, Verma P, Khode P, et al. Innovating supervised and semi supervised learning: Reducing label dependency, enhancing data efficiency and developing scalable, adaptive models for real time applications. *Int Conf Inf Commun Technol Compet Strateg*. 2024;1319:133–143.
16. Albalate A, Minker W. *Semi-Supervised and Unsupervised Machine Learning: Novel Strategies*. New York: John Wiley & Sons; 2013.
17. Yang X, Song Z, King I, et al. A survey on deep semi-supervised learning. *IEEE Trans Knowl Data Eng*. 2022;35:8934–8954.
18. Liu Y, Wei Q, Wang W, et al. Multiscale modeling and data-driven life prediction of kinematic interface behaviors in mechanical drive systems. *Coatings*. 2025;15:660.
19. Qureshi W, Curá F, Mura A. Principal component analysis for characterization of fretting wear experiments on spline couplings. *Procedia Eng*. 2015;109:73–79.
20. Shantal M, Othman Z, Bakar AA. A novel approach for data feature weighting using correlation coefficients and min–max normalization. *Symmetry*. 2023;15:2185.
21. Chawla NV, Bowyer KW, Hall LO, et al. SMOTE: Synthetic minority over-sampling technique. *J Artif Intell Res*. 2002;16:321–357.
22. Singh J, Singh S, Singh JP, et al. Case study on implementation of ANN technique for analysis of erosion of hard coatings in different slurries. *AIP Conf Proc*. 2024;2986:030105.
23. Camacho Olmedo MT, Paegelow M, Mas JF, et al. *Geomatic Approaches for Modeling Land Change Scenarios. An Introduction*. Cham, Switzerland: Springer Nature; 2018.
24. Bai C, Huang L, Pan X, et al. Optimization of deep convolutional neural network for large scale image retrieval. *Neurocomputing*. 2018;303:60–67. Available from: https://doi.org/10.1016/j.neucom.2018.04.034.
25. Horne BG. Progress in supervised neural networks. *Signal Process Mag IEEE*. 1993;10:8–39.
26. Kuo CCJ. Understanding convolutional neural networks with a mathematical model. *J Vis Commun Image Represent*. 2016;41:406–413.
27. Spanhol FA, Oliveira LS, Petitjean C, et al. Breast cancer histopathological image classification using convolutional neural networks. *Int Jt Conf Neural Networks (IJCNN 2016)*. 2017;2017:1868–1873.
28. Zhu J, Wang X, Kou L, et al. Prediction of control parameters corresponding to in-flight particles in atmospheric plasma spray employing convolutional neural networks. *Surf Coatings Technol*. 2020;394:125862.
29. Rumelhart DE, Hinton GE, Williams RJ. Learning representations by back-propagating errors. *Nature*. 1986;323:533–536.
30. Kingma DP, Ba J. Adam: A method for stochastic optimization. arXiv Prepr arXiv14126980. 2014.
31. Chen W-B, Standfield BN, Gao S, et al. A fully automated porosity measure for thermal barrier coating images. *Int J Multimed Data Eng Manag*. 2018;9:40–58.
32. Lu Y, Chen W-B, Wang X, et al. Deep learning-based models for porosity measurement in thermal barrier coating images. *Int J Multimed Data Eng Manag*. 2020;11:20–35.

33. Pan SJ, Yang Q. A survey on transfer learning. *IEEE Trans Knowl Data Eng.* 2009;22:1345.
34. Zhu J, Wang X, Kou L, et al. Application of combined transfer learning and convolutional neural networks to optimize plasma spraying. *Appl Surf Sci.* 2021;563:150098.
35. Singh S, Kasana SS. Quantitative estimation of soil properties using hybrid features and RNN variants. *Chemosphere.* 2022;287:131889. Available from: https://doi.org/10.1016/j.chemosphere.2021.131889.
36. Madhavan PV, Shahgaldi S, Li X. Long short-term memory time series modelling of pressure valves for hydrogen-powered vehicles and infrastructure. *Int J Hydrogen Energy.* 2025;124:67–83. Available from: https://doi.org/10.1016/j.ijhydene.2025.04.028.
37. Zhong K, Meng Q, Sun M, et al. Artificial neural network (ANN) modeling for predicting performance of SBS modified asphalt. *Materials.* 2022;15:8695.
38. Ikeuchi D, Vargas-Uscategui A, Wu X, et al. Data-efficient neural network for track profile modelling in cold spray additive manufacturing. *Appl Sci.* 2021;11:1654.
39. Demirkiran AS, Celik E, Yargan M, et al. Oxidation behaviour of functionally gradient coatings including different composition of cermets. *Surf Coatings Technol.* 2001;142:551–556.
40. Cetinel H, Öztürk H, Celik E, et al. Artificial neural network-based prediction technique for wear loss quantities in Mo coatings. *Wear.* 2006;261:1064–1068.
41. Singh J, Singh S. Neural network prediction of slurry erosion of heavy-duty pump impeller/casing materials 18Cr-8Ni, 16Cr-10Ni-2Mo, super duplex 24Cr-6Ni-3Mo-N, and grey cast iron. *Wear.* 2021;476:203741.
42. Satapathy A, Mishra SC, Das R, et al. Prediction of erosion behaviour of plasma sprayed fly ash coatings using neural network. *Int J Appl Eng Res.* 2006;10:3911–3914.
43. Pati PR, Satapathy A, Gupta G, et al. Optimizing wear analysis of plasma sprayed Linz-Donawitz slag-Al_2O_3 coatings using experimental design and neural network. *Proc Inst Mech Eng Part J J Eng Tribol.* 2022;236:1723–1736.
44. Rong Z, Li Y, Wu L, et al. An Advanced tool wear forecasting technique with uncertainty quantification using bayesian inference and support vector regression. *Sensors.* 2024;24:3394.
45. Singh J, Kumar S, Mohapatra SK. Tribological performance of Yttrium (III) and Zirconium (IV) ceramics reinforced WC–10Co4Cr cermet powder HVOF thermally sprayed on X2CrNiMo-17-12-2 steel. *Ceram Int.* 2019;45:23126–23142. Available from: https://doi.org/10.1016/j.ceramint.2019.08.007.
46. Singh J, Kumar S, Mohapatra SK. Erosion tribo-performance of HVOF deposited Stellite-6 and Colmonoy-88 micron layers on SS-316L. *Tribol Int.* 2020;147:105262. Available from: https://doi.org/10.1016/j.triboint.2018.06.004.

7 Applications of AI–Tribology in Aerospace and Renewable Energy

Amrinder Mehta, Brijesh Prasad, Hitesh Vasudev, and Abdul Wahab Hashmi

7.1 INTRODUCTION

Tribology has played its own role as a frontier discipline since the 1980s to conserve resources and prolong the service life of mechanical devices by means of the systematic study of friction, wear, and lubrication. This revealed that there were major losses applicable to the economy because of the poor knowledge of tribological scientific phenomena, which culminated in the official definition of the science of tribology [1]. Over the last several years, the number of research activities in the aerospace and renewable energy fields has been growing in scale and scope as the complexity of engineering systems exposed to extreme environments has increased. The potential of artificial intelligence (AI) and deep learning (DL) to achieve sustainability in different sectors is enormous. The advancement of AI and DL and their application in achieving sustainable development goals (SDGs), renewable energy, environmental health, and smart building energy management [2].

Figure 7.1 shows that AI and DL are revolutionizing renewable energy by enabling advanced data-driven solutions across the entire energy lifecycle. In energy prediction, AI classifiers provide precise short-term, long term, and multi-source energy projections to harmonize supply and demand. Anomaly detection systems use real-time, predictive, and postmortem analyses to identify anomalies in power networks before they become serious problems. The best energy management applications, including load balancing, demand response, and efficiency optimization, use resources in a cost-efficient manner. The advantages of AI-guided optimum siting, seamless integration, and storage configuration are a gain in the renewable system design to optimize performance. Battery management solutions are based on state-of-charge and state-of-health estimation, smart charging and discharging, and extension of the life cycle. Resilience modeling, cybersecurity threat detection, and fault detection are all means of enhancing grid stability and security. Condition-based and predictive maintenance use maintenance scheduling, failure prediction, and asset health to reduce downtime. Finally, price forecasting, market design algorithms, and automated trading systems are used in energy pricing and trading platforms to streamline market operations [3].

DOI: 10.1201/9781003624165-7

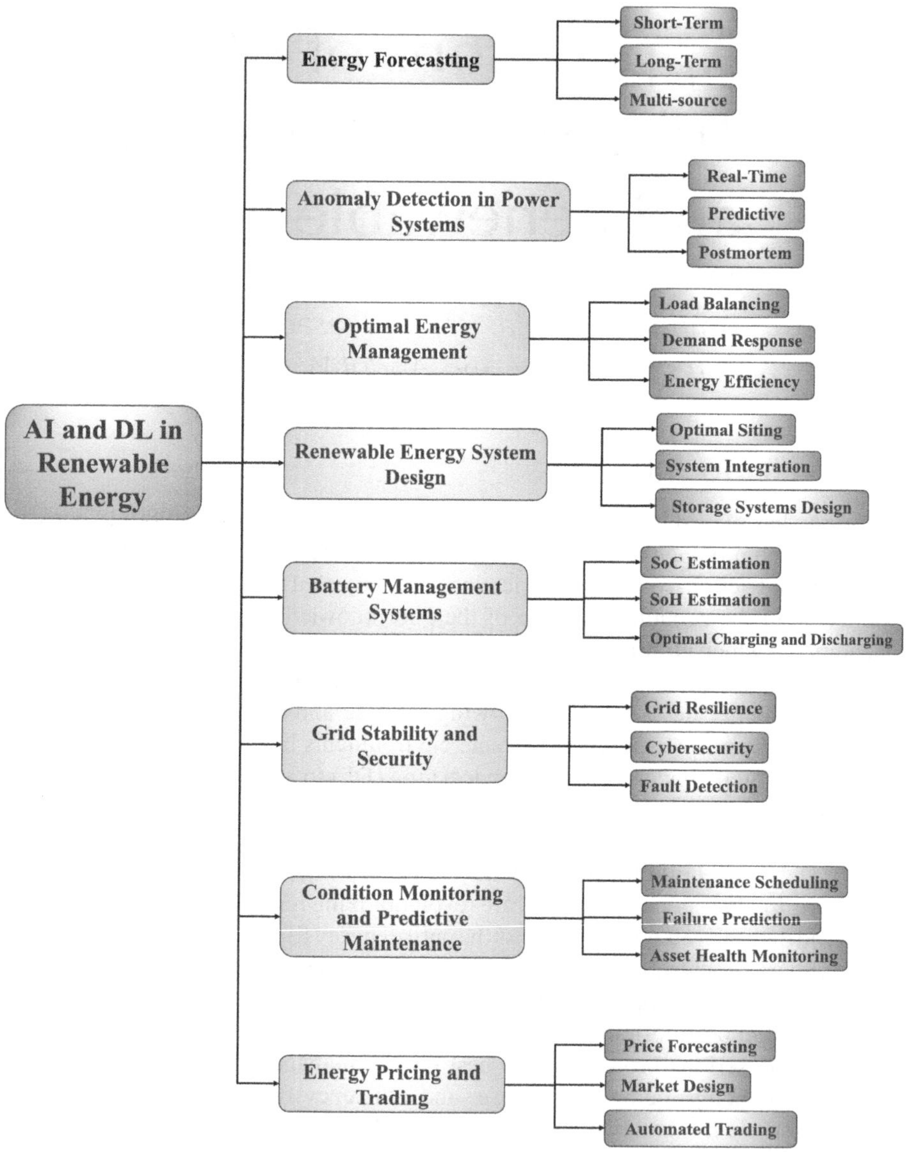

FIGURE 7.1 AI and DL in the field of renewable energy [3]. (Under CC 4.0 Attribution.)

With the development of engineering technology, the structures of industrial products and specialized equipment are becoming increasingly sophisticated, and the working conditions are becoming worse, including extreme temperature, pressure, and corrosion regimes. Simultaneously, operational reliability and stability pressures have become increasingly important, especially in high-processing and high-performance applications such as aircraft engines, wind turbine gearboxes, and space exploration equipment, where tribological failure can have disastrous effects.

This transformation has led to the creation of superior materials, coatings, and lubrication systems that can cope with severe working conditions while still providing peak operations over long periods of service [4].

Figure 7.2 shows the integration of AI within aerospace engineering, exemplified by the digital twin paradigm, where a physical aircraft's design, performance, and operational data flow continuously between real and virtual environments. Internet of Things (IoT) sensors on the airframe verify real-time telemetry and pass it on through cloud and edge computing platforms to AI-based analytics engines. These intelligent systems perform predictive simulations and machine–product modeling to predict component fatigue, optimize maintenance schedules, and increase the reliability. Using augmented and virtual reality interfaces, engineers can observe the complex behaviors of systems and rehearse their responses to anomalies within immersive environments. Data intelligence and performance-monitoring algorithms detect inefficiencies and recommend optimization solutions, and more advanced metaverse enterprises offer collaborative 3D virtual workspaces to remote workforces. Overall, this AI-based ecosystem can turn the existing engineering of airplane products into a dynamic process that can be iterated to provide a greater design speed, reduce downtime, and enhance safety and performance throughout the aerospace lifecycle [5].

Figure 7.3 shows that tribological systems begin with diverse input parameters, including the applied load, relative sliding velocity and time, contact geometry, friction mode, and lubrication conditions, which govern the interactions at the material interfaces. The system composition, which is an arrangement of friction pairs (the surfaces of contact), lubricants selected, and factors that control the ambient

FIGURE 7.2 AI in aerospace engineering [5]. (Under CC 4.0 Attribution.)

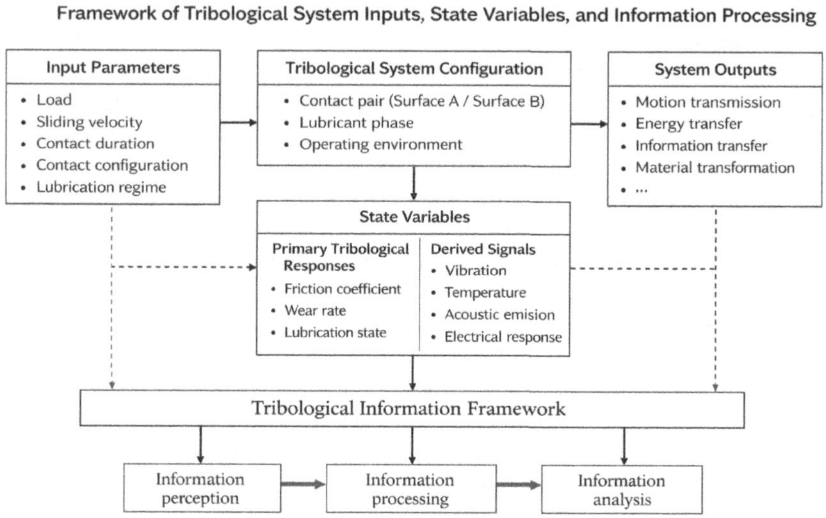

FIGURE 7.3 Shows the components of tribo-systems [6]. (Under CC 4.0 Attribution.)

environment, defined these inputs. The system creates state quantities as it runs, both direct tribological quantities (friction force, wear rate, and lubrication behavior) and derivative quantities (acoustic emissions, vibration pattern, temperature variation, imaging data, and electrical response). These abundant datasets represent the totality of tribology information, which circulates through three processing steps: information perception, where signals are gathered by sensors; information processing, which winnows, transforms, and fuses raw signal data into meaningful features; and information analysis, where machine learning (ML) and statistical models are used to extrapolate insights into the diagnosis of wear mechanisms, component-life prediction, and lubrication strategy optimization. Finally, this information-based method correlates basic inputs to operation outputs, such as efficient movement transmission, energy transfer, information exchange, and controlled material processing, to allow accurate monitoring and intelligent control of mechanical systems [6].

ML and AI are fast permeating tribology – AI and ML are quickly changing the way friction, wear, and lubrication phenomena are comprehended and controlled in aerospace and renewable energy applications. In aerospace, tribological models that use AI to extract sensor information to anticipate component wear and tear can be used to ensure proactive maintenance and reduce unplanned downtime [7]. Such models are also used to design bearings, seals, and coats to resist extreme temperatures, pressures, and loading conditions, thus greatly improving reliability and service. AI algorithms in the renewable energy sector process real-time wind turbine gearbox vibration, acoustic, and thermal measurements and share early fault indicators and optimized lubrication regimes. DL-based predictive maintenance systems can predict wear evolution and plan maintenance activities with reduced costs and energy waste. Moreover, AI-based adaptive control can be used to maintain the operational parameters at their optimal tribology, increasing the efficiency of the systems and enhancing the life of critical components in wind and solar installations [8].

Figure 7.4 shows that a comprehensive prognostics workflow begins with data capture and analysis, where automated thermography, other nondestructive testing (NDT) techniques, and advanced image processing feed into specialized software for initial defect detection. The monitored component, which is blanketed with in-service degradation, is evaluated using NDT to identify the present damage patterns, which are defined to quantify material degradation processes. These degradation data are used to inform the remaining useful life (RUL) estimation to combine empirical service knowledge, component design, and manufacturing parameters with predictive modeling to estimate future performance margins. The two-way arrow between NDT and RUL represents an iterative process of refinement: RUL predictions are used to determine the locations of specific NDT campaigns, and new inspection data are used to update the model. Finally, the combined degradation characterization and RUL projections allow a solid decision phase to be made; therefore, maintenance planners can plan the scheduled repairs and replacements or continue running the machine. The benefit of this closed-loop structure, lifecycle optimization, is achieved by associating real-time diagnostics to predictive analytics and informed maintenance measures [9].

Recent reviews have noted the rise of a cross-disciplinary field, known as tribo-informatics, which combines tribology and informatics to utilize AI and gain deeper insights into and control of friction, wear, and lubrication. Tribo-informatics

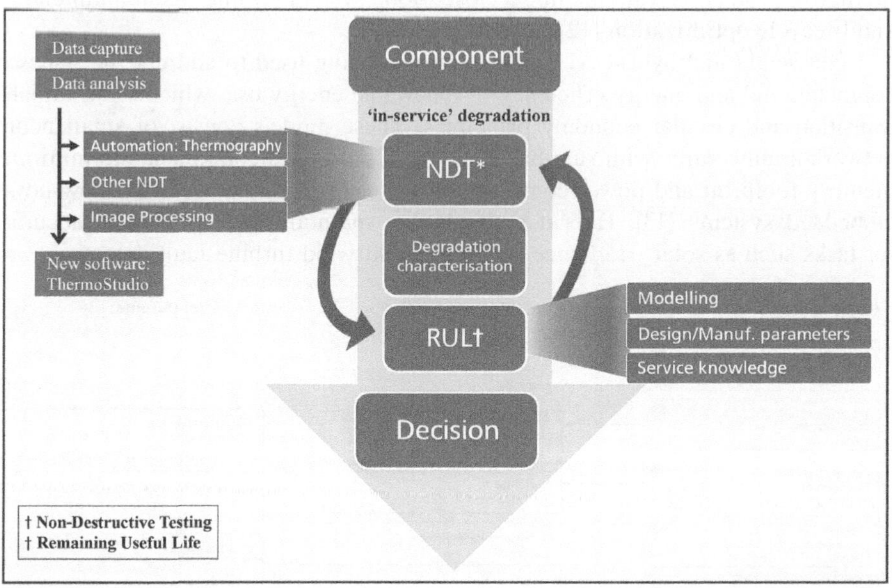

FIGURE 7.4 Framework showing an 'in-service' maintenance system that performs damage assessment through feedback for system design and manufacture [9]. (Under CC 4.0 Attribution.)

is a combination of real-time sensor measurements of vibration, acoustic emission, temperature, and surface images with AI-powered analytics to sense the state of the system and identify anomalies [10]. ML models then forecast the wear and frictional behavior of machines under different loads and environments and allow the dynamic optimization of lubrication strategies and component design. Tribo-informatics provides an effective platform for predictive maintenance, performance improvement, and resource optimization in all industries by combining data-centric informatics with conventional tribological experimentation [11].

Figure 7.5 shows that in tribological systems, a comprehensive information flow underpins status monitoring, fault diagnosis, and lifetime prediction. It starts with system inputs, including the lubrication mode, relative motion speed, and operation time, and with system-inherent data, including the tooling geometry and coating materials. During the operation of the machine, tribology state data (amount of wear, friction force, and wear form) and derived data (cutting forces, vibration, acoustic emissions, torque, and wear images) were continuously monitored. Simultaneously, system issues such as surface roughness, flatness deviation, and mechanical properties are benchmark performance indicators. All data obtained were processed using ML algorithms: regression models were used in measurement of tribological properties, in classification to identify different types of wear or faults, in clustering to discover previously unknown patterns in wear behavior, and in dimensionality reduction to simplify the high-dimensional data used in wear analysis. The resulting insights facilitate real-time monitoring of tool wear, accurate prediction of machining quality, and strong machine fault diagnosis and predictive protection, establishing a closed-loop system for intelligent maintenance and lifecycle optimization [12].

Lightweight and hybrid AI models are also being used to address the issues of computational and energy efficiency in renewable energy use, which fit sustainable transition and circular economy principles. These models consist of small neural network architectures with rule-based or physics-inspired components to minimize memory footprint and power consumption for use on edge devices and low-power embedded systems [13]. Hybrid methods can maintain high predictive accuracy for tasks such as solar irradiance prediction and wind turbine fault detection, with

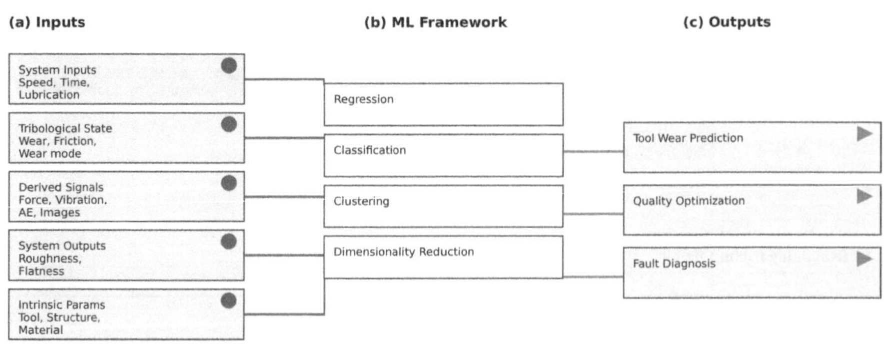

FIGURE 7.5 Information flow within the process of status monitoring, fault diagnosis, and life prediction of tribo-systems [12]. (Under CC 4.0 Attribution.)

minimum computational costs by optimizing model complexity and incorporating domain knowledge. This performance helps sustain off-grid and microgrid settings of constantly monitored and adaptively controlled systems, extends equipment service life, reduces electronic waste, and facilitates resource-efficient energy infrastructure [14].

Figure 7.6 shows that the iterative physics-informed neural network (PINN) approach for hydrodynamic contact begins with network initialization and an assumed film thickness profile. The PINN then estimates the pressure distribution PPP, which is assessed to balance the loads; in the case of failure to satisfy the balance condition, it is modified, and the process is repeated until convergence is achieved. This is a repeated process that provides the correct pressure and film thickness. By plotting the outlet film thickness versus sliding velocity, it was found that the PINN outcomes were closely comparable to those of both the finite-element method (FEM) simulations and experimental data at various velocities, indicating the predictive capability of this method. The fact that the PINN-predicted pressure contours closely resemble the traditional parabolic distribution of the FEM proves the three-dimensional visualizations of the pressure field, showing that the highest-pressure values are in the center of the contact and that there is a smooth gradient at the edges of the PINN. In general, the PINN framework offers a computationally effective, mesh-free variant of the conventional FEM with the highest levels of agreement with both experimental and standard numerical procedures [15].

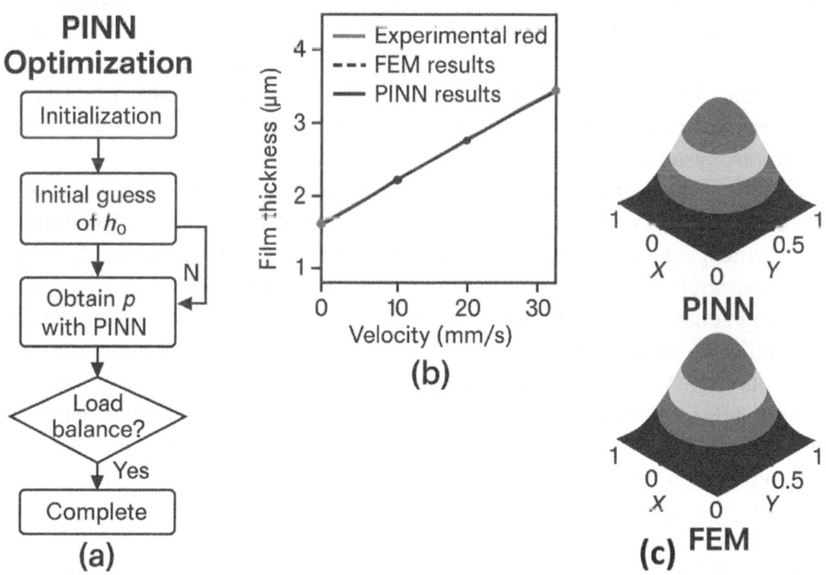

FIGURE 7.6 (a) Flowchart of the iterative PINN approach for hydrodynamic contact. (b) Outlet film thickness at different sliding velocities for the PINN method compared to FEM simulation and experimental results. (c) Pressure distribution predicted using the PINN (left) and FEM (right) [15]. (Under CC 4.0 Attribution.)

Nevertheless, the pace of development has led to numerous obstacles to AI–tribology adoption, such as the variability of data quality across nonhomogeneous sensors, the lack of model transparency because of the DL architecture, and the inability to integrate AI solutions with old mechanical and control systems. Moreover, data formatting, model validation, and performance benchmarking structures are not standardized, which hinders overall deployment [16]. This review unites modern studies of AI-based tribological systems in the aerospace and renewable energy industries, outlining the new directions of AI-based predictive maintenance and lightweight hybrid models, assessing their efficiency and dependability advantages, and critically analyzing the current constraints. This chapter closes with a description of future directions toward AI-tribology solutions that are robust, interoperable, and explainable.

7.2 AI–TRIBOLOGY IN AEROSPACE: APPLICATIONS AND TRENDS

The rapid development of AI in the aerospace industry has triggered a revolutionary shift in the entire industry, radically changing traditional operational paradigms. With AI-driven predictive maintenance algorithms, it is now possible to detect faults early, predict the likelihood of equipment failure prior to its occurrence, and reduce the number of unscheduled maintenance events by as much as 30%, not to mention the extent to which aircraft downtimes can be reduced [17]. ML and computer vision technologies have been combined, contributing to improved safety systems by providing real-time surveillance, automated quality control in the manufacturing process, and intelligent analysis of large amounts of operational data. AI has also facilitated the optimization of the supply chain, efficiency in flight paths, and advanced automation in factory and assembly line activities. All these technological developments have increased operational efficiency, lowered costs, improved safety standards, and placed the aerospace industry in a position to further innovate in autonomous systems and clever aircraft design [18].

Aerospace maintenance is changing owing to AI and ML, which make it possible to monitor and identify faults in the most critical tribological systems (bearings, engines, and airframes) in a predictive manner. ML models identify slight anomalies, such as vibrations, temperature, and audio emissions, by constantly measuring sensor data to monitor wear and tear or to alert for upcoming failures [19]. Predictive algorithms can predict the remaining useful life, and maintenance can be planned to minimize unplanned downtime and maximize the service life of components. System optimization is also made more effective using optimization techniques that suggest lubrication regimes, modify operating parameters, and refine design tolerances. This is an empirical method that enhances the safety, reliability, and cost-effectiveness of aerospace assets by transforming maintenance practices into condition-based practices [20].

Figure 7.7 demonstrates that AI in the aerospace industry represents a set of technologies that are advancing by passing through consecutive ages. In its most general sense, AI is a system that emulates human reasoning and decision-making, starting with simple rule-based expert systems of the 1950s to current-day cognitive architectures that can autonomously control flight. ML as part of AI Golden parachute ML

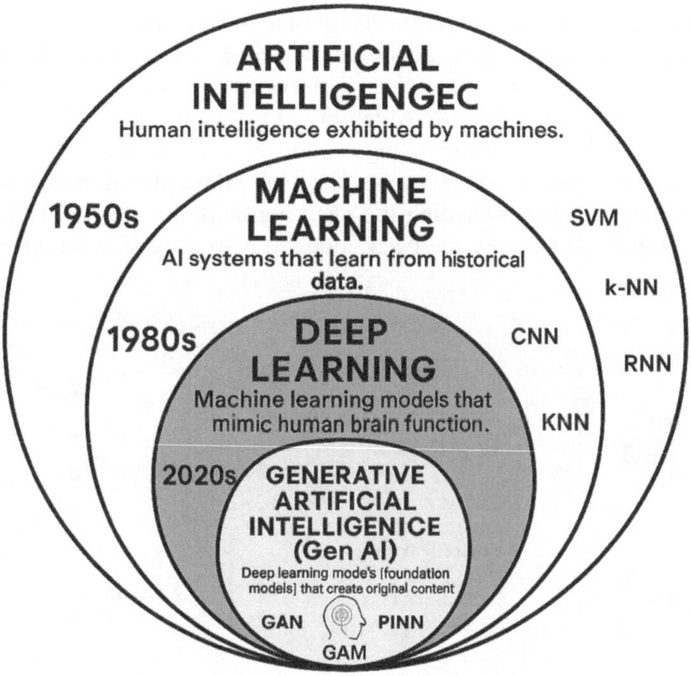

FIGURE 7.7 Schematic representation of AI, ML, DL, and GenAI in the aerospace context [21]. (Under CC 4.0 Attribution.)

emerged in the 1980s as data-driven algorithms, including support vector machines, k-nearest neighbors, decision trees, and gradient boosting machines, that are trained on historical flight, maintenance, and sensor data to optimize performance and failure prediction. DL, which became prominent in the 2010s, uses artificial neural networks, such as convolutional neural networks (CNNs) and recurrent neural networks (RNNs) and long short-term memory networks, to operate on high-dimensional observations, such as radar images, engine vibration measurements, and human-like aerodynamic simulator responses. Finally, the generative AI of the current decade includes foundation models such as generative adversarial networks, deep generative models, PINNs, and hybrid AI systems that generate novel design prototypes, synthesize training data, and simulate system behavior, which propel the fast innovation of aerospace design, digital twins, and mission planning [21].

DL and hybrid AI models enable real-time monitoring and diagnostics by combining high-frequency sensor data with physics-based knowledge. CNNs and RNNs are used to process streaming data (vibration spectra, thermal patterns, and acoustic emissions) and detect anomalies and predict fault types in real time. Hybrid frameworks combine domain knowledge, lowering false alarm rates and improving interpretability, and lightweight architectures imply on-device inferencing with very low latency [22]. This round-the-clock monitoring allows the detection of the arising problems in real time, which provides an automatic alarm and diagnostic suggestions. These systems enhance operational safety, increase the efficiency of maintenance interventions, and reduce unplanned downtime through the accurate identification of the mechanisms of deterioration and prediction of the succession of failures, eventually increasing the availability of assets and cost-effectiveness [23].

Figure 7.8 CNNs excel at extracting spatial features by applying learned filters across input grids, such as images or sensor maps, through alternating convolution and pooling layers before flattening the resulting feature maps for classification or regression outputs. RNNs use feedback loops to store hidden state information on

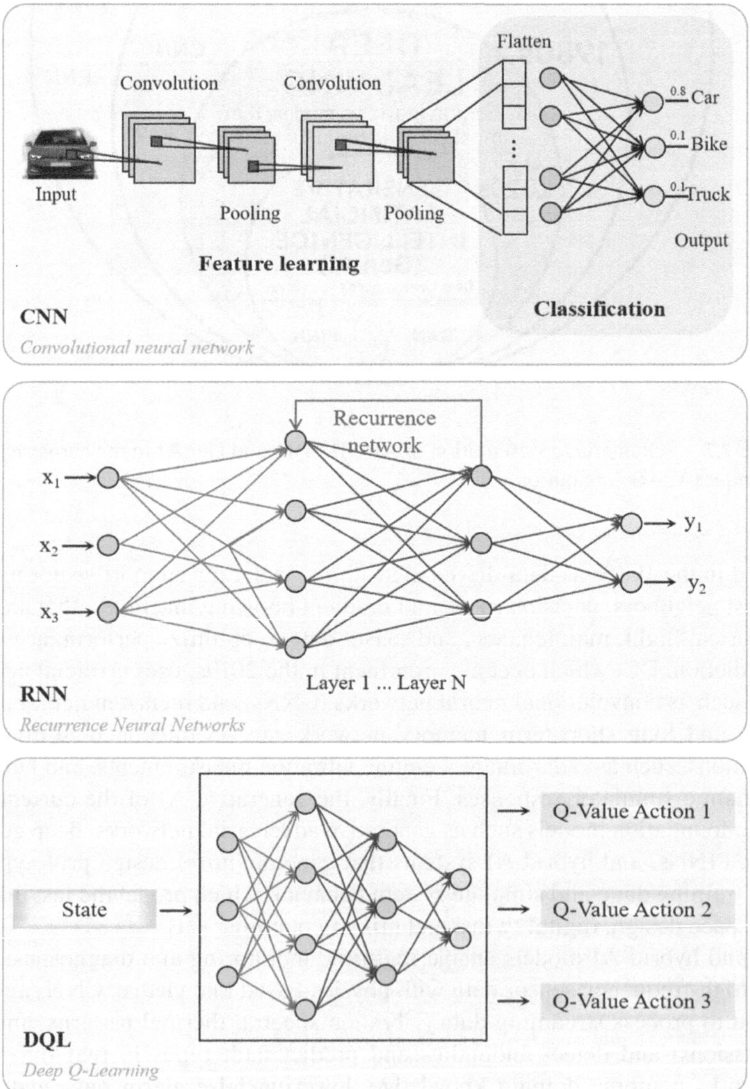

FIGURE 7.8 DL models including CNN, RNN, and DQL [24]. (Under CC 4.0 Attribution.)

a sequence of inputs to allow them to learn temporal dependencies in a time series, text, or signal sequence by repeatedly applying past information through a sequence of layers. Deep Q-learning (DQL) is a strategy that integrates deep neural networks with reinforcement learning to predict the expected cumulative reward of a sequence of actions in each state. A network estimates the Q-value function; agents update the network weights by minimizing the temporal difference error between the predicted and target Q-values based on the observed transitions. All these architectures support the capabilities of modern AI: CNNs to recognize spatial patterns, RNNs to model sequences, and DQL to make decisions in changing environments [24].

Design optimization and simulation are evolving with the use of generative AI and digital twins to generate virtual models of complex systems and realize countless design variations in an automated manner. Generative models present new geometries and material patterns that are optimized by performance-related parameters such as weight, durability, and efficiency. Digital twins reflect real-world properties in real time by combining sensor streams and physics-based simulations to assess what-if conditions, accelerate prototyping, and test operational plans. With the maturity of these technologies, ethics have taken center-stage: establishing transparency in AI-generated designs, avoiding data-driven bias in the results of a simulation, and maintaining human control over the process to make autonomous decisions safer and fairer [25].

7.3 AI-DRIVEN TRIBOLOGY IN RENEWABLE ENERGY SYSTEMS

With mounting environmental crises and security issues in energy supply, the need for sustainable development and intense efforts to adopt sustainable energy has become a priority in the modern global world. This paradigm shift is a life-saving reaction to the pressing necessity of solving the problems of climate change and resource shortages that undermine global stability. Renewable energy solutions, including solar, wind, hydropower, nuclear, hydrogen, and geothermal technologies, are becoming central players in ensuring that more energy sources are clean and efficient to sustain the increasing demand for cleaner and efficient energy sources [26]. These systems provide sustainable alternatives to fossil fuels, which currently supply 80% of the global energy, and play a significant role in limiting the emission of greenhouse gases and curbing the effects of climate change. The rapid development of renewable technologies, especially solar and wind energy, has resulted in these becoming cost-competitive with the use of fossil fuels, allowing deployment rates and capacity increases like never before. AI–tribology plays a central role in renewable energy systems, such as wind and solar [27].

Figure 7.9 demonstrates that a solar power forecasting workflow starts with the collection of important input variables that affect photovoltaic (PV) output: the global horizontal irradiance (GHI) measures the amount of sunlight available, the surface temperature of the PV module influences the conversion rate, prediction timeframes (1 hour in advance, next day) are defined by forecast horizons, and weather type (clear, cloudy and partly cloudy) is illustrated by weather classification. These were concurrently inserted into AI models, a multilayer neural network, and a random forest ensemble. In a neural network, features of the input are passed across

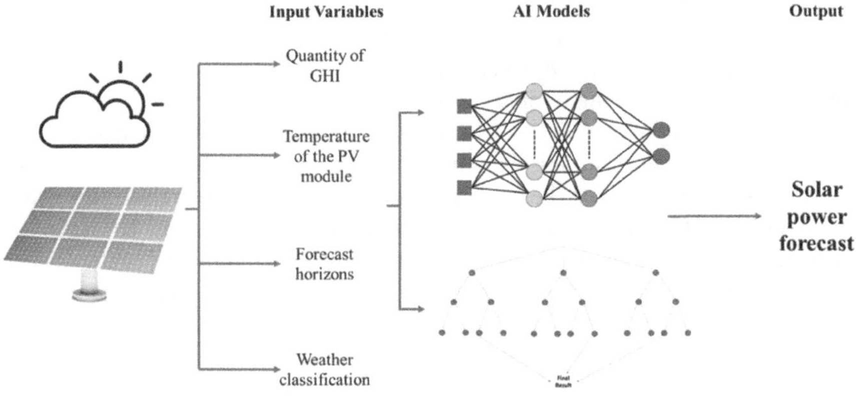

FIGURE 7.9 Schematic showing the process of input variable feeding into AI models (e.g., neural network and RF) to obtain solar power forecasts [28]. (Under CC 4.0 Attribution.)

interconnected hidden layers, in which weighted summations and nonlinear activation functions are learned to learn the interaction of complex features. The random forest model builds several decision trees that divide the input space based on feature thresholds to compute individual estimates of power, which are combined to produce a strong forecast. Lastly, both models generate a solar power forecast that numerically predicts future PV generation, which guides grid operators and energy managers for optimal dispatch, load balance, and storage [28].

AI-based predictive maintenance has revolutionized the operation of wind farms, allowing the detection of faults with a high degree of accuracy and simplifying the inspection process. ML uses sensor data (vibration, acoustic emissions, and temperature) to indicate anomalies, making them more specific and less time-consuming for human inspectors. Power output forecasting, maintenance scheduling, and real-time condition monitoring were performed. DL models are widely used in variable renewable energy systems to maintain balance in generation and ensure the dependable performance of assets [29]. In addition to turbine health, AI streamlines grid workflows through demand prediction, dispatching of energy storage, and coordination of distributed energy to stabilize the grid. Modern studies focus on lightweight AI models to reduce computation and energy costs, which is in line with the principles of a sustainable and circular economy. Additional predictive accuracy and adaptation to dynamic environmental conditions can be obtained using hybrid modeling methods that combine data-driven learning with physics-based mechanisms. Simultaneously, the developing field of tribo-informatics combines the data of friction, wear, and lubrication with AI analytics to provide comprehensive knowledge of the system for use in maintenance optimization and prolonged equipment life [30].

7.4 CHALLENGES, LIMITATIONS, AND FUTURE DIRECTIONS

The critical issues of AI–tribology integration include data quality problems due to inconsistent sensor calibration and mixed data formats and interoperability issues that prevent unproblematic communication between different monitoring systems and platforms. The issue of model transparency is that DL algorithms are complex black boxes that do not allow for trust and regulatory acceptance. Connectivity with legacy systems is technologically challenging because of the compatibility of the protocols and the old infrastructure [31]. Security weaknesses expose critical systems to cyberattacks, and privacy issues arise due to the sensitive data collection of operations. Moral aspects take center-stage in aerospace and critical infrastructure applications, where the decisions made by AI directly affect safety and general welfare and thus require strong governance mechanisms, interpretable algorithms, and elaborate risk assessment protocols to guarantee responsible usage [32].

The transfer learning workflow within the Mask R-CNN framework is depicted in Figure 7.10. A nontrained Mask R-CNN model, whose convolutional and RoIAlign layers are initialized randomly, cannot detect or segment objects meaningfully. During the training stage, this initial model was presented with the large-scale COCO dataset, which allowed it to learn generic feature representations, region proposal methods, and class-specific masks for a wide range of object types. The pretrained Mask R-CNN results in powerful backbone weights and head parameters, which are applicable in general semantic segmentation. This pretrained network is then retrained on a domain-specific dataset, such as high-resolution aircraft images, by modifying only the classification and mask heads (and potentially some convolutional layers) to extract domain-specific features and object categories. This results in a trained Mask R-CNN with the same general segmentation capability as that of COCO but optimized to detect and generate masks of target objects most accurately in the new dataset, with a significantly reduced training time and higher accuracy compared to training on the new dataset [33].

Standardized architectures are required to align the data types, validation standards, and performance thresholds to develop and test AI–tribology solutions consistently. Elucidating AI methods will need to be embraced to clarify the direction of

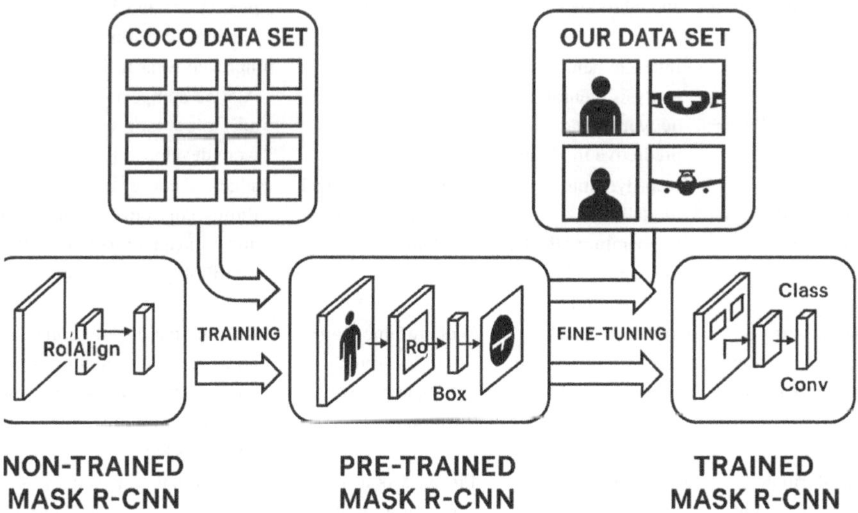

FIGURE 7.10 Transfer learning applied to the MASK R-CNN framework [33]. (Under CC 4.0 Attribution.)

model decisions, which will help clear stakeholder suspicion and streamline compliance with the regulations. Industry consortia, academic institutions, and government agencies can create collaborative policies to determine the best practices in data governance, cybersecurity, and ethical supervision. This type of coordinated work will ensure that AI-based tribological systems harmonize with the existing infrastructure, focus on safety and privacy, and meet sustainability goals. By coordinating technical standards with policy regulations and transparency demands, the field can be deployed in a scalable, responsible, and effective manner in the aerospace and renewable energy fields [34].

Table 7.1 demonstrates the revolutionary role of AI in tribological problems in aerospace and renewable energy, where ML systems improve the prediction of tribological behavior and system performance optimization at levels not previously observed. This integration is also a paradigm shift from the older methods of experimentation to data-driven predictive techniques, which are much less costly and time-consuming than the old systems to develop and are more reliable and efficient.

The application of AI to tribology in aerospace and renewable energy is driving incredible advances in reliability, efficiency, and sustainability. According to state-of-the-art

TABLE 7.1
Impact of AI on Tribological Challenges in Aerospace and Renewable Energy Sectors

Domain Focus	Methodology	Sample/ Application	Key Results
Material discovery and tribological property prediction	LSBoost algorithm, random forest, SVM, and neural networks	Various engineering materials (metals, ceramics, and polymers)	LSBoost achieved R^2 >0.95 for friction coefficient and wear rate predictions and 30%–40% faster material screening [35]
Aerospace composite materials and wear modeling	Finite-element analysis, neural networks, and predictive modeling	Al5052 composites with cenosphere particle reinforcement	Improved wear resistance in Al5052 composites, and accurate wear prediction models were developed [36]
Wind turbine gearbox tribology and energy efficiency	Tribo-dynamics modeling, temperature effects, and shape modification	Wind turbine helical gears and bearing systems	Enhanced energy efficiency through optimized gear meshing and reduced temperature effects on tribological performance [37]
Wind turbine bearing tribology and lubrication	Fault tree analysis (FTA), experimental tribology, and lubrication analysis	Wind turbine main shaft, gearbox, pitch, and yaw bearings	Identified major failure modes: abrasion, micro-etching, and white etched cracks; developed lubrication optimization strategies [38]
General tribology applications with AI/ML	Artificial neural networks, DL, and pattern recognition	Pin-on-disk tests, fretting wear and various tribological systems	ANN models achieved 95%+ accuracy in tribological property prediction and eliminated the need for extensive experimental testing [39]

reviews and experimental studies, AI-based predictive maintenance and diagnostic systems can significantly decrease unplanned downtimes; increase the lifespan of bearings, seals, and turbine parts; and decrease the total cost of maintenance. Tribo-informatics has been explored and proposed, blending friction, wear, and lubrication measurements with high-level analytics to provide insights into the system [40]. Hybrid models, which blend data-driven algorithms with physics-based models, provide more accurate predictions and higher flexibility when operating under dynamically changing conditions. Slim AI designs also enable scalable deployment through reduced computational and energy requirements, enabling real-time monitoring at the edge. Although this has been done, there are still critical issues, such as obtaining high-quality, interoperable datasets, creating explainable AI methods to build trust and regulatory compliance, and creating effective security systems that can be used to safeguard critical infrastructure. To overcome these problems, engineers, data scientists, and policymakers must work together with interdisciplinarity and develop common protocols and supporting regulations that will make AI–tribology integration ethical, responsible, and effective [41].

7.5 CONCLUSION

Tribological material testing determines important friction, wear, and lubrication characteristics in diverse environments. AI–tribology combines experimental datasets with ML and analytics to streamline the performance of a system by applying patterns to tribological behavior. In aerospace and renewable energy use cases, AI-based models use this combination of data to facilitate predictive maintenance, identify emerging failures, and plan interventions ahead of time. Using recurrent sensor input analyses and lab-based material properties, AI–tribology improves component dependability, prolongs service life, and reduces unexpected downtime by making prudent and information-driven choices. Digital twins and generative AI in aerospace can be used to explore the design space through automation and simulate complex artificial environments, which accelerates prototyping and validation. In renewable energy, AI algorithms can provide better fault detection by vibration and thermal sensors on wind turbines, better forecasting of power output by advanced pattern recognition, and optimization of maintenance schedules to reduce downtime. The use of lightweight AI models will further improve computational efficiency and energy consumption, allowing real-time analytics on edge devices and facilitating sustainability objectives across the distributed energy systems.

The need to address the issue of data collection, which is a challenge due to the sheer amount of data, tends to drive the data collection of AI applications. A broad-based data governance framework is critical for solving the data quality problems of AI. The challenges posed by the rapid advancement of AI technologies require technical expertise in AI to solve them. Researchers and policymakers could resolve these issues through interdisciplinary cooperation, education and training investment, and encouraging the creation of more ethical, secure, and human-oriented AI systems, which would open up the path for further progress in AI. Responsible AI–tribology deployment requires standardized architectures, explainable AI, and cooperative policies.

REFERENCES

1. Nosonovsky M, Bhushan B. *Green Tribology: Principles, Research Areas and Challenges*. London: The Royal Society Publishing; 2010. pp. 4677–94.
2. Aydin F, Durgut R, Mustu M, Demir B. Prediction of wear performance of ZK60/CeO_2 composites using machine learning models. *Tribology International*. 2023;177:107945.
3. Fan Z, Yan Z, Wen S. Deep learning and artificial intelligence in sustainability: A review of SDGs, renewable energy, and environmental health. *Sustainability*. 2023;15:13493.
4. Al-Haddad LA, Al-Muslim YM, Hammood AS, Al-Zubaidi AA, Khalil AM, Ibraheem Y, et al. Enhancing building sustainability through aerodynamic shading devices: An integrated design methodology using finite element analysis and optimized neural networks. *Asian Journal of Civil Engineering*. 2024;25:4281–94.
5. Shenwai PG, Choudhary A, Pokuri T, Basak A, Manikandan M, Singh B. On the role of artificial intelligence in aerospace engineering: Current state of the art and future trajectories. *The Aeronautical Journal*. 2025;17:1–27.
6. Yin N, Xing Z, He K, Zhang Z. Tribo-informatics approaches in tribology research: A review. *Friction*. 2023;11:1–22.
7. Al-Lami A, Hilmer P, Sinapius M. Eco-efficiency assessment of manufacturing carbon fiber reinforced polymers (CFRP) in aerospace industry. *Aerospace Science and Technology*. 2018;79:669–78.
8. Atescan-Yuksek Y, Mills A, Ayre D, Koziol K, Salonitis K. Comparative life cycle assessment of aluminium and CFRP composites: The case of aerospace manufacturing. *The International Journal of Advanced Manufacturing Technology*. 2024;131:4345–57.
9. Addepalli S, Roy R, Axinte D, Mehnen J. 'In-situ'inspection technologies: Trends in degradation assessment and associated technologies. *Procedia Cirp*. 2017;59:35–40.
10. Basheer AA. Advances in the smart materials applications in the aerospace industries. *Aircraft Engineering Aerospace Technology*. 2020;92:1027–35.
11. Balaji D, Ranga J, Bhuvaneswari V, Arulmurugan B, Rajeshkumar L, Manimohan MP, et al. Additive manufacturing for aerospace from inception to certification. *Journal of Nanomaterials*. 2022;6:1–18.
12. Yin N, Yang P, Liu S, Pan S, Zhang Z. AI for tribology: Present and future. *Friction*. 2024;12:1060–97.
13. Bharadwaj KA. Emergence of neuromorphic chips to simulate human brain. *3rd International conference on Electronics, Communication and Aerospace Technology (ICECA): IEEE*; Coimbatore, India, 2019. pp. 1331–4.
14. Bhattacharya S, Agarwal AK, Prakash O, Singh S, Pandey M, Kant R. Introduction to sensors for aerospace and automotive applications. *Sensors for Automotiveaerospace Applications*. 2019;264:1–6.
15. Zhao Y, Guo L, Wong PPL. Application of physics-informed neural network in the analysis of hydrodynamic lubrication. *Friction*. 2023;11:1253–64.
16. Chinchanikar S, Shaikh AA. A review on machine learning, big data analytics, and design for additive manufacturing for aerospace applications. *Journal of Materials Engineering Performance*. 2022;31:6112–30.
17. Costanza G, Tata ME. Shape memory alloys for aerospace, recent developments, and new applications: A short review. *Materials*. 2020;13:1856.
18. Daniyan I-A, Muvunzi R, Fameso F, Ndambuki J, Kupolati W, Snyman J. Application of multi-criteria decision and simulation approaches to selection of additive manufacturing technology for aerospace application. *Computers, Materials and Continua*. 2025;83:1623–48.

19. Huang X, Tepylo N, Pommier-Budinger V, Budinger M, Bonaccurso E, Villedieu P, et al. A survey of icephobic coatings and their potential use in a hybrid coating/active ice protection system for aerospace applications. *Progress in Aerospace Sciences*. 2019;105:74–97.
20. Filippatos A, Markatos D, Tzortzinis G, Abhyankar K, Malefaki S, Gude M, et al. Sustainability-driven design of aircraft composite components. *Aerospace*. 2024;11:86.
21. Derrick M, David S, Frederic M, Tajebe B. A review on aerospace-AI, with ethics and implications. *Journal of Civil, Construction and Environmental Engineering*. 2025;10:60–74.
22. Karagiannis D, Stamatelos D, Kappatos V, Spathopoulos T. An investigation of shape memory alloys as actuating elements in aerospace morphing applications. *Mechanics of Advanced Materials Structures*. 2017;24:647–57.
23. Li SS, Yue X, Li QY, Peng HL, Dong BX, Liu TS, et al. Development and applications of aluminum alloys for aerospace industry. *Journal of Materials Research and Technology*. 2023;27:944–83.
24. Wang Q, Li Y, Li R. Integrating artificial intelligence in energy transition: A comprehensive review. *Energy Strategy Reviews*. 2025;57:101600.
25. Madhavadas V, Srivastava D, Chadha U, Raj SA, Sultan MTH, Shahar FS, et al. A review on metal additive manufacturing for intricately shaped aerospace components. *CIRP Journal of Manufacturing Science Technology*. 2022;39:18–36.
26. Mohanavel V, Ali KA, Ranganathan K, Jeffrey JA, Ravikumar M, Rajkumar S. The roles and applications of additive manufacturing in the aerospace and automobile sector. *Materials Today: Proceedings*. 2021;47:405–9.
27. Mohd Yusuf S, Cutler S, Gao N. The impact of metal additive manufacturing on the aerospace industry. *Metals*. 2019;9:1286.
28. Yousef LA, Yousef H, Rocha-Meneses L. Artificial intelligence for management of variable renewable energy systems: A review of current status and future directions. *Energies*. 2023;16:8057.
29. Cai M, Zhang P, Xiong Q, Cai Z, Luo S, Gu L, et al. Finite element simulation of fretting wear behaviors under the ball-on-flat contact configuration. *Tribology International*. 2023;177:107930.
30. Dai Y, Li K, Xiang Q, Ou M, Yang F, Liu J. Microstructure and tribology behaviors of WC coating fabricated by surface mechanical composite strengthening. *Applied Surface Science*. 2023;619:156759.
31. Deliwala AA, Dubey K, Yerramalli CS. Predicting the erosion rate of uni-directional glass fiber reinforced polymer composites using machine-learning algorithms. *Journal of Tribology*. 2022;144:091707.
32. Mahapatra SK, Satapathy A. Development of machine learning models for the prediction of erosion wear of hybrid composites. *Polymer Composites*. 2024;45:7950–66.
33. Doğru A, Bouarfa S, Arizar R, Aydoğan R. Using convolutional neural networks to automate aircraft maintenance visual inspection. *Aerospace*. 2020;7:171.
34. Marian M, Tremmel S. Physics-informed machine learning—An emerging trend in tribology. *Lubricants*. 2023;11:463.
35. Jatavallabhula JK, Shabana S, Pappula B. Development and evaluation of machine learning based predictive models for tribological properties of blended coatings at elevated temperature. *Journal of Bio-and Tribo-Corrosion*. 2025;11:25.
36. Sheikh KA, Khan MM, Roga S, Qureshi T. Advanced tribological simulations and predictive modeling of wear behavior in Al5052/cenosphere composites using machine learning. *Journal of Tribology*. 2026;148:011403.
37. Yin J, Meng X, Cheng S. Enhancing wind turbine energy efficiency: Tribo-dynamics modeling and shape modification. *Renewable and Sustainable Energy Reviews*. 2025;208:115071.

38. Peng H, Zhang H, Fan Y, Shangguan L, Yang Y. A review of research on wind turbine bearings' failure analysis and fault diagnosis. *Lubricants*. 2022;11:14.
39. Rouf S, Raina A, Ul Haq MI, Naveed N. Sensors and tribological systems: Applications for industry 4.0. *Industrial Robot: The International Journal of Robotics Research and Application*. 2022;49:442–60.
40. Neupane D, Bouadjenek MR, Dazeley R, Aryal S. Data-driven machinery fault detection: A comprehensive review. *Neurocomputing*. 2024;627:129588.
41. Ng WL, Goh GL, Goh GD, Ten JSJ, Yeong WY. Progress and opportunities for machine learning in materials and processes of additive manufacturing. *Advanced Materials*. 2024;36:2310006.

8 Sustainability through Artificial Intelligence in Tribology

Jyoti Bhattacharjee, Subhasis Roy, and Hyun Suk Jung

8.1 INTRODUCTION

In the field of tribology, which is the nature of friction, wear, and lubrication of surfaces in motion, artificial intelligence (AI) [1] and machine learning (ML) are emerging as impactful and powerful technologies. AI and tribology integration has truly helped and ultimately accelerated our appreciation of complex tribological data, while also illuminating unseen mechanisms and honing system performance in previously unfathomable ways. Tribological phenomena are complex and multifaceted systems involving interactions of many types and at many different scales – often in time and space. This also necessitates advanced and extended approaches to characterise and understand tribological systems [2]. Indeed, AI is capable of addressing the significant amounts of data produced by experiments, simulations, and historical data at a scale and resolution where human analysis would be pointless. Practically, for example, ML models that have been trained by using high-quality annotated data can accurately predict wear rates, friction coefficients, and lubricant performance under different operating conditions, while expending fewer physical experiments. Bayesian optimisation loops serve as a great example of how AI can accelerate the decision-making process in planning tribology experiments by maximising past knowledge and iteratively suggesting, due to data constraints, the most informative experiments to complete, while effectively saving time and money [3].

AI-driven predictive modelling can enhance researchers' understanding of the physics involved in tribology and wear and better inform them in the design of application-specific materials and lubricants [4–6]. AI could also aid in the discovery of new tribological materials and additives, in part by rapidly screening candidate compounds to evaluate their performance against a range of criteria. AI methods would allow for not only the optimisation of materials but also the online monitoring and control of tribological systems in the context of industrial machinery to create a mechanism of responsive maintenance to avert failures and extend asset life. The successful application of AI in this area has required high-quality, well-annotated, and interoperable data sources, which may be somewhat limited in the area of tribology due to the interdisciplinary nature of the work. However, the integration of AI with tribology represents an opportunity to improve the efficiency, reliability, and

DOI: 10.1201/9781003624165-8

sustainability of many types of mechanical equipment and provides stimulation for a flow of innovations from fundamental science to industrial applications [7].

8.1.1 Role of Sustainability in Contemporary Tribological Systems

The current idea of sustainability is a transcendent aspect of modern tribological systems, which is required to mitigate the environmental impact, reduce energy consumption, and improve resource efficiency globally. The two largest issues confronting tribology, wear and friction, are the most significant contributors to energy loss and consumptive materials. Roughly 23% of global consumption can be linked with frictional inefficiencies associated with tribological interactions, which contribute about 20% of total consumption. The management of friction and wear indicates that various amounts of energy can be saved while also reducing greenhouse gas emissions. The service life of mechanical components can be extended, leading to less consumption of raw materials and emission of waste. In this sense, the idea of Green Tribology has gained visibility and relevance in relation to tribological performance and environmental responsibility, represented in Figure 8.1 [8,9].

Green Tribology is a branch of tribology that focuses on manufacturing environmentally safe lubricants, implantable and biomimetic materials, and eco-friendly tribological processes that produce power. The goal is to not just making the environment less polluted but also mitigating heat and energy loss to friction. Such a plan coincides with the basic requirements of sustainability, which are longer product life, less transportation and manufacturing efficiency, and closure of the loop with better recyclability and resource utilisation in the circular economy. The automotive,

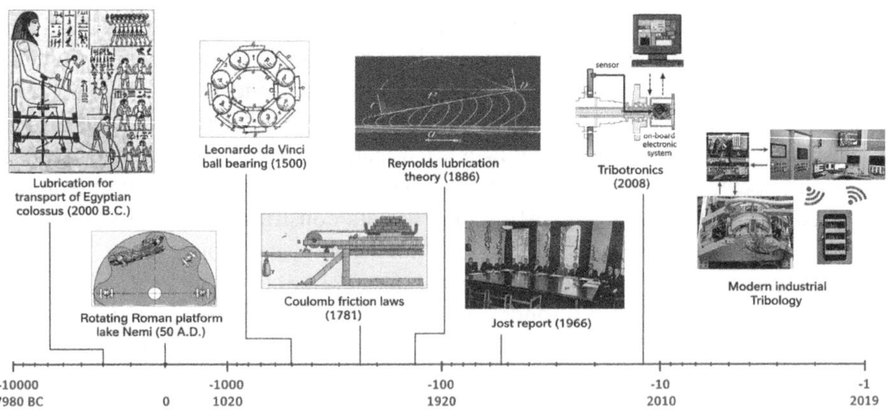

FIGURE 8.1 The timeline graphically illustrates the development of tribology with the following events in time: moving large structures in ancient Egypt with lubricated channels (2000 B.C.); the Lake Nemi rotating platform (50 A.D.); da Vinci's bearings (1500); the work of Coulomb (on friction) (1781); and Reynolds on lubrication (1886). The evolution continues more contemporarily with the Jost Report, enabling modern tribology (1966); Tribotronics (2008); and smart industrial tribological systems with sensors and observations for digital monitoring and predictive maintenance [9]. (Reproduced with permission.)

aerospace, power generation, and manufacturing industries could be the biggest beneficiaries of sustainable tribology, where the developments in lubrication, surface engineering, and materials science are directly related to both economic and environmental gains [10–12].

8.1.2 Current Challenges Facing Sustainable Tribological Performance

Despite significant advancements and increased attention to sustainability, sustainable tribological performance still faces many difficulties from the technical, environmental, and economic perspectives. One of the most important issues to resolve is finding the right mix between the highest possible tribological performance and the environment, health, and safety regulations. Usually, high-performance lubricants and coatings are made from very rare, toxic, or nonrenewable materials that not only pollute the Earth but are also difficult to recycle or dispose of safely. The use of bio-based or biodegradable lubricants seems promising, but their performance still falls short of that of traditional lubricants, and more research is needed to develop formulations and additives that meet both operational and sustainability requirements [13]. The complexity of the tribological system, however, where multiscale interactions occur in changing conditions, does not lend itself to predictive modelling and field testing. A lack of data combined with the inaccessibility of shared databases, restricts the capability of AI and ML to expand existing knowledge and engineer sustainable solutions. It requires cooperation and collaboration to develop large interoperable datasets from a variety of materials, conditions, and operating parameters [13].

One more significant obstacle we have to face is the uneven distribution of sustainability issues that have to be solved throughout the entire lifecycle or lifespan of tribological components, starting from the selection and manufacturing of materials to end-of-life waste management. Life-cycle analysis is a must; however, it is usually hard to conduct because of the intricate impacts and the particular requirements of the industry involved. Moreover, the economic drivers for the transition to greener tribological technologies are very faint, except in the case of the absence of the above-mentioned conditions or if the related costs and benefits are not clearly analysed, and this negatively impacts the adoption. Severe working conditions, like those in the aerospace or renewable energy sectors, pose further tribological difficulties that, among other things, constitute thermal, electrical, and mechanical stresses that accelerate the wear and lubrication problem. The invention of new materials and lubricants that can operate under these conditions and still provide good performance demands revolutionary advancements in the fields of materials science and engineering [14].

8.2 OVERVIEW OF AI TECHNIQUES FOR TRIBOLOGY

The application of ML and AI in tribology is one of the popular areas of technology as these systems deal with the intricate problems and multiparameter issues that would be otherwise left to conventional methods. Despite this, the main AI approaches employ supervised and unsupervised learning algorithms, Bayesian optimisation, and deep learning models such as convolutional neural networks (CNN). Supervised learning algorithms consider a range of input parameters related to lubricants, load, temperature, and material types (e.g., composition) to predict tribological properties from input parameters such as friction coefficients and wear rates. Examples include Gaussian Process Regression (GPR), Random Forest (RF), Support Vector Machines (SVM), and Extreme Gradient Boosting (XGB), as shown in Figure 8.2 [15].

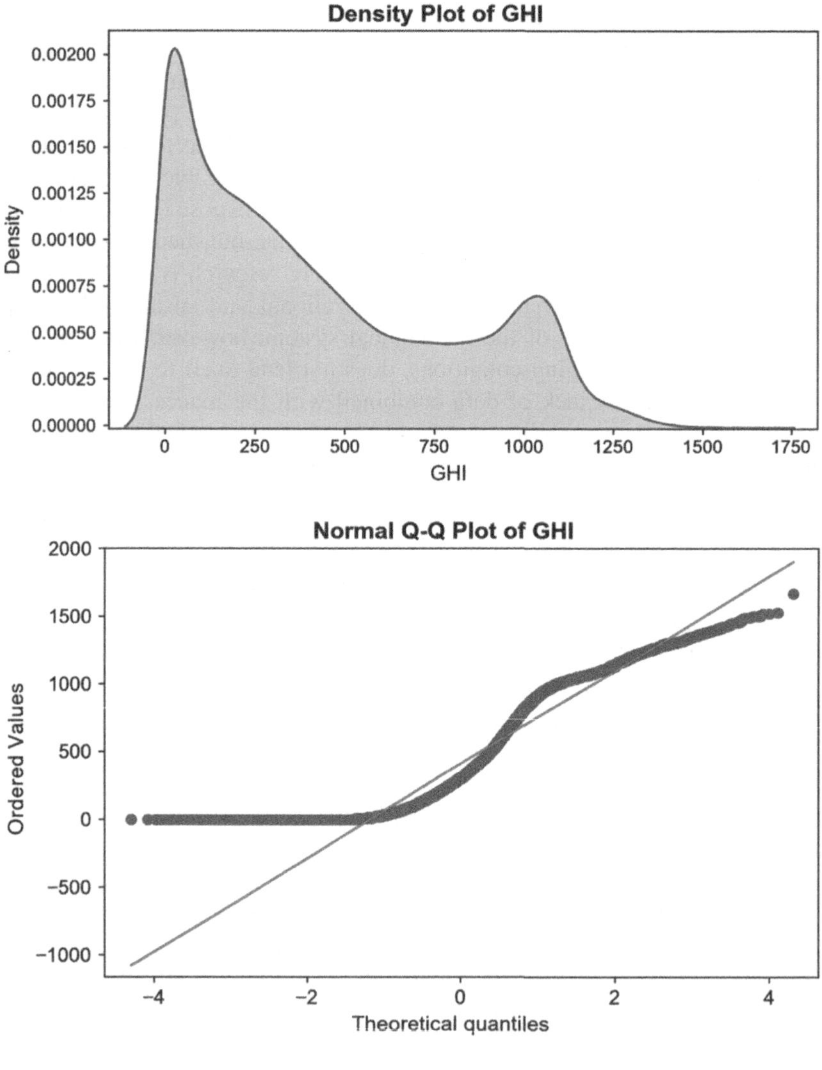

FIGURE 8.2 The density plot vividly showed a distribution that was skewed to the right, and GHI values were predominantly between 0 and 200 W/m^2, whereas, on the other side, there were just extremely few cases of GHI values above a thousand, recorded over a thousand W/m^2. In addition, the Q–Q plot confirmed the non-normality, bringing to light a strong skewness and possible outliers. These visual aids backed the EDA that GHI data are not normally distributed [16]. (Reproduced with permission.)

Clustering and dimensionality reduction, which are the major unsupervised learning methods, can sometimes lead to the establishment of hidden patterns in the enormous data, such as combining the wear mechanisms with similar behaviour together or detecting the anomalies in the operational data [16,17].

8.2.1 Data-Driven Decision-Making: From Intuition to Precision

Data-driven decision-making (DDDM) in tribology signifies a shift from the use of experience and intuition to leveraging structured data and analytical models for decision-making. DDDM employs quantitative data such as friction measurements, wear rates, surface topography, and operating conditions [18] that inform the design, optimisation, and maintenance of tribological systems. With large datasets generated from experiments, simulations, and real-world monitoring, engineers are capable of discovering relationships and causal connections that were previously beyond the scope of traditional analysis. The key aspect of DDDM is the quality and availability of data. Tribological data should conform to the FAIR (Findable, Accessible, Interoperable, and Reusable) data principles: Findable, Accessible, Interoperable, and Reusable—to permit new models to be built on complete and consistent data. High-quality datasets allow for the development of reliable predictive models that can, in many situations, accurately predict how a system has been operating under a range of conditions, facilitating virtual prototyping and scenario trials before the design is physically implemented [18]. In general, DDDM reduces development time, cost, and waste of material. In addition, DDDM enables proactive decision-making by establishing trends and anticipating failures before they happen. For example, real-time monitoring of machinery has been generating data that can serve as training input for ML models and report early signs of wear, breakdown of lubrication, or another problem before equipment failure [19,20].

8.2.2 Machine Learning's Role in Predictive Maintenance and Optimisation

Within tribological applications, ML has primarily aided in predictive maintenance and system optimisation. The use of ML algorithms on known historical and real-time performance data allows for forecasting of asset deterioration, predicting the remaining useful life, and providing maintenance recommendations before catastrophic failure. The use of ML is proving to be very beneficial in many industries, such as automotive, aerospace, and manufacturing, in which unexpected or uncontrolled applications of machines lead to very high costs and safety risks [21]. ML workshops, such as RF, K-Nearest Neighbours (KNN), and Long Short-term Memory (LSTM), are technologies that capture the time evolution of friction and wear rates to help with performance prediction patterns. Vibration signals, temperature profiles, and lubricant condition could be included utilising classification models and feature extraction methods to indicate early signs of impending faults within bearings, gears, or clutches. These algorithms are learning each time they are used, and their diagnostic performance should improve as they are trained with more data. ML can support maintenance activities, as well as offer tribological system optimisation through identification of the best material, coating, and lubricant combinations for a given application. The intersection of ML and experimental design allows for rapid development of suitable candidate materials, especially showing promise as low-friction and wear resistance materials to consider and greatly compress the development cycle. Data-driven optimisation can provide improved outcomes, improve sustainability, and enhance service intervals through energy reduction loss and improved service intervals. Industry 4.0 and machine connectivity, along with advanced ML supporting tribological systems, have taken us into an era of sustainability and efficiency [22].

8.3 SUPERVISED AND UNSUPERVISED LEARNING-BASED PREDICTIVE MODELS OF WEAR

The use of ML for predictive modelling of wear behaviour involves two main approaches: supervised and unsupervised learning. Supervised learning utilises labelled datasets where input variables (such as, temperature, sliding velocity, and load) are associated with known outputs (such as wear volume or wear rate) [23,24]. The main objective is to build models that understand the functional link between operating parameters and wear and accurately predict wear under new circumstances. The conventional supervised methods used are regression models, decision trees, support vector machines, and artificial neural networks [23].

Supervised learning is good for cases when large amounts of labelled wear datasets exist; experimental tribological data, however, are typically expensive or time-consuming to obtain, limiting dataset size. This chapter describes the algorithms for unsupervised learning that deal with unlabelled data, and they are capable of finding out the existing patterns and groupings in the data without needing the outputs to be given. Clustering and dimension reduction can be used to find out the different wear modes, to identify the operating conditions that influence wear, or to uncover the characteristics of wear that are beyond the normal wear behaviour. The approach of using unsupervised techniques is mainly meant for data exploration and hypothesis creation, which can eventually lead to the informing of the targeted supervised modelling. The combination of the supervised and the unsupervised learning methods has been increasingly used in the area of wear forecasting [24,25]. One of the examples can be that unsupervised clustering might divide the wear data into different regimes, and then the specific supervised models for each such cluster can be developed further to enhance the accuracy of prediction. The combination of the unsupervised learning's ability to recognise patterns with the supervised techniques' ability for precise outcome modelling best describes this hybrid method. The combination of different datasets, like that of sensor readings, surface examination, and environmental factors, also makes it possible to create more accurate predictive wear. The wear rate prediction done by AI models is more accurate than that done by empirical models, which are based on the characterisation of complex nonlinear dependencies among multiple factors. Moreover, AI techniques also enable real-time updating of the model based on the incoming new data, thus allowing for adaptive maintenance planning and optimal tribological design [24].

8.3.1 Neural Network Methods for Coefficient of Friction Estimation

Neural networks (NNs), a family of ML models inspired by the analysis of biological NNs, are well-suited to map challenging, nonlinear relationships in tribology, such as the estimation of the coefficient of friction (COF), as shown in Figure 8.3.

The COF is a function of several factors, including surface roughness, material, temperature, lubrication, and operating speed, making analytical prediction via traditional means challenging. NNs, particularly deep structures and recurrent patterns, can capture intricate dependencies in data without relying on prior physical rules [25, 28].

Convolutional neural networks (CNNs) have been successful in classifying tribofilm images to distinguish high- and low-friction scenarios based on spatial features related to surface films formed during sliding. Recurrent neural networks (RNNs), and especially those utilising Gated Recurrent Units (GRUs), have, in addition to CNNs, also been successful in capturing temporal shifts in contribution to frictional action when varied conditions are present and successfully identifying dynamic friction effects such as velocity-dependent changes and state progression at the interface. NN-based models can be advantageous compared to traditional phenomenological laws for friction based on their ability to learn from experimental or simulated data and represent fine-scale phenomena that can be overshadowed, such as microscopic

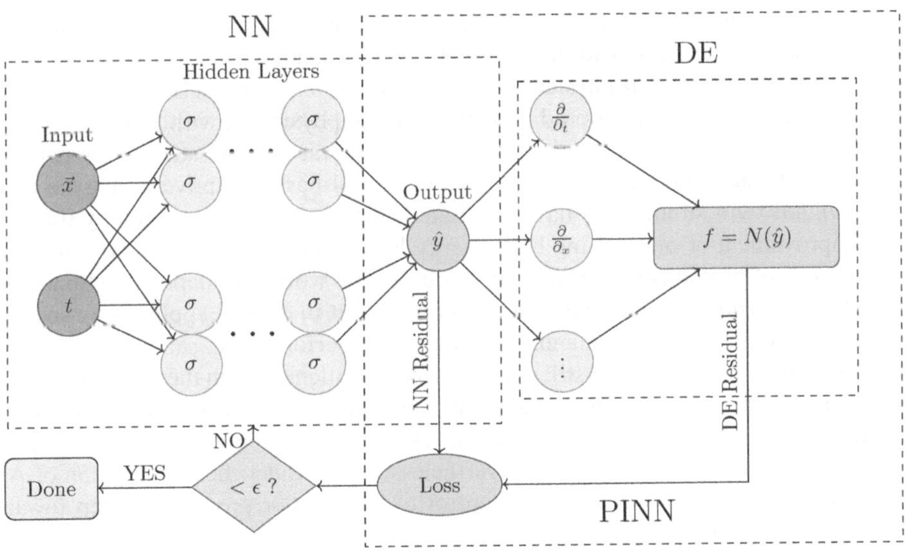

FIGURE 8.3 An example of the physics-informed neural network (PINN) framework in which the residuals of the differential equations (DEs) with the dynamic elements N are mixed into the overall loss function. The neural network finds and tunes the optimal parameters such that the total loss is minimum, along with the simultaneous satisfaction of the initial and boundary conditions (IC/BCs). As an extra measure, a cost function is used, which confirms the physical validity of the DEs, thus transferring some domain-specific knowledge to the neural architecture that finally increases both the accuracy and stability of the predictions [26]. (Reproduced with permission.)

healing or stick–slip switching. Nonetheless, NNs often require the dataset to be carefully curated and can particularly benefit from physics-informed training regimes, which include known frictional physics into loss functions to increase generalisation and interpretability. Furthermore, they can be employed for time-synced prediction of friction in-operational systems, where rapid and accurate COF prediction could be used to improve control strategies and component protection [26,27].

8.3.2 Real-Time Monitoring and Adaptive Control Systems

AI real-time monitoring and adaptive control systems constitute a large step forward for tribology, enabling zeroing and continuous testing and optimisation of friction, wear, and lubrication performance under operational conditions. The combination of Internet of Things (IoT) sensors with AI analytics in real time has provided applications for real-time measurement of friction, with applications spanning from development to end-user experiences and operational systems.

According to the original content, adaptive control systems utilising AI algorithms can adaptively modify operating parameters such as load, speed, or quality rate immediately after the system detects the tribological condition. Reinforcement learning (RL)-based structures and other AI systems are constantly learning the optimum control policy, which can simultaneously improve system stability, efficiency, and durability. Domains of AI-based real-time monitoring are enabled for predictive maintenance by detecting wear or lubrication faults in a real-time fashion to prevent unexpected downtime and costly repairs. Digital twin technology creates virtual representations of physical systems based upon AI models that so they can test changes *in silico*, as opposed to applying them physically, with simulated tribological behaviour initially resulting in the current state of operation. This permits improved decision-making in operation within an AI-based adaptive strategy or control. Digital twin simulations have facilitated further sustainable tribology through the improvement of minimising both energy loss and mechanical wear through the better control of contact interfaces. In addition, the worth of adaptive technologies should be considered, with potential differences felt in critical applications such as aerospace, automotive, and manufacturing, where performance, safety, and sustainability goals converge. There still exists inherent challenges with the design of robust versions that can tolerate noise, uncertainty, and varying operating environments and therefore matching them with any AI control implementation into current industrial system application infrastructure. Nevertheless, we consider the combination of AI, real-time monitoring, and adaptive control to be a major progressive step toward intelligent, sustainable systems [29–31].

8.4 GENERATIVE ADVERSARIAL NETWORKS (GANs)

GANs are made up of a pair of NNs known as a generator and a discriminator. The generator creates surface texture patterns, while the discriminator judges their authenticity and usefulness, which leads to the generator's constant self-updation. This adversarial training approach allows GANs to produce very creative and lifelike texture patterns that would be very difficult to produce through conventional computational or experimental methods [32].

Surface texture is a dominant parameter in tribology to control friction, wear, lubrication, and hydrophobicity [33]. The ability of GANs to produce rich yet functionally optimal texture libraries accelerates material surface engineering and enables access to nature-inspired surface-based design. Globally prominent studies highlight that the use of GANs in tribological surface engineering is capable of generating textures with a level of roughness, depth, and geometry dynamically optimised to a certain lubrication regime or wear situation. Such a benefit can complement other types of AI, like RL and topology optimisation, to push the boundaries of surface engineering towards highly efficient, tailored tribological solutions to cover areas ranging from biomedical implants to industrial machinery [34].

8.4.1 Simulation of Shark-Skin–Inspired Surfaces for Drag Reduction

Shark skin has led to a new generation of drag-reducing surface technologies, thanks to its hierarchical microgrooved structure, known as denticles. The denticles create riblets that disrupt the boundary layer flow, significantly decreasing turbulent drag forces. Simulations of shark skin-inspired surfaces are using computational fluid dynamics (CFD) and fluid–structure interaction models to investigate and optimise the above parameters: riblet shape, spacing, and aspect ratio. The researchers have verified that synthetic riblet surfaces designed after shark skin can cut the drag in turbulent flow by 6%–10%, thus making it possible to save significant energy in the case of aircraft, marine, and pipelines. CFD analysis indicates that riblet patterns generate favourable flow features like streamwise vortices and recirculation zones that bring back momentum close to the surface and, hence, reducing skin friction drag [35].

The simulated riblet geometry optimisation includes different working conditions like flow speed, pressure gradient, and surface elasticity. The latest research suggested the use of AI-based parametric models for the optimisation of shark-skin surface patterns through the dragging performance mapping of riblet geometry. AI simulations are powerful enough to reduce the number of design cycles to a minimum and at the same time to produce nature-inspired surfaces by applying cutting-edge manufacturing methods like 3D printing or laser texturing. In short, the simulations of shark-skin-mimetic surfaces are a reflection of the future where biomimicry intertwined with AI has been gradually replacing the traditional method of drag reduction with environment-friendly and energy-efficient solutions across various sectors [35].

8.4.2 Fabrication of Sustainable Textured Coatings Using AI-Assisted Technology

AI is creating a new frontier in the production of sustainable textured coatings by integrating real-time data analysis, process refinements, and quality assurance into manufacturing processes. Sustainable coatings are coated surfaces that incorporate eco-friendly materials that have superior durability, corrosion resistance, and tribological performance. AI platforms can optimise a range of critical fabrication parameters such as spray rates during application, optimal curing times, deposition thickness, and texturing parameters for laser systems to produce uniform coatings and surface morphologies mimicking bio-inspired structures or engineered designs [36]. Predictive models can supply operational insights relating to the predicted response of a coating based on real-time process data to optimise variable adjustments in environmental conditions and minimise energy consumption if drying or curing times are being utilised. AI-enhanced design software can even enable users to incorporate bespoke changes to textured coatings to improve specific applications (i.e., self-cleaning, corrosion prevention, or friction minimisation). Sustainable coatings, therefore, deliver longer lifecycles and less maintenance, which contribute to sustainability by lowering resource and pollution impact [37]. The convergence of AI, therefore, introduces consistency in batch production processes by ensuring repeatability, which is vital for the durability and reliability of protective coatings for time-sensitive and hostile environments.

8.5 OPTIMISATION OF MACHINING AND FINISHING PROCESSES THROUGH AI

AI is revolutionising the optimisation of machining and finishing processes using historic levels of precision, efficiency, and sustainability. The traditional machining parameters, such as cutting speed, feed rate, and depth of cut, have long been determined by empirical experience or conservative heuristics, which, in most instances, have led to inefficient rates of material removal as well as excessive tool wear. The AI-based approaches, particularly RL and ML models, set these parameters optimally in a dynamic fashion by learning progressively from the process feedback [38].

RL methods view machining as a sequential decision problem, where the control parameters are adjusted to optimise an objective function that balances productivity (maximising metal removal) with quality (minimising surface damage and tool wear). This adaptive method significantly improves tool life, reduces energy usage, and refines surface finishes. For instance, AI algorithms avoid thermal damage by gradually reducing velocities when tool wear reaches its critical stage, while operating at high velocities initially to maximise efficiency [38].

In addition, AI integrates process monitoring data, such as vibration, acoustic emission, and temperature data, to predict machining conditions and anticipatorily control parameters. This enhances the quality of the product and reduces scrap rates while lowering energy usage by minimising waste in cycles and eliminating the need for tool replacement. AI-facilitated process optimisation enables sustainable manufacturing goals by utilising materials with maximum efficiency, extending tooling life, and reducing operational carbon footprints. A number of articles record that machining processes enabled by AI can achieve over a 50% increase in material removal rates without any compromise or even improvement in surface integrity, emphasising the transformative capability of AI to drive eco-friendly manufacturing, as shown in Figure 8.4.

8.5.1 Data Mining and ML Algorithm-aided Sustainable Material Discovery

Sustainable materials discovery is continually improving with the integration of ML and data mining techniques, enabling the rapid discovery of materials with enhanced environmental and performance characteristics [40,41]. Furthermore, natural

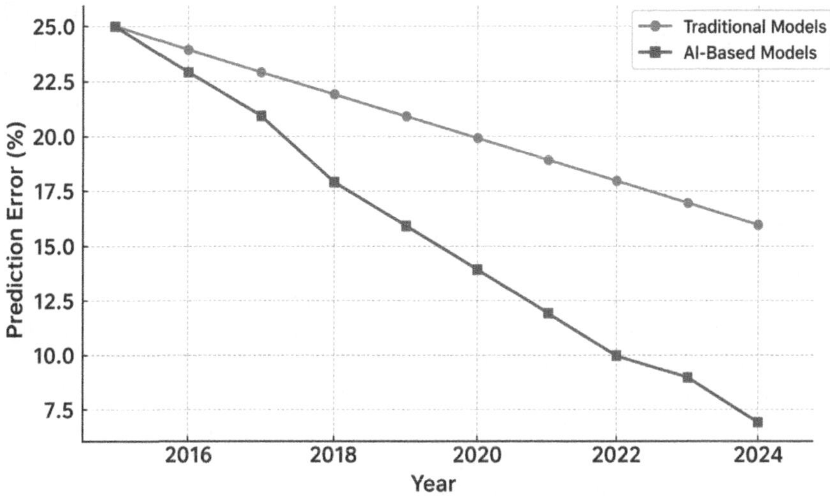

FIGURE 8.4 This chapter clearly shows how much better the predictions of hurricane's paths were with the use of DL-based weather forecasting models. The improvement was remarkable: 15% increase in the accuracy of forecasts over both traditional numerical weather prediction and regular meteorological models, thus indicating the betterment of AI use in tropical cyclone forecasting through better accuracy and reliability [39]. (Reproduced with permission.)

language processing (NLP) computer programmes extract information from millions of scientific articles, enabling the discovery of novel compounds and the optimisation of synthesis pathways in relation to sustainability metrics. Such ML-based discovery pipelines accelerate the discovery of bio-based polymers, low-environmental footprint high-entropy alloys, and materials with fewer rare elements that require fewer catalysts. This synergy is instrumental in achieving global sustainability goals in high-resource-intensive industries, such as construction, transportation, energy, and electronics [41].

8.5.2 Reducing Use of Synthetic Additives and Non-Renewable Materials

Decreasing reliance on synthetic additives and nonrenewable materials is an urgent topic within sustainability that AI can assist with through material engineering. Synthetic additives, such as lubricants, plasticisers, and stabilisers, are the primary components of hazardous, nonbiodegradable, or hard-to-source raw materials that have significant human and environmental impacts [42]. Using generative ML models, scientists can generate optimised molecular scaffolds that lower human and environmental impacts while ensuring compatibility and performance. These models can reduce vast chemical space to produce viable hit candidates while meeting green chemistry principles (i.e., low toxicity, construed to be found as biodegradable, and relatively simple to create). Additionally, ML models create the potential to predict important physicochemical properties, stability, inventiveness, and recyclability in material testing much faster than traditional experimental testing pathways. While processing and curb concerns (i.e., eco-friendly practices, reducing material, and energy and emissions) have been trending and AI techniques have emerged with new processing strategies, processing ranges have been demonstrated to be used in additive manufacturing as well as chemical synthetic processes. AI has the ability to learn the best processing conditions from its modelling reaction kinetics and processing different information where AI can offset the negative impacts caused by synthesis (i.e., detrimental by-products, rare catalyst, or ionic solvent) for which AI-driven retrosynthetic aid approaches have been developed [43, 44].

There is hope that the future use of AI in the industry has enabled reworking of products recalibrated to sustainable additives that enhance compliance, reduction of costs, and marketability. This complete stream approach to AI from a molecular design solution to a process optimisation solution enables fundamental shifts to circular economy systems, as well as green supply chains by reducing environmental impacts with sustainability of product performance or improved performance, which is a key development towards sustainable manufacturing and material development.

AI is disrupting machining and finishing with dynamic process parameter optimisation, enhancing efficiency and sustainability. Traditional machining utilises static settings, e.g., cutting speed and feed rate, that deliver poor tool life and huge amounts of energy across almost all use cases. AI-complemented approaches, especially RL and ML, synergistically accept real-time responses and realisation feeds to modify parameters adaptively while enhancing metal removal rates, surface finishes, and tool wear consideration. Data-driven machining serves to balance doing machining work more effectively while still benchmarking against goals associated with green manufacturing, as shown in Figure 8.5.

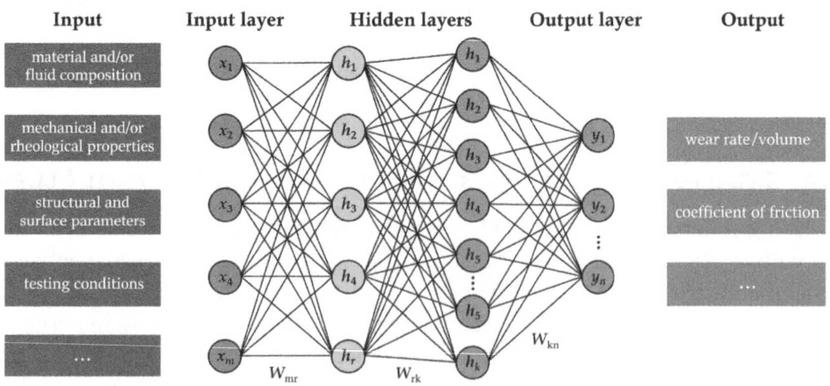

FIGURE 8.5 Artificial neural networks have been applied to discover the relationship between material characteristics and testing environments. The model captures the nonlinear interactions among the input variables, like material mixture, surface texture, weight, and climate conditions, and thus it can very accurately predict the friction and wear under different tribological test conditions. (Reproduced from [45].)

8.6 DATA MINING AND ML ALGORITHM-DRIVEN SUSTAINABLE MATERIAL DISCOVERY

ML and data mining accelerate sustainable material discovery by forecasting environmental and functional characteristics from huge computational and experimental databases. Unlike traditional trial-and-error, density functional theory (DFT) calculations and ML models trained in materials databases allow for virtual high-throughput screening to rapidly identify materials with optimised recyclability, low toxicity, and energy-efficient synthesis. Transfer and active learning strategies for small data improve predictability even with limited data [46]. Physics-informed ML uses domain knowledge for enhanced explainability and resilience. This AI-assisted discovery accelerates the development of biopolymers, green catalysts, and low-carbon, environmentally friendly alloys with reduced environmental footprints. In this way, AI supports a design decision that ensures performance while considering circular economy principles, vital in addressing sustainability issues facing construction, transportation, and electronics.

AI has reduced the use of synthetic additives and nonrenewable resources through novel molecular substitution design and optimisation. Most synthetic additives, for example, plasticisers or stabilisers, are either toxic or in short supply. Generative ML algorithms have generated candidate molecules that display the same performance but that also fall within green chemistry guidelines, such as compostable, low toxicity, stability, and being renewable. Predictive models based on chemical physicochemical and environmental profiles have quickly filtered candidate molecules—faster and less expensive than traditional experimentation. AI has also provided feedback to improve production processes, reducing emissions and waste, using simulation of reaction kinetics, and selecting sustainable synthetic pathways. Also, AI provides believable future biobased additives from renewable starting materials through AI-aided retrosynthesis, reducing the use of fossil fuels. AI-enabled reformulation of industrial applications reduces regulatory burden, saves money, and increases sustainable supply chains. This integrated framework from molecular design to optimisation enables and supports circular economy models that provide a lower ecological footprint plant for high-performance products [47].

8.7 INTEGRATION OF GREEN AI MODELS TOWARDS GLOBAL SUSTAINABILITY GOALS

The forward-looking approach to tribology has been primarily based upon sustainable AI models tailored towards optimising global sustainability goals like carbon neutrality and efficiency of resources. These models are marked by energy-efficient algorithms, low computational demands, and associated with a low carbon footprint during training. The advancement of lightweight, interpretable AI models specifically for tribological-related actions has enabled methodologies that monitor and optimise lubrication, wear, and friction in real time in a capacity that is affordable in terms of energy usage [48].

The integration of AI has influenced a whole new sphere of carbon-neutral compositions and tribological materials by exploring environmentally friendly formulations, assessing their lifecycle implications, or directing low-energy manufacturing. For instance, using AI-backed formulations to optimise the biodegradable lubricant production environment has reduced embodied energy while also providing mechanical performance. In industrial environments, AI systems have been optimised in real time to reduce energy consumption and emissions, while also extending the lifespan of tribological components, providing increased carbon neutrality at the nexus of industry and transportation. AI models have also brought collective agency together by organising widespread efforts around the world to collect, share, and analyse data across multiple disciplines and organisations. This unified action has been able to use AI to inform global best practices, collaborative sustainability metrics, and create an increased speed of carbon-neutral solutions worldwide [49].

8.7.1 Open Challenges and Ethical Implications of AI Use

Although AI holds promise for enhancing sustainability and equity in tribology, two issues related to the events must still be tackled: large-scale open problems and ethical issues. Some large-scale technical problems are as follows: data quality and scarcity, especially for rare or complex tribological phenomena; generating robustness, explainability, and usability for AI models across many operating conditions; and developing standardised measures of sustainability. Ethical problems focus on bias, transparency, and accountability. Typically (if not always), AI models are "black boxes", which raises questions of accountability with respect to safety and environmental footprint and implications for economic justice. For instance, biases in the training data (such as, geographic area or the material being over-represented) may induce biased decision-making and potential harm to a community or industry [50]. Responsible AI development often relates to providing transparency into the algorithm, appropriate data diversity, and stakeholder engagement. Additionally, misuse of AI or too much dependency on the AI technology can lead to diminishing trust, affecting human responsibility and accountability and reinforcing digital inequality. Ultimately, enveloping AI into tribological practice and research means use of supportive frameworks for responsible governance and ongoing stakeholder engagement, meanwhile honouring commitments to ethical principles, such as inclusivity and equity [50,51].

8.7.2 AI-Enabled Circular Tribology Vision

The future vision includes AI-powered circular tribology—a space where lubricants, parts, and materials are designed, used, and recycled within an ongoing closed-loop system, generating no waste or environmental impacts. AI has enabled that transition by providing intelligent material selection, predictive maintenance, and adaptive surface engineering to maximise part life while encouraging recycling [52].

AI-powered simulation software has provided the best surface finish, coating, and lubricant formulation to maximise durability while allowing for the recovery of wear particles and debris. ML algorithms have predicted end-of-life paths for parts and recommended maintenance or refurbishment at appropriate times to avoid unnecessary replacement. Digital twins, virtual replicas of physical tribological systems, have also monitored current conditions, predicted the end of life, and controlled the recycling or remanufacturing processes with very little human intervention, facilitating the creation of self-repairing and nature-inspired tribological surfaces that are capable of renewing their operating life by either repairing the wear damage on their own or self-modulating according to the changing operating conditions. Besides, such surfaces have been prolonging their cycle of use through self-repair or self-modulation, thus minimising the impact of tribological systems on the environment through wear, dependence on raw materials, and so on. Moreover, sustainability, or even more, ecological friendliness, has been the principal aspect of the whole development of tribological systems [53–55]. There is a need for extensive interdisciplinary cooperation, sharing of data, and development of standard, transparent models to make the transition to AI-based circular tribology. The global AI-based framework has played a major role in the achievement of a future for tribology and other technologies that is scalable, sustainable, and resource-efficient.

8.8 CONCLUSIONS

The direction of tribology in the future is towards the development of smart, autonomous systems that have been be able to feed themselves and supervise themselves, understand the changing conditions, and do the self-optimisation throughout their life without human intervention. These systems are combined with AI, ML, and sensor technologies to perform real-time diagnostics, preventive maintenance, and friction, wear, and lubrication control, using the devices' reliability, energy efficiency, and material life self-sustainability factor that limits human intervention and surprise failures. Self-sustainable tribology has been using digital twins, where the physical systems are represented by computer models, to constantly learn and through simulation-driven optimisation. The AI embedded in the system has been controlling the lubrication with precision, dynamically adjusting to the surface conditions, and sharing loads in a way that reduces material wear and energy losses. The predictive and preventive anticipation of wear or friction irregularities has been prolonging the asset lifespan and minimising resource utilisation, and thus, the indirect implication of human environmental sustainability. The convergence of nanotech innovations with AI-as-a-service has the potential of granting such systems the capabilities of self-healing surfaces and bioinspired adaptive materials that change their properties in response to the environment. This smart interaction between materials and AI-driven processes represents a transition from traditional reactive maintenance to proactive, sustainable, design. The hope is for tribological solutions that have not only been high-performing but also a source of regeneration and responsive to the environment, therefore in line with the global sustainability aspirations.

ACKNOWLEDEGEMENT

The author (S. Roy) would like to acknowledge 'Scheme for Transformational and Advanced Research in Sciences (STARS)' (MoE-STARS/STARS-2/2023-0175) by the Ministry of Education for promoting translational India-centric research in sciences implemented and managed by the Indian Institute of Science (IISc), Bangalore, for their support.

REFERENCES

1. Kumar, A., R. Gupta, & G. Mani. The DMSO-catalyzed CO_2 reduction with 9-BBN as a source for the selective N-formylation, room temperature N-methylation of amines, N-hydroxymethylation of heterocycles, and C–N couplings. *The Journal of Organic Chemistry* 2025, 90(29), 10316–10331. https://doi.org/10.1021/acs.joc.5c00965
2. Wang, Z.-F., Y. Fei, A. Qin, S. Zhang, & X. Zhang. Robust fluorine-decorated {Yb4}-organic framework for C_2H_6 capture and efficient catalytic performance on CO_2-epoxide cycloaddition. *ACS Applied Materials & Interfaces* 2025, 17(20), 29788–29798. https://doi.org/10.1021/acsami.5c07205
3. Wang, X., T. Hu, L. Fan, Q.-P. Qin, & X. Zhang. A precise preparation strategy for 2D nanoporous thulium-organic framework: High catalytic performance in CO_2-epoxide cycloaddition and knoevenagel condensation. *Inorganic Chemistry* 2025, 64(9), 4461–4471. https://doi.org/10.1021/acs.inorgchem.4c05273
4. Yao, Y., B. Tian, M. Liu, F. Xu, K. Cui, P. Wen, R. Dong, & M. Fan. An ecofriendly design for superhydrophilic natural deep eutectic solvent lubricants and their excellent interfacial behavior. *ACS Sustainable Chemistry & Engineering* 2024, 12(46), 16934–16948. https://doi.org/10.1021/acssuschemeng.4c06944
5. Yan, Y., X. Zhang, Y. Wang, S. Xue, S. Liu, Q. Ye, F. Zhou, & W. Liu. Grafting poly(ionic liquid)s on covalent organic frameworks via surface-initiated ring-opening metathesis polymerization for enhancing lubrication performance. *ACS Applied Polymer Materials* 2024, 6(16), 10072–10082. https://doi.org/10.1021/acsapm.4c01654
6. Yang, C., G. Wang, S. Guo, & J. Sun. Potential and progress of two-dimensional nanomaterials in oil-based lubrication. *Chinese Chemical Letters* 2025, 36(9), 111178. https://doi.org/10.1016/j.cclet.2025.111178
7. Li, Y., X. Kong, D. Zhu, T. Guo, X. Wu, W. Yu, & L. Wang. Development of the DES-contained reduced graphene oxide system with efficient CO_2 adsorption and photothermal desorption for pre-gas purification in AEMFCs. *Separation and Purification Technology* 2025, 360, 131193. https://doi.org/10.1016/j.seppur.2024.131193
8. Ren, M., B. Zhao, C. Li, Y. Fei, X. Wang, L. Fan, T. Hu, & X. Zhang. Defect-engineered indium–organic framework displays the higher CO_2 adsorption and more excellent catalytic performance on the cycloaddition of CO_2 with epoxides under mild conditions. *Molecular Diversity* 2025, 29(3), 2017–2031. https://doi.org/10.1007/s11030-024-10956-z
9. Ciulli, E. Vastness of tribology research fields and their contribution to sustainable development. *Lubricants* 2024, 12(2), 33. https://doi.org/10.3390/lubricants12020033
10. Hua, K., Y. Shi, M. Tang, X. Zuo, J. Gao, & X. Zhang. In-situ surface modified water-soluble Cu nanoparticles as lubrication additives in water-based cutting fluids. *Colloids and Surfaces A: Physicochemical and Engineering Aspects* 2024, 698, 134605. https://doi.org/10.1016/j.colsurfa.2024.134605
11. Yang, C., J. Sun, Q. Chang, D. Su, Y. Xu, G. Wang, & S. Guo. Tribo-induced interfacial lubrication performance and self-repair mechanism of oleylamine-grafted chlorinated graphene. *Langmuir* 2025, 41(36), 24754–24764. https://doi.org/10.1021/acs.langmuir.5c03049
12. Bhattacharjee, J., & S. Roy. A review on photocatalysis and nanocatalysts for advanced organic synthesis. *Hybrid Advances* 2024, 6, 100268. https://doi.org/10.1016/j.hybadv.2024.100268
13. Liu, X., Z. Zheng, J. Xiang, H. Wang, H. Wang, D. Feng, D. Qiao, & H. Li. Perspective of tribological mechanisms for α-alkene molecules with different chain lengths from interface behavior. *Langmuir* 2024, 40(52), 27364–27376. https://doi.org/10.1021/acs.langmuir.4c03592

14. Yu, H., X. Zuo, X. Zhang, & W. Liu. CO_2 capture and conversion into carbon-based tribofilm by in situ tribochemical reaction for green lubrication. *ACS Sustainable Chemistry & Engineering* 2024, 12(10), 4070–4080. https://doi.org/10.1021/acssuschemeng.3c07283
15. Guo, Y., S. Yang, P. Ren, H. Liu, X. Miao, Z. Wu, Y. Lv, J. ShangGuan, & Y. Bai. Interfacial self-healing lubrication in pen nib steel: Water-based composite nanoparticle lubricants for friction reduction. *Colloids and Surfaces A: Physicochemical and Engineering Aspects* 2025, 720, 137087. https://doi.org/10.1016/j.colsurfa.2025.137087
16. Mugware, F. W., T. Ravele, & C. Sigauke. Short-term predictions of global horizontal irradiance using recurrent neural networks, support vector regression, gradient boosting random forest and advanced stacking ensemble approaches. *Computation* 2025, 13(3), 72. https://doi.org/10.3390/computation13030072
17. Shaikh, A. A., J. Bhattacharjee, P. Datta, & S. Roy. A comprehensive review of the oxidation states of molybdenum oxides and their diverse applications. *Sustainable Chemistry for the Environment* 2024, 7, 100125. https://doi.org/10.1016/j.scenv.2024.100125
18. Li, S., C. Gao, C. Duan, S. Zhang, P. Zhang, & Z. Zhang. Achieving oil-based superlubricity with near-zero wear via synergistic effect between PEEK-PTFE and PAO40 containing DDP-Cu nanoparticles. *Tribology International* 2025, 208, 110645. https://doi.org/10.1016/j.triboint.2025.110645
19. Tian, B., K. Cui, P. Wen, R. Dong, & M. Fan. Reinforcement of tribological performance assisted by in-situ carbon layer from the release, capture and tribo-chemistry of ionic liquid hydrotalcite composite lubricating materials. *Carbon* 2025, 243, 120532. https://doi.org/10.1016/j.carbon.2025.120532
20. Li, Z., T. Wang, J. Sun, X. Li, & J. Zhao. Tribological behavior of graphene reinforced ceramics. *Science China Materials* 2025, 68(8), 2636–2669. https://doi.org/10.1007/s40843-025-3478-4
21. Fan, J., Y. Shen, Y. Liu, J. Qu, J. Liu, B. Yu, & J. Xu. Improved friction and wear performance utilized with aminoguanidine based ionic liquid over wide temperature range for reciprocating frictional contact surface. *Tribology Letters* 2025, 73(2), 6. https://doi.org/10.1007/s11249-025-01973-6
22. Wang, J., C. Gao, C. Duan, G. Yang, J. Ma, L. Yu, S. Zhang, P. Zhang, & Z. Zhang. Whole course oil-based superlubricity for polyetheretherketone-bearing steel sliding pair by introducing tribocatalytically active dialkyl-dithiophosphoric-modified copper nanoparticles in poly-alpha olefin oil. *Friction* 2025, 13(5), 9440954. https://doi.org/10.26599/FRICT.2025.9440954
23. Li, H., X. Sui, P. Ayala, E. Marquis, H. Rabl, A. Ertl, P. Bilotto, Y. Shang, J. Li, L. Xu, M. C. Righi, D. Eder, & C. Gachot. Advanced solid lubrication with COK-47: Mechanistic insights on the role of water and performance evaluation. *Advanced Science* 2025, 12(17), 15268. https://doi.org/10.1002/advs.202415268
24. Yu, H., X. Zuo, X. Zhang, X. Wang, & F. Zhou. Bioinspired carbon quantum dots nanofluid fabricated by a facile two-step synthesis toward high performance lubrication. *Chemical Engineering Journal* 2025, 510, 161810. https://doi.org/10.1016/j.cej.2025.161810
25. Wang, Z., J. Wang, R. Huang, J. Wang, B. Zhang, K. Grinkevych, & J. Xu. Tribological performance of lubricant with high concentration graphene additive dispersed by ultrasonic technology. *Tribology International* 2025, 202, 110354. https://doi.org/10.1016/j.triboint.2024.110354
26. Farea, A., O. Yli-Harja, & F. Emmert-Streib. Understanding physics-informed neural networks: Techniques, applications, trends, and challenges. *AI* 2024, 5(3), 1534–1557. https://doi.org/10.3390/ai5030074

27. Huynh, K. K., A. K. Tieu, C. Lu, L. Smillie, C. Nguyen, & S. T. Pham. In-situ engineering catalytically active surfaces for tribocatalysis with layered double hydroxide nanoparticles. *Carbon* 2024, 228, 119324. https://doi.org/10.1016/j.carbon.2024.119324
28. Wang, Z., R. Huang, Z. Zhao, H. Qiang, J. Wang, K. Grinkevych, & J. Xu. Preparation and tribological properties of the friction induced Mo-based film. *Tribology International* 2024, 195, 109633. https://doi.org/10.1016/j.triboint.2024.109633
29. Zhang, Y., Q. Niu, Y. Xu, J. Chen, C. Zhai, U. Azhar, Y. Lu, H. Li, & C. Zong. Synthesis of long straight-chain alkane substituted ferrocene with ultra-robust oil solubility as multifunctional organometallic additive. *Applied Organometallic Chemistry* 2024, 38(6), 7482. https://doi.org/10.1002/aoc.7482
30. Peng, F., S. Fan, N. Song, C. Gao, S. Zhang, & Y. Zhang. Effect of nickel acetyl acetonate as lubricant additive in base oils with different molecular structure on in-situ formation and tribomechanism of carbon-based tribofilms of steel-steel sliding pair. *Tribology Letters* 2024, 72(2), 1859. https://doi.org/10.1007/s11249-024-01859-z
31. Bhattacharjee, J., S. Roy, M. Mandal, & A. Mallik. Critical review of biomass and nanotech synergies using Artificial Intelligence (AI) process. *Energy* 2025, 360(4), 100038. https://doi.org/10.1016/j.energ.2025.100038
32. Huynh, K. K., S. T. Pham, K. A. Tieu, & S. Wan. Tribocatalysis induced carbon-based tribofilms—An emerging tribological approach for sustainable lubrications. *Lubricants* 2023, 11(8), 327. https://doi.org/10.3390/lubricants11080327
33. Zhang, L., Z. Zhang, X. Gao, S. J. Matlan, & N. A. Taha. The preparations of fluorographene nanosheets and research in tribological properties in high vacuum. *Materials* 2023, 16(11), 3929. https://doi.org/10.3390/ma16113929
34. Jyoti Bhattacharjee, S. R., A. A. Shaikh, & P. Datta. Efficient fabrication and characterization of doped nanocomposites for thermoelectric materials. *Nano Trends* 2025, 10, 100109. https://doi.org/10.1016/j.nwnano.2025.100109
35. Li, D., R. Lv, G. Si, & Y. You. Hybrid neural network-based prediction model for tribological properties of polyamide6-based friction materials. *Polymer Composites* 2017, 38, 1705–1711.
36. Sardar, S., S. Dey, & D. Das. Modelling of tribological responses of composites using integrated ANN-GA technique. *Journal of Composite Materials* 2020, 55, 75.
37. Vinoth, A., & S. Datta. Design of the ultrahigh molecular weight polyethylene composites with multiple nanoparticles: An artificial intelligence approach. *Journal of Composite Materials* 2020, 54, 179–192.
38. Aleksendrić, D., & D. C. Barton. Neural network prediction of disc brake performance. *Tribology International* 2009, 42, 1074–1080.
39. Leong, W. Y. Artificial intelligence-driven analytics for monitoring and mitigating climate change impacts. *Engineering Proceedings* 2025, 108(1), 7. https://doi.org/10.3390/engproc2025108007
40. Senatore, A., & S. Ciortan. An application of artificial neural networks to piston ring friction losses prediction. *Mechanical Testing and Diagnostics* 2011, 1, 7–14.
41. Bhattacharjee, J., & S. Roy. A new approach: Spinel nano-ferrites for wastewater-based microplastics. In: Thakur, A., & P. Thakur (Eds) *Applications of Spinel Nano-Ferrites in Health, Environmental Sustainability, and Safety*; CRC Press: Boca Raton, FL, 2025; pp. 171–183.
42. Roy, S., & J. Bhattacharjee. Nano-scale computational modeling: Challenges, limitations, and emerging trends. In: Behera, A., & J. K. Patra (Eds) *Advanced Nanomaterials in Biomedical Implants: Processing, Structures, Properties and, Applications*; Elsevier: Oxford, UK, 2024; pp. 471–494. https://doi.org/10.1016/B978-0-443-27378-0.00018-0
43. Echávarri Otero, J., E. De La Guerra Ochoa, E. Chacón Tanarro, P. Lafont Morgado, A. Díaz Lantada, J. M. Munoz-Guijosa, & J. L. Muñoz Sanz. Artificial neural network approach to predict the lubricated friction coefficient. *Lubrication Science* 2014, 26, 141–162.

44. Bhakta, D., D. Gayen, J. Bhattacharjee, R. Chatterjee, P. Dastidar, & S. Roy. Repurposing industrial and agricultural wastes as sustainable alternatives in photovoltaic cells and dye degradation. *Chemistry Select* 2024, 9, 38.
45. Rosenkranz A., M. Marian, F. J. Profito, N. Aragon, & R. Shah. The use of artificial intelligence in tribology—A perspective. *Lubricants* 2021, 9(1), 2. https://doi.org/10.3390/lubricants9010002
46. Marian, M., P. Grützmacher, A. Rosenkranz, S. Tremmel, F. Mücklich, & S. Wartzack. Designing surface textures for EHL point-contacts—Transient 3D simulations, meta-modeling and experimental validation. *Tribology International* 2019, 137, 152–163.
47. Boidi, G., M. Rodrigues da Silva, F. J. J. Profito, & I. F. Machado. Using machine learning radial basis function (RBF) method for predicting lubricated friction on textured and porous surfaces. *Surface Topography: Metrology and Properties* 2020, 8, 044002.
48. Kügler, P., M. Marian, B. Schleich, S. Tremmel, & S. Wartzack. tribAIn—Towards an Explicit specification of shared tribological understanding. *Applied Sciences* 2020, 10, 4421.
49. Senatore, A., V. D'Agostino, R. Di Giuda, & V. Petrone. Experimental investigation and neural network prediction of brakes and clutch material frictional behaviour considering the sliding acceleration influence. *Tribology International* 2011, 44, 1199–1207.
50. Marian, M., M. Bartz, S. Wartzack, & A. Rosenkranz. Non-dimensional groups, film thickness equations and correction factors for elastohydrodynamic lubrication: A review. *Lubricants* 2020, 8, 95.
51. Alambeigi, F., S. M. Khadem; H. Khorsand, & E. Mirza Seied Hasan. A comparison of performance of artificial intelligence methods in prediction of dry sliding wear behavior. *The International Journal of Advanced Manufacturing Technology* 2016, 84, 1981–1994.
52. Bhattacharjee, J., S. Roy, & A. A Shaikh. Artificial intelligence and machine learning for the optimization of photocatalytic performance. *Energy* 2025, 360(3), 100027. https://doi.org/10.1016/j.energ.2025.100027
53. Dai, X., J. He, Y. Chen, A. Fu, W. Zhang, & Q. Huang. Self-lubrication and enhanced wear and corrosion resistance of calcium silicate hydrate nanostructures on Ti-based micro-arc oxidized coatings. *Smart Materials in Manufacturing* 2025, 3, 100081. https://doi.org/10.1016/j.smmf.2025.100081
54. Busse, M., & A. K. Schlarb. A novel neural network approach for modeling tribological properties of polyphenylene sulfide reinforced on different scales. In: Busse, M., & A. K. Schlarb (Eds) *Tribology of Polymeric Nanocomposites*; Elsevier: Oxford, UK, 2013; pp. 779–793.
55. Mandal, D., J. Bhattacharjee, D. Mitra, & S. Roy. Aspects of energy efficiency and the execution of modern technologies for achieving net zero. *Graduate Journal of Interdisciplinary Research, Reports and Reviews* 2025, 2(03), 150–163. https://doi.org/10.5281/ZENODO.15087049

9 Artificial Intelligence-Based Emerging Tools and Techniques in Tribology

Praveen Kumar Verma and Hitesh Vasudev

9.1 INTRODUCTION

One interpretation of the Greek tribos 'rubbing', tribology is the crossing of disciplines that probes the fundamental physicochemical and mechanical interfacial phenomena of bodies in relative motion. Incorporating the friction and wear of lubricants, it has grown from ancient empirical practices to a fully integrated discipline that includes mechanical engineering, surface science, materials chemistry, condensed matter physics, and biomechanics [1]. The tribological system designs and their operation complexities arise from the nonlinear, nonstationary, and multiscale problems of friction, lubrication, and wear [2]. System response at the atomic level, say, of surface asperities, is influenced by the mesoscale structural and the macroscale mechanical system response. Linear, ideal, and stochastic assumptions that permeate analytical and numerical techniques fail to predict the system behavior that is predominantly history-dependent [3]. Hence, it remains highly challenging to predict, *a priori* and accurately, wear rate, friction response in different operating conditions, and the stability of lubricants in a given service. In contrast to siloed methodologies of the past, tribo-informatics facilitates cross-domain correlation mining, high-fidelity predictive modeling, and design–optimization loops in shorter cycles, while unveiling hidden structure–property–performance coupling elusive to standard experimental–theoretical strategies [4].

Predictive modeling of wear and friction is likely to be the most rewarding application of artificial intelligence (AI) in tribology. Current models, such as Archard's wear law or Hertzian contact theory, are useful in practice but suffer from their inability to handle complex surface interactions under varying load, velocity, temperature, and lubrication conditions. AI models such as machine learning (ML) and deep learning (DL) can be trained from huge experimental data to learn the nonlinear relationships and to produce very precise predictions for tribological performance. Such models can be easily trained for novel materials, coatings, or lubricating regimes, thus eliminating the need for repeated and expensive experimental campaigns. AI can analyze these data in real time to identify anomalies, forecast failure, and suggest remedial measures before failures become catastrophic breakdowns.

 DOI: 10.1201/9781003624165-9

In the aerospace and automotive industries, where safety and efficiency are at a premium, AI-based condition monitoring is soon becoming de rigueur [5]. Figure 9.1 shows a conceptual overview of how AI, via ML, DL, the Internet of Things, remote sensing (RS), and hybrid models, facilitates decision-making, prediction, and optimization in real-time monitoring, infrastructure, hydrological, and hybrid modeling systems.

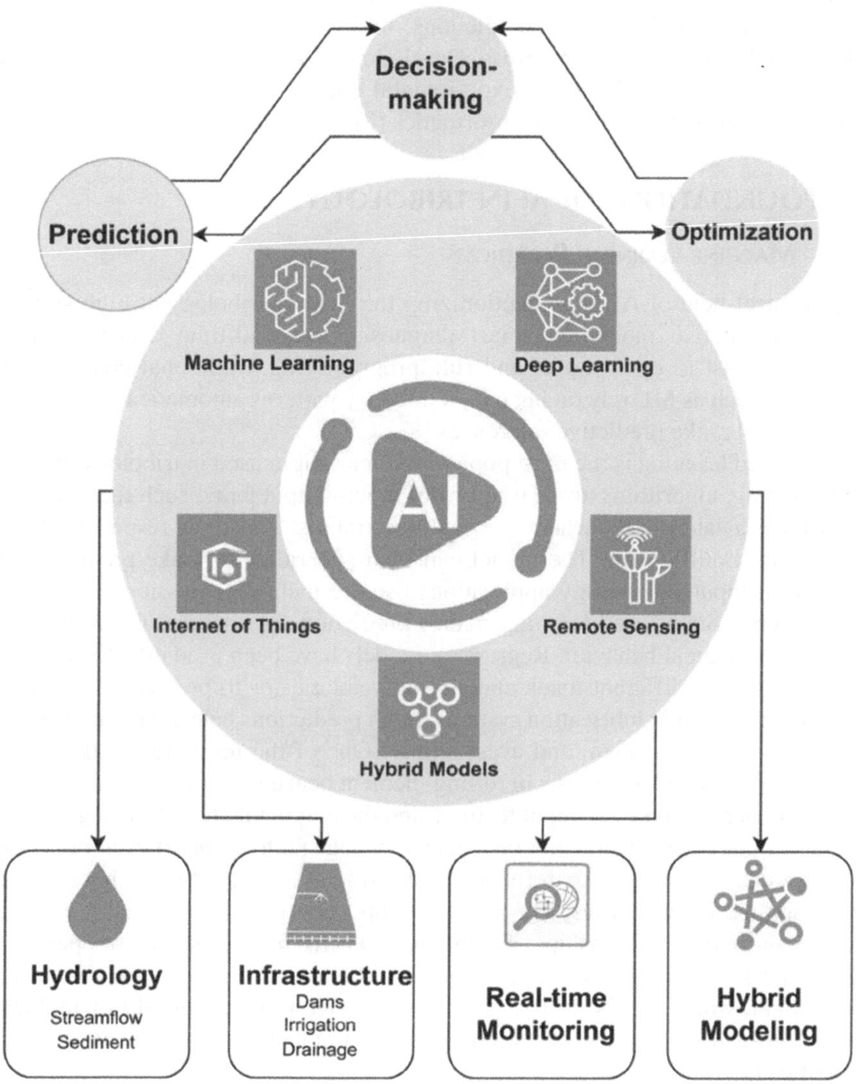

FIGURE 9.1 A conceptual framework of AI-assisted prediction, optimization, and decision support in hydrology, infrastructure, real-time monitoring, and hybrid model systems based on machine learning (ML), deep learning (DL), the Internet of Things (IoT), remote sensing (RS), and hybrid models [6]. (CC Attribution by 4.0.)

Applications of AI in tribology extend beyond utilitarian purposes to the quest for fundamental understanding. With unsupervised learning techniques, AI can classify data to distinguish previously unrecognized wear mechanisms or systematically classify surface topologies. Reinforcement learning (RL) approaches have even been proposed for the generation of adaptive tribology systems with the ability to modify lubrication tactics dynamically based on the operating conditions. However, it is not without its challenges. Lack of data and inconsistency in data from one laboratory to another can affect the creation of sound AI models. Interpretability is another issue: whereas AI may make precise predictions, understanding the underlying physics will be crucial for scientific progress. Development of open-access tribo-databases of reproducible and well-curated experimental data is another potential avenue to promote the development of tribo-informatics [7].

9.2 FOUNDATIONS OF AI IN TRIBOLOGY

9.2.1 Machine Learning Paradigms

ML, a central field of AI, is revolutionizing the field of tribology at high speed by allowing predictive modeling, smart diagnosis, and real-time decision-making. In stark contrast to open model and rule programming, traditional computational approaches such as ML rely on big data to identify patterns automatically, build relationships, and make predictive inferences [8].

Supervised learning is the most popular ML paradigm used in tribology. In supervised learning, algorithms are learned from input–output pairs, such that the model learns from a labeled set where the features (inputs) and their respective targets (outputs) are both known. The model can then generalize to make predictions for new unseen inputs. Tribology applications usually include regression or classification problems. Supervised learning makes predictions of continuous variables that describe tribological behavior. Regression models have been used effectively to predict wear rate for different loads and sliding velocities or to predict oil film thickness in hydrodynamic lubrication systems. Such predictions help create maintenance schedules, optimize design, and avert failure. One of the most frequently encountered examples is fault diagnosis in rolling-element bearings, with vibration or acoustic emission signals taken as input features and the output taken as the category of the fault (e.g., inner race defect, outer race defect, or cage fault). Support vector machines (SVMs), decision trees, random forests, and artificial neural networks have all been widely applied in such cases. The power of supervised learning is that it can learn direct input–output relationships of complex inputs from experimental or operational data. Its performance, however, relies heavily on the presence of good-to-excellent labeled data, which can be challenging or expensive to be acquired in tribological studies because of the cost of experimental equipment or long-term wear testing necessities [9].

Unsupervised algorithms, on the other hand, deal with unlabeled data. Rather than relating inputs to familiar outputs, these algorithms seek to reveal hidden structures, patterns, or clusters within the data. Unsupervised methods are especially useful in tribology for exploratory analysis, anomaly detection, and dimensionality reduction.

One example that comes to mind is wear particle analysis of lubricating oils. Wear particles are generated during operation, and their form, size, and distribution contain vital information on the status of contacting surfaces. Manual analysis of these particles can be subjective and labor-intensive. Unsupervised clustering algorithms, such as k-means or hierarchical clustering, can automatically group wear particles into categories (e.g., abrasive wear, adhesive wear, and fatigue wear), providing insights into dominant wear mechanisms. Principal component analysis (PCA) and other dimensionality reduction techniques can convert high-dimensional sensor data into lower-dimensional space, making visualization, interpretation, and detection of normal behavior deviations simpler. Another application of unsupervised learning is anomaly detection, which is rapidly gaining significance in predictive maintenance. By acquiring the typical patterns of operation of a tribological system, unsupervised models can highlight anomalous conditions, like initial-stage bearing faults or non-normal lubrication states, even before failure is catastrophic. This improves system reliability and minimizes downtime in industrial systems [10]. Figure 9.2 shows the schematic representation that highlights the uses of key ML methods and algorithms in data analysis, prediction, and optimization by classifying them into supervised, unsupervised, RL, and DL categories.

RL is a new and developing paradigm in ML. In contrast to supervised and unsupervised learning, both of which are concerned with static data collections, RL deals with dynamic interactions between an agent and the environment. An agent performs actions in a state, gets feedback through rewards or penalties, and learns an optimal policy in terms of maximizing cumulative rewards over time. Despite being yet in its early stages in tribology, RL is extremely promising for adaptive and intelligent lubrication approaches. For example, in real-time lubrication control, an RL agent may vary lubricant delivery rates, viscosity modifiers, or additive levels in response to constantly varying operating conditions like load, speed, or temperature. In the long run, the system would learn to reduce friction and wear to a minimum while maximizing energy efficiency and component life. Similarly, RL paradigms can be

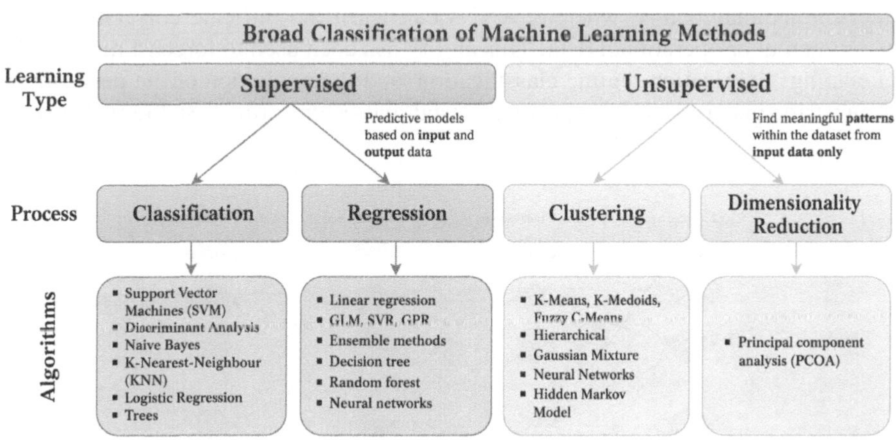

FIGURE 9.2 A diagram generally classifying existing ML methods and algorithms [11]. (CC Attribution by 4.0.)

used in adaptive control of sophisticated tribotesting machines so that the system adjusts test parameters automatically to mimic real wear scenarios or speed up material screening. The foremost strength of RL is that it learns best actions without any predefined labeled datasets. Still, its application in tribology is hampered by the necessity of real-world simulation environments, lengthy training times, and cautious exploration strategies so as not to destroy costly equipment in the learning process. Still, with increasing adoption of digital twins and high-fidelity tribological simulations, RL will find more and more applications in the future [12].

9.2.2 Neural Networks and Deep Learning

Artificial neural networks (ANNs) are one of the most powerful ML algorithm classes, based on the biological neural networks in the human brain. They are composed of interconnected nodes, or 'neurons', in layers that take input data, transfer information across hidden layers, and produce outputs. Each of the connections among neurons is linked with a weight that is computed iteratively during training to allow the network to learn intricate relationships in datasets. For tribology, in which the behavior of surface interactions is ruled by nonlinear effects, stochastic fluctuations, and dependencies at multiple scales, ANNs provide an invaluable modeling tool. Unlike traditional regression models and other statistical techniques, ANNs can address nonlinear functions, learn in uncertain conditions, and provide generalized output even if multiple interdependent variables, such as load, sliding velocity, lubrication condition, surface roughness, and material properties, change. The use of neural networks is most apparent in predictive modeling of tribological attributes such as friction coefficient, wear rate, and lubricant film thickness. For instance, in most tribological tests, the output parameter is affected by multiple interrelated input parameters. An apparent modification in sliding speed or contact temperature can engage with surface chemistry and microstructural characteristics, which results in nonlinear behaviors. Conventional modeling methods struggle to delineate such interactions analytically, whereas ANNs can learn the intricate correlations from experimental or operational data. This ability has been used to forecast the wear life of coatings, lubrication regime classification, and the prediction of the performance of new tribo materials without prolonged trial-and-error testing [13]. Figure 9.3 shows

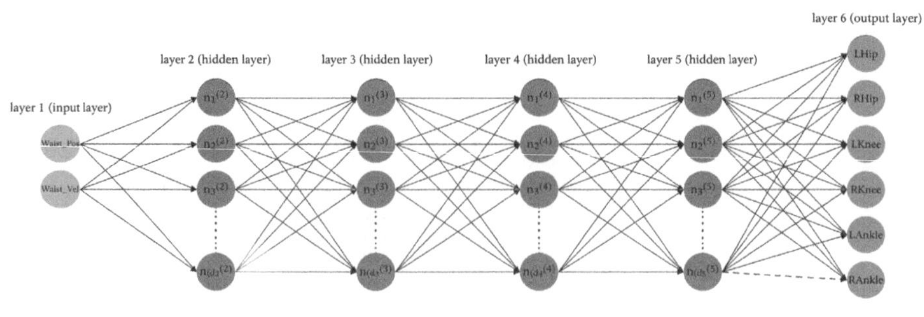

FIGURE 9.3 The deep neural network structure of the simplified model [14]. (CC Attribution by 4.0.)

the deep neural network structure of the simplified model, illustrating interconnected input, hidden, and output layers used for learning complex patterns and improving predictive accuracy in tribological data analysis.

DL, as an area of ML developed from ANNs, builds upon these abilities even further with the use of multilayered network architectures. In DL algorithms, several layers of hidden layers are cascaded in between the input and the output, allowing the system to learn data's hierarchical representations automatically. This hierarchical learning enables the model to learn low-level features in the earlier layers (for example, edges or texture in a picture of a wear scar) and gradually construct increasingly abstract representations in the later layers (for example, wear type classification or surface degradation severity prediction). Such feature extraction is done automatically, lessening the workload of preprocessing or hand-engineered feature engineering, which is commonly needed in traditional ML methods [14,15]. This is especially useful for predictive maintenance in those industries where tribological failures can cause expensive downtime, like aerospace, automotive, and energy industries [16]. Figure 9.4 shows the general machining tool highlighting key aspects discussed in this review, including tribological properties, surface interactions, and associated problems influencing tool performance and material removal processes.

Also, DL frameworks enable multimodal data fusion. Tribological systems tend to generate heterogeneous data types, such as force measurement, temperature traces, lubrication film thickness, images, and acoustic signals. Fusing these multimodal heterogeneous data types into one comprehensive predictive model is difficult for conventional models. But DL models, with their ability to process multisource data, are capable of combining information across modalities to give overall insight. For example, the fusion of surface images and vibration data allows stronger forecasts of wear evolution than the separate analysis of one or both datasets. This multimodal learning is in line with the increasing interest in digital twins of tribological systems,

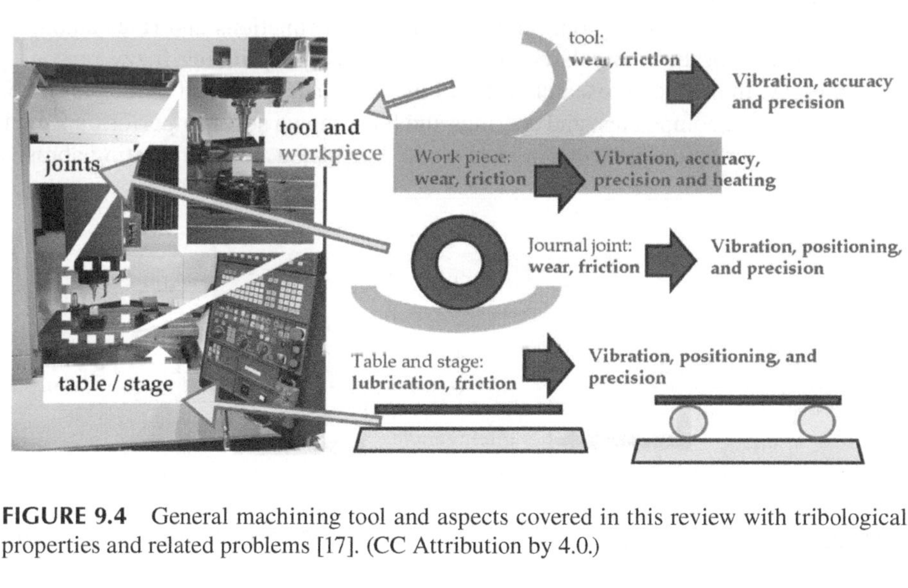

FIGURE 9.4 General machining tool and aspects covered in this review with tribological properties and related problems [17]. (CC Attribution by 4.0.)

where real-time streams of information are incorporated to form virtual copies that mimic and optimize system behavior [18,19].

9.2.3 Fuzzy Logic and Hybrid Systems

Fuzzy logic, originally created by Lotfi Zadeh during the 1960s, is a mathematical method of representing uncertainty, imprecision, and vagueness in decision-making. In contrast to classical binary logic, which places variables into definite sets of 'true' and 'false', fuzzy logic allows variables to be members of multiple sets with partial membership grade. This approach is closer to human thought, where boundaries are typically imprecise and conclusions are based on linguistic rules such as low, medium, or high rather than on definite numerical values. For tribology, where test conditions are not fixed, operating parameters vary, and material behaviors are nonlinear, fuzzy logic presents an effective strategy for modeling complex systems and predicting response with more flexibility and interpretability [20]. The fuzzification, rule base, inference engine, and defuzzification modules utilized to define uncertainty and predict tribological reactions are illustrated as building blocks of a fuzzy logic system (FLS) in Figure 9.5.

Fuzzy logic can be applied when one has to take into account wear, friction, and lubrication phenomena depending on interacting parameters such as load, speed, surface hardness, temperature, and lubrication conditions. The interactions among the above parameters are not mathematically well described but may be represented using experience-based knowledge and heuristic rules. For example, fuzzy systems have been applied to estimate wear rates for composite and steel materials under various operating conditions. One of the common uses of fuzzy logic in tribology is the Adaptive Neuro-Fuzzy Inference System (ANFIS). ANFIS unites human-like inference of fuzzy inference systems (FIS) and adaptability and self-learners in ANNs. While fuzzy systems rely on expert-defined membership functions and rule bases, ANNs learn to recognize patterns in data well but become uninterpretable. ANFIS combines these strengths by optimizing membership functions and rules automatically with training data to enhance prediction without sacrificing fuzzy reasoning transparency. In tribology research, ANFIS models have been used in the prediction of wear of steels, composites, and coatings and have shown superior performance relative to fuzzy or neural network models alone. For instance, in forecasting abrasive

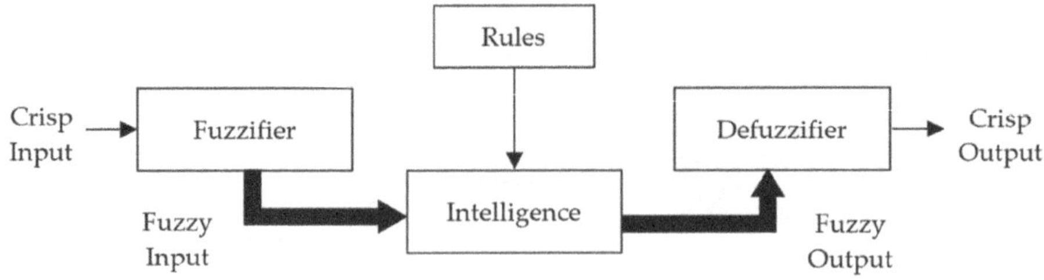

FIGURE 9.5 The components of a fuzzy logic system (FLS) [21]. (CC Attribution by 4.0.)

wear in fiber-reinforced composites, ANFIS has been demonstrated to be capable of capturing material response nonlinearities while yielding explainable rules to relate wear mechanisms to operating conditions [22]. Hybrid systems go further than ANFIS by combining fuzzy logic with other types of computational intelligence approaches such as evolutionary algorithms, SVMs, and DL architectures [23].

FLS and hybrid systems, however, also present some challenges. Fuzzy model design is usually dependent on expert knowledge to specify membership functions and rule bases, which can bring in subjectivity. ANFIS alleviates this by learning from data but is sensitive to training dataset quality and size, typically lacking in tribology research. Hybrid systems are powerful but also add complexity. Because of this, poorly executed hybrid systems are not ideal for real-time constrained applications. Optimizing algorithms will help these systems. Using evolutionary algorithms to adjust fuzzy rule bases, incorporating transfer learning to reduce data dependency, and using cloud systems for computational overload are possible solutions [24]. In the context of fuzzy logic and hybrid systems, FLS are hybrid systems to deal with the uncertainty, imprecision, and nonlinearity of tribological systems. The fuzzy logic inference and the hybrid systems, such as ANFIS and fuzzy-neural hybrids, help gain precision and flexibility. Systems used for wear prediction, lubrication control, and fault diagnosis show the applied research for these in the industry and labs. With AI-based predictive maintenance and a digital twin, FLS and hybrid systems will be used to develop clever, explainable, and robust solutions that restore the balance of precision and reasoning to systems [25].

9.2.4 Tribo-Informatics

Developments in the complexity of tribological systems and the speed of advancements in AI have created yet another paradigm, namely, tribo-informatics. The power of informatics tools in the management and smart analysis of data is utilized in the fusion of science with tribology. Instead of putting together an unfortunate system of information in an *ad hoc* method, as is the case with traditional systems which have standalone interpretations of different tribological measurements, tribo-informatics permits an integrated organized information system. This system is made up of inputs, system characteristics, system outputs, and tribological and derived states, which assist in comprehending system behavior and describing it in a systematic manner. Inputs refer to the system's external load conditions, sliding speed, lubrication, and temperature. Ambient conditions, as well as inherent material and design characteristics including hardness, surface and microstructure, microstructure, and coatings, are described in inherent system properties [26].

Bearings and gears use acoustic and vibration analysis and oil condition monitoring to prevent costly breakdowns. Saving costs by preventing costly breakdowns is a form of predictive maintenance. With AI, predictive maintenance is very efficient. The predictive maintenance module in the tribo-informatics system reviews archived data to analyze and estimate wear evolution and remaining useful life to plan schedules to maintain the system. This leads to prior maintenance, which is a huge contrast to the breakdown maintenance. This predictive maintenance approach addresses the operational downtime and costs associated with breakdown maintenance. Minimizing frictional losses, extending the life of the component, and saving lubricants are the goals of the tribo-informatics system in optimizing parameters, and the various systems mentioned are the different applications in which the energy efficiency, system life, and process performance improvements are realized [27].

In spite of its potential, tribo-informatics also suffers from a set of critical challenges, primarily the unavailability of large, high-quality, standardized datasets because tribological tests are frequently costly, time-consuming, and arduous to reproduce in multiple laboratories. Furthermore, blending heterogeneous streams of data also presents technical challenges as sensor signals, material parameters, and image datasets tend to be different in format, scale, and sampling frequency, necessitating strong data fusion techniques and standardized procedures. Another key challenge is explainability; although AI models are capable of making precise predictions, engineers and practitioners need to know the rationale behind the predictions in order for it to be trusted, verified, and practically applied, which has ignited interest in explainable AI solutions specific to tribological application [28,29].

9.3 AI FOR TRIBOLOGICAL DATA ANALYSIS

Tribological information is inherently multiscale, from atomic-scale contacts to system-level macroscopic performance, over more than one domain such as mechanical, thermal, and chemical behavior and more than one kind of data such as time-series sensor data, microscopy images, spectroscopy information, and numerical measurements [30]. Such models are capable of expressing nonlinear relationships that are hard to formulate analytically, allowing precise prediction of system performance for various operating regimes. Aside from regression, classification models are especially appropriate in classifying tribological events into different classes, e.g., discriminating between adhesive, abrasive, and fatigue wear mechanisms or recognizing lubrication conditions as boundary, mixed, or hydrodynamic. Over the last few years, convolutional neural networks (CNNs), a specific type of DL models, have been used to classify wear scar images from optical or scanning electron microscopy with high accuracy and reliability versus human specialists [31]. Figure 9.6 shows

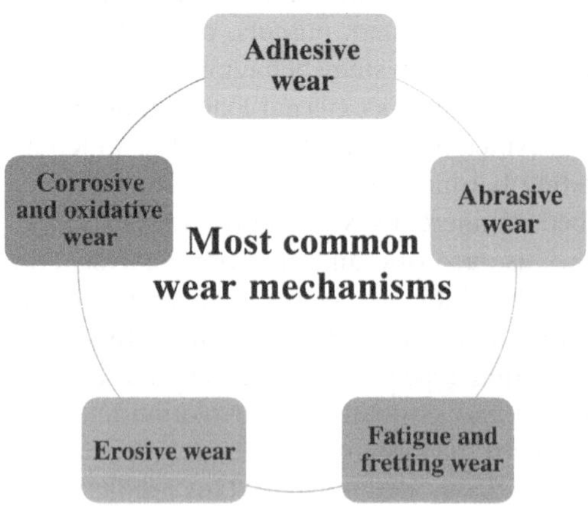

FIGURE 9.6 Major wear mechanisms [32]. (CC Attribution by 4.0.)

the main wear mechanisms that affect material deterioration and tribological performance under various operating situations. These mechanisms include adhesive, abrasive, fatigue, and corrosive wear.

Automated classification of images is a means of expediting the analysis of images of wear features, while at the same time reducing the subjectivity of interpretation. In a similar fashion, classification models aid predictive maintenance in fault detection by processing acoustic emissions and vibration signals from defected bearings and gears, classifying either the defect type or lubrication failure. In contrast, clustering techniques address situations in the absence of labeled data. Unsupervised clustering techniques, such as k-means and hierarchical clustering, provide significant insights into wear process evolution and condition monitoring by clustering wear particles and surfaces, morphologies, and lubrication states based on the aforementioned similarities. For instance, liberal use of clustering techniques on wear debris particles extracted from lubricants may highlight dominant wear mechanisms and detect transitions from mild to heavy wear, all without defining specific boundaries, thus providing a warning of impending failure of the component. The second significant category of AI techniques to address tribological data is the use of various dimension reduction techniques to deal with the high dimensionality of tribological datasets [33].

Dimensionality reduction techniques like principal component analysis (PCA) and autoencoders compress large multivariant data sets into lower-dimensional representations while retaining the most important aspects. This affords visualization and interpretability while also streamlining subsequent ML operations by reducing superfluous and noisy attributes. In vibration analysis, for instance, the PCA technique facilitates the reduction of hundreds of signal attributes into a few representative components, thereby enhancing fault detection, which becomes both robust and interpretable. Autoencoders, a more advanced DL approach to dimensionality reduction, learn nonlinear mappings to capture more complex relationships in tribological data, thereby helping in the analysis of large, real-time sensor datasets for data created in real time [34].

This accelerates innovation in the areas of coatings, lubricants, and composite materials, where the combinations of parameters are enormous, making exhaustive research through traditional methods impractical. In comparison to the fields of computer vision and natural language processing, tribology lacks extensive standardized datasets and metadata repositories. This limitation makes it difficult to formulate generalizable and robust AI systems. The absence of consolidated benchmarks further inhibits the assessment and comparison of different algorithms' efficiencies across various research efforts. Hence, setting tribological datasets' conditions, structuring datasets, and defining metadata should become a priority to leverage AI and communal knowledge. A fruitful future research direction is the unification of disparate datasets, integrated signals, and systems of images and material parameters to form a singular predictive model. This approach offers a more comprehensive understanding of tribological systems compared to the traditional, fragmented examination where individual datasets are analyzed in isolation [35]. The use of AI approaches in regression, classification, clustering, and dimensionality reduction is revolutionizing the processing of tribological data and shifting the discipline from traditional methodologies to cutting-edge, data-centric approaches. Coupled with high-throughput testing and supported by extensive databases, AI can tremendously increase the pace of discovery, improve predictive capabilities, and enable dynamic optimization of tribological systems across various applications in the automotive, aerospace, manufacturing, and energy sectors. This will greatly improve systematic, scalable, and intelligent tribology [36].

9.4 AI IN TRIBOLOGICAL SYSTEM PREDICTION AND DESIGN

One of the most powerful and groundbreaking applications of tribology is the use of AI to design systems, materials and lubricants, predictive maintenance systems, and predictive maintenance systems with remarkable accuracy and predictive power. Verbal and estimation analytic equations, as well as other analytical approaches, are often insufficient for tribology performance predicting because processes relating to friction, wear, and lubrication are nonlinear, random, and multiscale systems. In contrast, AI systems effectively handle nonlinearities, high-dimensional data spaces, and complex interdependencies, thus showing great value for predictive maintenance systems in tribological systems. One of the most well-known applications in the literature is the prediction of friction and wear. In the case of predictive maintenance, neural networks, and specifically ANNs, have outperformed traditional regression methods in predicting wear rates and friction coefficients of cutting tools, bearings, and composite materials, which is due to their ability to learn complex and nonlinear interdependencies of control variables such as load, sliding velocity, hardness, microstructure, and other variables intrinsic to the systems. An example is the estimation of the tribological properties of sintered steels, wherein ANNs demonstrated their superior ability to simulate the nonlinear behavior of steel wear, thus allowing for enhanced predictions of the lifetime of the components at varying operational conditions. Neural network models, beyond the field of metals, have been deployed in predicting the wear resistance and frictional stability of polymer composites and ceramics under varying conditions for informing material selection and design [37]. ML models have extensively been used to map real-time signals like vibration, acoustic emission, and temperature data to lubrication regimes like boundary, mixed, and hydrodynamic lubrication. Figure 9.7 shows key avenues for the advancement of AI

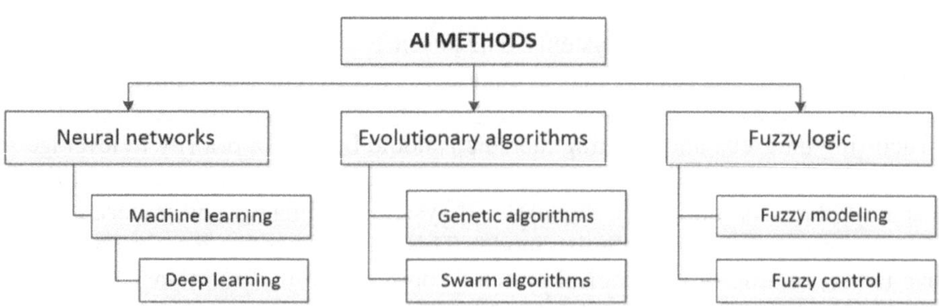

FIGURE 9.7 Main directions of AI technique development in hydraulic system design [38]. (CC Attribution by 4.0.)

techniques in hydraulic system design, emphasizing their use in condition monitoring, predictive modeling, fault diagnosis, optimization, and integration with digital twins and intelligent control frameworks.

9.5 AI FOR ONLINE CONDITION MONITORING AND FAULT DIAGNOSIS

AI has been shown to be a groundbreaking instrument in online condition monitoring and fault diagnosis, particularly in tribology and manufacturing systems where wear, lubrication, and fatigue failure remain key concerns [39]. Condition monitoring refers to the preventative maintenance strategy of constantly observing the status of machines and components to detect early symptoms of deterioration before total collapse. The integration of AI has rectified much of this by enabling the extraction of latent patterns, adaptive learning from real-time data streams, and intelligent decisions [40,41]. CNNs' superior feature learning of hierarchical features from raw inputs such as time-series signals and spectrograms has been employed to automate anomaly detection without significant human feature engineering. These technologies have enabled the transition from offline diagnosis to online monitoring, with AI algorithms working on on-the-fly high-dimensional data streams to provide real-time feedback with respect to machine health. ML algorithms based on vibration, force, and sound signals can forecast the development of wear to enable tool replacement in advance of deterioration as an expensive failure [42].

AI contributes significantly to detecting unusual lubrication conditions, starvation, contamination, or over-viscosity change, which have a direct impact on friction, heat generation, and material removal processes. ML algorithms, especially using recurrent neural networks (RNNs) and long short-term memory (LSTM) networks, can model temporal dependencies in sensor readings and are, therefore, promising for RUL prediction. The increasing use of the Industrial Internet of Things (IIoT) also reinforces AI applications by furnishing a large network of sensors that produce real-time streams of data across remote assets, enabling AI algorithms to function at a systems level. AI enables predictive maintenance and minimizes downtime, which enhances operating reliability and improves safety. Thus, AI not only transforms industrial practice but also paves the way for smart, resilient, and sustainable manufacturing ecosystems. [43].

9.6 HYBRID AI–TRIBOLOGY APPROACHES IN SURFACE CHARACTERIZATION

However, there is excessive use of pin-on-disk wear testing, scanning electron microscopy, atomic force microscopy, X-ray diffraction, microhardness testing, and experimental tribology surface analyses. While these approaches provide good information regarding wear and surface texture, they tend to be time-consuming, expensive, and only provide limited predictive capabilities regarding potential behavior across various operational conditions. The predictive capabilities of AI on data can provide newer approaches via experimentation to address nonlinear correlation mapping between the material properties, the process inputs, and the output performances [44,45].

When trained with enough datasets consisting of surface images and the corresponding tribological parameters, ML algorithms can identify concealed aspects and interrelations not easily distinguishable through manual analysis. This facilitates the acceleration of root-cause analysis of wear failures and the design of composites and coatings with surface characteristics according to specific needs. The significant advantage of hybrid AI–tribology systems lies in the reduction of time and costs associated with experiments. Traditional wear testing requires numerous trials under various loads, sliding speeds, and environmental conditions to build comprehensive robust datasets. AI systems can predict untested conditions using limited datasets, eliminating the need for extensive trials. This targeted approach saves materials, reduces the number of required testing cycles, and improves the effectiveness of R&D. Moreover, techniques such as GPR and DL provide uncertainty estimates in addition to predictions, enabling researchers to gauge confidence and refine testing in more advanced ways [46].

The hybrid strategy advances eco-friendly design by minimizing the reinforcement content, conserving resources, and prolonging the lives of structural components. For instance, predictive models can suggest strategically optimized loadings of nanoparticles that balance the mechanical performance of a composite with cost constraints and avoid unnecessary reinforcement that may hinder ductility or complicate machinability. Predictive models of wear can also suggest design provisions for lightweight, wear-resistant fabrics used in high-performance aerospace, automotive, and biomedical engineering applications, where both performance and eco-friendliness are critical [47].

9.7 EMERGING FIELDS OF AI IN TRIBOLOGY

AI is really changing the future of tribology, the study of friction, wear, lubrication, and the results of these interactions and isn't just focused on the old ways of diagnosing faults and wear monitoring. It is branching out into specialized areas and addressing current engineering challenges. AI's flexibility is evident in its application to various tribology areas in advanced manufacturing, biomedical engineering, sustainability, aerospace, and even outer space. All of these areas, together, mark the beginning of intelligent, adaptive, and predictive systems in the practice and development of science tribology [48]. Intelligent Tribology is using AI to develop monitoring systems that detect anomalies and in real-time automatically adjust lubrication and other parameters within a predictable lid so that the system is optimized. Conventional automated lubrication systems rely on predefined schedules and manual inputs, resulting in inefficiencies. AI-enabled systems integrate sensors and adaptive control algorithms to optimize lubrication, enhancing efficiency, reducing wear, and improving reliability [49,50]. There has been an increase in the relevance of sustainable Green Tribology. The use of AI in developing and improving lubricants that are environmentally friendly (such as biodegradable lubricants), ionic liquids, and aqueous solutions has been invaluable. AI programs conduct extensive reviews of the formulated lubricants, additives, tribology performance and chemistry and then identify eco-friendly alternatives that will perform comparably. Innovative AI programs also perform a life-cycle analysis of tribological systems, enabling industries to achieve sustainable reliability and efficiency [51].

9.8 FUTURE PERSPECTIVES

The integration of AI tribology and the development of adaptive systems that tackle real-world engineering problems more effectively are on the horizon. AI-enabled systems that are adaptable and self-learning to assist with real-world engineering problems are on the horizon. Building tribo-databases is one of the key avenues to pursue. These databases will contain standardized tribological data types including wear rates, friction coefficients, lubrication profiles, and surface characterization databases. Insufficient standardized datasets have consistently hindered the advancement of powerful AI and ML models within tribology. The development of these systems, which academia, industry and research bodies contribute to, will significantly enhance the pace of model development and assist in cross-system condition collaboration and benchmarking [52]. Digital twins monitor systems in real time, predict system faults in the future, and optimize tribological systems without stopping physical processes, which is a game changer for many industries. The aerospace, automotive, and energy industries, where system reliability is key and system downtime is costly, are transformed by these systems [53].

The growing focus on explainable artificial intelligence (XAI) in the discipline of tribology is also remarkable as it focuses on one of the primary barriers to industrial uptake—trust. AI may be able to make highly accurate predictions, but engineers struggle to understand the rationale of decisions owing to the opaque nature of black-box models. XAI tries to mitigate this issue by highlighting the most influential variables used to make a prediction, visualizing the paths taken to reach a decision, and correlating the results with known tribological attributes. Such clarity improves user trust and eases the regulatory approval processes for industrial use. The integration of the most promising emerging components—tribo-databases, physics-informed AI, RL, digital twins, and XAI—will transform tribology into a predictive, smart, and sustainable discipline that addresses complex advanced manufacturing, biomedical, and extreme environmental applications with reliability and efficiency that is unprecedented in scope and magnitude [54].

9.9 CONCLUSION

The use of predictive, monitoring, and designing capabilities of AI is changing the discipline of tribology AI. Starting with the use of neural networks for predicting wear to the use of complex hybrid AI for characterizing nanocomposites, the discipline of tribology AI is achieving novel and advanced applications. The rise of tribo-informatics is indicative of the comprehensive assimilation of data and AI, while hybrid techniques showcase the benefits of juxtaposing empirical research with computational analysis. Data standardization, industrial validation, and interpretability of AI remain as major obstacles; however, research in AI-mediated tribology is expected to become a primary constituent of engineering systems. Integration of AI for design, monitoring, and incorporating feedback loops for optimization in tribology will transform the discipline from an empirical science to a predictive field to achieve global efficiency, sustainability, and reliability.

REFERENCES

1. Yin N, Yang P, Liu S, Pan S, Zhang Z. AI for tribology: Present and future. *Friction* 2024;12(6):1060–1097.
2. Liu Y, Wei Q, Wang W, Zhao L, Hu N. Multiscale Modeling and Data-Driven Life Prediction of Kinematic Interface Behaviors in Mechanical Drive Systems. Coatings. 2025 May 30;15(6):660.
3. To AC, Liu WK, Olson GB, Belytschko T, Chen W, Shephard MS, Chung YW, Ghanem R, Voorhees PW, Seidman DN, Wolverton C. Materials integrity in microsystems: a framework for a petascale predictive-science-based multiscale modeling and simulation system. *Computational Mechanics*. 2008;42(4):485–510.
4. Marian M, Tremmel S. Current trends and applications of machine learning in tribology—A review. *Lubricants*. 2021;9(9):86.
5. Muniyappan A, Muthuvel PA. Predictive Modelling of Wear and Friction in Surface Textured TiAlN Coated Ti6Al4V Alloy Using Artificial Neural Networks. In *International Conference on Advanced Materials Manufacturing and Structures*. 2024, February (pp. 131–142). Cham: Springer Nature Switzerland.
6. Gacu JG, Monjardin CE, Mangulabnan RG, Pugat GC, Solmerin JG. Artificial intelligence (AI) in surface water management: A comprehensive review of methods, applications, and challenges. *Water*. 2025 Jun 4;17(11):1707.
7. Tian Y, et al. Real-time prediction of wear morphology and coefficient of friction via deep learning combining acoustic signals and FEM. *Processes*. 2025;13(6):1762.
8. Pantić M, Jovanović S, Djordjevic A, Petrović Savić S, Radenković M, Šarkoćević Ž. Predicting wear rate and friction coefficient of $Li_2Si_2O_5$ dental ceramic using optimized artificial neural networks. *Applied Sciences*. 2025;15(4):1789.
9. Marian M, et al. Physics-informed machine learning—An emerging trend in tribology. *Lubricants*. 2023;11(11):463. DOI:10.3390/lubricants11110463
10. Sose AT, Joshi SY, Kunche LK, Wang F, Deshmukh SA. A review of recent advances and applications of machine learning in tribology. *Physical Chemistry Chemical Physics*. 2023;25:4408–4443.
11. Rosenkranz A, Marian M, Profito FJ, Aragon N, Shah R. The use of artificial intelligence in tribology—A perspective. *Lubricants*. 2020 Dec 26;9(1):2.
12. Zhang Z, Pan S, Raeymaekers B. Guest editorial: AI & emerging computational approaches for tribology. *Friction*. 2024;12:1057–1059.
13. Paturi UM, Palakurthy ST, Reddy NS. The role of machine learning in tribology: A systematic review. *Archives of Computational Methods in Engineering*. 2023 Mar;30(2):1345–1397.
14. Wang Y, Shan G, Li H, Wang L. A wearable-sensor system with AI technology for real-time biomechanical feedback training in hammer throw. *Sensors*. 2022 Dec 30;23(1):425.
15. Nsor M. Predictive maintenance using machine learning for engineering systems through real-time sensor data and anomaly detection models. *International Journal of Research Publication and Reviews*. 2024;5(10):5167–5183.
16. Khakbiz M, Akhlaghi F. A hybrid artificial intelligence and tribological approaches for surface characterization of nanocomposites. *Tribology International*. 2025 Aug 18;214:111105.
17. Kano S. A short review: Tribology in machining to understand conventional and latest modeling methods with machine learning. *Machines*. 2025 Jan 23;13(2):81.
18. Johns-Rahnejat PM, Rahmani R, Rahnejat H. Current and future trends in tribological research. *Lubricants*. 2023 Sep 11;11(9):391.
19. Zhang Z, Yin N. *Tribo-Informatics: The Systematic Fusion of AI and Tribology*. Boca Raton, FL: CRC Press; 2024 Apr 11.

20. Vinoth A, Nirmal KN, Khedar R, Datta S. Optimizing the tribological properties of uhmwpe nanocomposites—An artificial intelligence based approach. In *Trends in Mechanical and Biomedical Design: Select Proceedings of ICMechD 2019*. 2020 Aug 21 (pp. 831–843). Singapore: Springer.
21. Azizan FL, Sathasivam S, Ali MK, Roslan N, Feng C. Hybridised network of fuzzy logic and a genetic algorithm in solving 3-satisfiability hopfield neural networks. *Axioms*. 2023 Mar 1;12(3):250.
22. Drees D, Lopes L, Baião P, Koutsomichalis A, Georgiou E. Solving real-life wear problems: Streamlined lab tests for diverse industrial applications. Can AI help?. *Wear*. 2025 Feb 3;571:205853.
23. Shabana S, Jatavallabhula JK, Kumar RN, Chinthalapudi R, Pappula B. Towards intelligent tribology: Predicting wear and friction in WC–Co coatings with machine learning. *Journal of Bio-and Tribo-Corrosion*. 2025 Sep;11(3):88.
24. Sharma P, Said Z, Tiwari AK. Soft computing tool (intelligent techniques) for nanorefrigerants and nanolubricants. In: Zafar S, Kumar A (Eds) *Tiwari Nano-Refrigerants and Nano-Lubricants*; 2024 Jan 1 (pp. 297–314). Amsterdam, Netherlands: Elsevier.
25. Herwig N, Peng Z, Borghesani P. Bridging the trust gap: Evaluating feature relevance in neural network-based gear wear mechanism analysis with explainable AI. *Tribology International*. 2023 Sep 1;187:108670.
26. Ghosh S, Knoblauch R, El Mansori M, Corleto C. Towards AI driven surface roughness evaluation in manufacturing: A prospective study. *Journal of Intelligent Manufacturing*. 2024 Oct 3;36:1–30.
27. Sieberg PM, Hanke S. Challenges and potentials in the classification of wear mechanisms by artificial intelligence. *Wear*. 2023 Jun 1;522:204725.
28. Pawar DM, Jagtap HP. Development of thermographic image-based tool wear monitoring system: A literature review. *Journal of Mines, Metals & Fuels*. 2025 Mar 1;73(3):47464.
29. Sieberg PM, Kurtulan D, Hanke S. Wear mechanism classification using artificial intelligence. *Materials*. 2022 Mar 22;15(7):2358.
30. Wilińska I, Wilkanowicz S. Advancements in environmentally friendly lubricant technologies: Towards sustainable performance and efficiency. *Energies*. 2025 Jul 28;18(15):4006.
31. Soares G, Chavarette FR, Gonçalves AC, Faria HA, Outa R, Mishra VN. Optimizing the transition: Replacing conventional lubricants with biological alternatives through artificial intelligence. *Journal of Applied and Computational Mechanics*. 2025 Apr 1;11(2):294–302.
32. Morshed-Behbahani K, Farhat Z, Nasiri A. Effect of surface nanocrystallization on wear behavior of steels: A review. *Materials*. 2024 Apr 1;17(7):1618.
33. Perčić M, Zelenika S, Mezić I. Artificial intelligence-based predictive model of nanoscale friction using experimental data. *Friction*. 2021 Dec;9(6):1726–1748.
34. Shah R, Pai N, Thomas G, Jha S, Mittal V, Shirvni K, Liang H. Machine learning in wear prediction. *Journal of Tribology*. 2025 Apr 1;147(4):040801.
35. Desai PS, Granja V, Higgs III CF. Lifetime prediction using a tribology-aware, deep learning-based digital twin of ball bearing-like tribosystems in oil and gas. *Processes*. 2021 May 24;9(6):922.
36. Sadeghi B, Cavaliere P, Shabani A, Pruncu CI, Lamberti L. Nano-scale wear: A critical review on its measuring methods and parameters affecting nano-tribology. *Proceedings of the Institution of Mechanical Engineers, Part J: Journal of Engineering Tribology*. 2024 Feb;238(2):125–155.
37. Karthick NK, Bharathiraja M. Ai-optimized Al6061-T6 hybrid metal matrix composites reinforced with Al_2O_3, graphite, and bauxite residue (SiO_2) for advanced brake hub applications: Tribological and machining performance. *Journal of Polymer & Composites*. 2025;13(6):182–198.

38. Filo G. Artificial intelligence methods in hydraulic system design. *Energies*. 2023 Apr 7;16(8):3320.
39. Deshwal M, Kumar P. AI-driven prediction and experimental validation of mechanical and microstructural performance in Al7075-B4C-Si3N4 hybrid composites. *Engineering Research Express*. 2025 Jul 30;7(3):035533.
40. Kaufmann T, Sahay S, Niemietz P, Trauth D, Maas W, Bergs T. AI-based framework for deep learning applications in grinding. In *2020 IEEE 18th World Symposium on Applied Machine Intelligence and Informatics (SAMI)* 2020 Jan 23 (pp. 195–200). New York: IEEE.
41. Ammisetti DK, Kruthiventi SH. Experimental investigation of the influence of various wear parameters on the tribological characteristics of AZ91 hybrid composites and their machine learning modeling. *Journal of Tribology*. 2024 May 1;146(5):051704.
42. Deepthi YP, Kalaga P, Sahu SK, Jacob JJ, S KP, Ma Q. AI-based machine learning prediction for optimization of copper coating process on graphite powder for green composite fabrication. *International Journal on Interactive Design and Manufacturing (IJIDeM)*. 2025 Jun;19(6):4123–4130.
43. Batu T, Lemu HG, Shimels H. Application of artificial intelligence for surface roughness prediction of additively manufactured components. *Materials*. 2023 Sep 18;16(18):6266.
44. He JH. AI-based nanotechnology: Breakthroughs, applications, challenges, and the road ahead. *Recent Patents on Nanotechnology*. 2025;20(1):1–9.
45. Nandipati M, Fatoki O, Desai S. Bridging nanomanufacturing and artificial intelligence—A comprehensive review. *Materials*. 2024 Apr 2;17(7):1621.
46. Subramani R, Leon RR, Nageswaren R, Rusho MA, Shankar KV. Tribological performance enhancement in FDM and SLA additive manufacturing: materials, mechanisms, surface engineering, and hybrid strategies—A holistic review. *Lubricants*. 2025 Jul 7;13(7):298.
47. Ko JH, Yin C. A review of artificial intelligence application for machining surface quality prediction: From key factors to model development. *Journal of Intelligent Manufacturing*. 2025 Feb 10;37:1–24.
48. Díaz Lantada A, Franco-Martínez F, Hengsbach S, Rupp F, Thelen R, Bade K. Artificial intelligence aided design of microtextured surfaces: Application to controlling wettability. *Nanomaterials*. 2020 Nov 18;10(11):2287.
49. Jha R, Jha BK. *Artificial Intelligence-aided Materials Design: AI-algorithms and Case Studies on Alloys and Metallurgical Processes*. Boca Raton, FL: CRC Press; 2022 Mar 15.
50. Shah Z, Albasheir NA, Raja MA, Almazah MM, Al Samman FM, Talha M, Jamil A, Waqas M. Deep learning multilayer stochastic intelligent computing for the analysis of irregular heat source of Carreau nanofluid within the vicinity of an exponentially expanding cylinder. *Tribology International*. 2025 Mar 1;203:110389.
51. Ramachandran RA, Barão VA, Ozevin D, Sukotjo C, Mathew M. Early predicting tribocorrosion rate of dental implant titanium materials using random forest machine learning models. *Tribology International*. 2023 Sep 1;187:108735.
52. Siddique A, Yadava GS, Singh B. Applications of artificial intelligence techniques for induction machine stator fault diagnostics. In *4th IEEE International Symposium on Diagnostics for Electric Machines, Power Electronics and Drives, 2003. SDEMPED 2003*. 2003 Aug 24 (pp. 29–34). New York: IEEE.
53. Zahmatkesh S, Esmaeili H, Valerio AT, Rizvi R. Predicting abrasion resistance in thermoplastic polyurethanes using machine learning. *Journal of Applied Polymer Science*. 2025;142:e57619.
54. Dhande D, Navale S, Gaikwad DP, Wadkar DV, Patil SR, Nalawade RD, Ghazali MJ, Nik WW. Prediction of abrasive wear behavior of the poly-tetra-fluoroethylene material using adaptive neuro fuzzy interface systems. *Journal of Tribology*. 2023;38:141–159.

10 Case Study on Artificial Intelligence in Automotive Tribology

Praveen Kumar Verma, Hitesh Vasudev, and Shivam Kumar

10.1 INTRODUCTION

Tribology remains an outstanding field with almost universal relevance today [1]. The friction, lubrication, and wear of materials are enabled and integrated within the systems consuming the majority of the energy worldwide, and the global energy consumption is indicative of the wear systems worldwide. Knowledge of tribology can open the doors to innovative solutions for future technical problems. Underpinning all progress are multitudes of accurate experiments and sophisticated computer simulations on various scales and in several physical fields [2]. As for tribology 4.0 or triboinformatics, sophisticated data processing, analysis, and learning techniques can be constructed based on this firm and data-rich foundation and utilized to increase the current knowledge. Furthermore, tribology is also distinguished by the fact that it is still not possible to describe underlying processes fully in mathematical terms, e.g., by differential equations [3]. Thus, contemporary machine learning (ML) or artificial intelligence (AI) techniques offer chances to investigate the intricate processes in tribosystems and to classify or quantify their behavior in an effective or even real time manner [4].

Tribology has generated numerous multidimensional and multidisciplinary studies because of its distinct system reliance, multidisciplinary connection, and temporal dependence.

Multistructural research data serve as the foundation for use of AI in tribology. As a result, the thorough integration of AI techniques with tribology will unavoidably enhance the effectiveness of knowledge collection, integration, analysis, and research in tribology and foster the field's growth. Early detection of mechanical wear is made possible, for instance, by the strong data and pattern recognition capabilities of AI, which may create correlations between wear and other signals (such as vibration, acoustics, electrical, and sound pressure) [5]. Automotive tribology holds a uniquely central place at the intersection of mechanical engineering, materials science, and energy systems, where the complex interrelations of friction, wear, and lubrication determine the running reliability, functional lifetime, and thermomechanical efficiency of vehicular subassemblies. From combustion engines and transmission sets to braking units and rolling element bearings, the tribological behavior

DOI: 10.1201/9781003624165-10

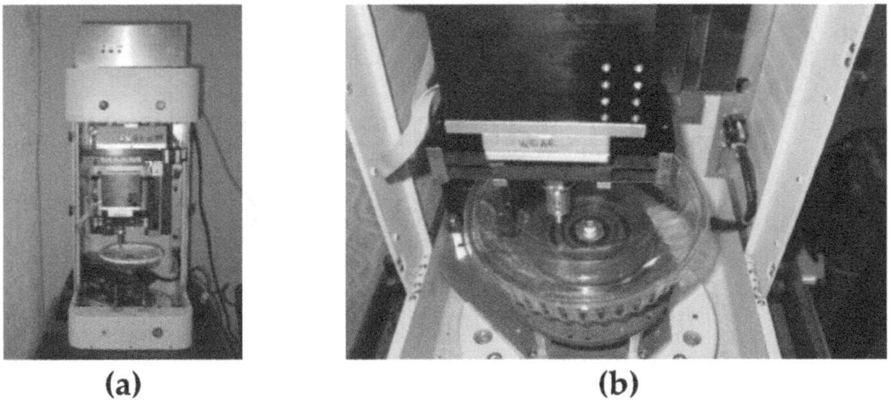

FIGURE 10.1 Tribo-test setup: (a) UMT-2 tribometer; (b) pin-on-disk module [7]. (CC 4.0 Attribution.)

of components directly influences energy dissipation, surface destruction kinetics, and catastrophic failure limits [6]. Figure 10.1 shows the pin-on-disk machine.

At the same time, strong difficult compliance and due attention to the environment aggravate the necessity of advancing the tribological optimization for friction and wear losses, which constitute about a third of the total primary global energy use and a huge share of the automotive transportation networks. More tribologically efficient systems, as a result of the combination of AI predictive maintenance, smart lubricant design, and adaptive surface technology, will lead to disproportionate energy and carbon emission savings. Recent positive AI research shows that properly trained predictive AI models using a fusion of multimodal sensors, which include vibration, acoustic emission, and thermal imaging, and oil debris analysis can diagnose emerging faults, predict wear in advance, and propose action plans prior to macroscopic failures. The aggressive and thermally unstable environment of motor vehicles makes the accuracy of these predictions highly questionable. Predictive maintenance for an uncontrolled mechanical failure seeks the question of safe operational boundaries and AI control mechanisms that should be applied in a safety-critical automotive environment [8]. In this case study, interface AI technology and automotive tribology have been used to describe emerging possibilities and analyze systemic limitations and epistemological contradictions. To clarify notable advances as distinct from fleeting rhetoric, this chapter contextualizes the history and the science of tribology technology. It also analyzes trends to construct hybrid models and, with the necessary discipline, anchors a realistic and sustainable agenda for the incorporation of AI into tribology engineering within the automotive industry.

10.2 BACKGROUND OF AI IN ENGINEERING TRIBOLOGY

The increasing influence of AI in comprehending and formulating the approaches of engineering tribology requires a transition from eliminativist mechanistic assumptions. Toward the understanding of algorithmically organized, data-driven epistemologies, which interrogate and predict multifarious interfacial complexities, are the shifting epistemological landscapes. Canonical implementations of AI—ML, deep learning (DL), Expert Systems (ES), fuzzy logic (FL), and evolutionary algorithms (EA)—are epistemic prostheses to help transcend the cognitive and computational intractabilities inherent in tribological systems, whose interfacial morphodynamics are governed by nonlinear, multiscale, and nonstationary couplings of surface topography, lubrication rheology, material microstructure, and dissipative thermomechanical fields. Ontological indeterminacy of such phenomena renders formalisms of classical deterministic form inappropriate, thereby mandating algorithmic forms that can fuse heterogenous streams of data, learn latent hierarchies of features, and extrapolate to experimentally intractable regimes [9,10]. Figure 10.2 shows the schematic of ML/AI model training steps and parameters for predicting lubrication performance from experimental and simulation data.

ML has been used to disentangle high-dimensional correlations hidden in noisy tribological data above the inferential power of routine regressionist practice. Supervised models—support vector regressors, random forests, and gradient boosting machines—have demonstrated predictive merit in estimating tribo-contact responses such as wear morphologies, coefficient-of-friction transients, and residual service lives under changing kinematic and thermodynamic conditions. Unsupervised clustering modalities have also been directed toward ontological wear debris morphologies and failure mode categorization without any *a priori* labeling. DL amplifies such epistemic potential through representational hierarchies implemented in convolutional, recurrent, and hybrid neural substrates that support the extraction of

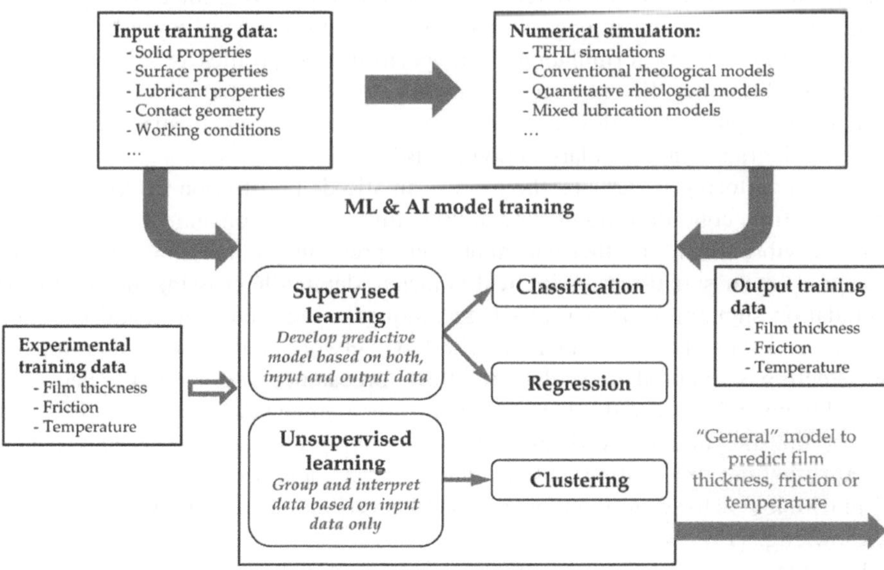

FIGURE 10.2 Schematic showing the main steps and parameters associated with the training process of ML/AI models to predict the lubrication performance from experimental and simulation datasets [11]. (CC 4.0 Attribution.)

invariant features from microscopy images, acoustic emission spectra, or vibration signatures, respectively, that would otherwise remain epistemologically out of reach for human experts. The capacity of DL models to encode approximations of nonlinear manifolds within high-dimensional spaces makes them centrally significant to the task of interpreting the emergent phenomenology of asperity-level contact evolution. Their logical chain-formulae provide prescriptive guidance in lubricant mix design, failure diagnosis, and proactive maintenance as knowledge bases of acquired experiential knowledge over time. FL architectures, however, realize the non-crisp semantics common in tribological linguistics. These structures have been well suited to controlling lubricant film thickness and viscosity adaptively *in situ* to mitigate instabilities resulting from dynamically varying loads and temperatures [12].

Data-driven approximations of first-principles strategies (e.g., density functional theory and molecular dynamics) have reduced in half the daunting computational costs traditionally associated with hardness, toughness, or tribo-corrosion resistance of novel alloys and coatings. Convolutional architectures trained on scanning electron or transmission electron micrographs self-segment and recognize microstructural objects—grain interfaces, porosity patterns, and second-phase particles—whose morphologies incontrovertibly define the interfacial response. ML surrogate models are increasingly being asked to speed up the identification of tribologically resilient structures within vast compositional spaces, including high-entropy alloys, bio-inspired films, and hybrid nanocomposites. In rheology, AI has been an incredibly useful proxy for experimental rheological measurement, aggregating gargantuan viscosity–temperature–pressure datasets into prognostic machines capable of extrapolating lubricant behavior over previously unexplored shear-rate and thermal regimes. Hybrid DL–FL systems have been required to predict synergistic and antagonistic interactions among lubricant additives, thus permitting rational design of green, low-toxicity formulations with performance profiles extending beyond mineral-oil baselines. Evolutionary algorithms, when integrated with tribochemical simulation outputs, also allow the optimization of additive chemistries to suppress oxidative degradation and volatilization [13,14].

Wear monitoring is perhaps the most dramatic demonstration of AI for tribology. Real-time condition-monitoring architectures, which consume high-frequency acoustic, vibrational, and thermal input, were previously constrained by the inadequacy of linear signal-processing techniques to demuddle crossing, nonstationary degradation signatures. Temporal DL techniques using recurrent neural networks (RNNs) and long short-term memory (LSTM) networks help resolve this impasse by detecting sequential dependencies, thus forecasting degradation trajectories. Hybrid fuzzy inference and ML classifiers further augment and strengthen resilience against missing and noisy datasets in industrial prognostic-health-management (PHM) pipelines. Integrating these techniques contributes extensively to the field of AI in tribology. However, the most profound contributions remain within the order of knowledge [15].

Introducing AI to engineering tribology is a pivotal change. This is not a peripheral addition. DL and ML technologies provide the most powerful data-hungry prognostic capabilities and the ability to hierarchize features, while expert systems and FL offer heuristic knowledge and uncertainty management. The ability to self-recalibrate in real time and the unsupervised learning potential of predictive maintenance and industrial PHM systems virtually redefine the ontological scope of tribological research. This integration indicates the intelligent and self-adaptive tribological systems and their ability to navigate complex operational ecologies of tribological systems, thereby redefining the scope of research [16].

10.3 AUTOMOTIVE TRIBOLOGY

Predicting wear is an equally daunting task, considering that material removal mechanisms in automotive systems are stochastic and path-dependent. Wear does not appear as a uniform wear but rather as a temporally and spatially heterogeneous process modulated by contact pressure distributions, lubrication starvation, microstructural anisotropies, and tribochemical reactions. In engines, the piston–ring–cylinder–liner interface is most vulnerable to adhesive and abrasive wear, exacerbated by elevated combustion pressures and transient lubrication conditions during cold starts. Micropitting, scuffing, and rolling-contact fatigue caused by cyclic stress and lubricant film failure assail gears. Brake systems, subject to intense thermomechanical loads, experience wear, which concurrently controls safety performance and particulate emissions, the latter now identified as a key contributor to non-exhaust urban air pollution [17]. Tires are intricate viscoelastic composites, and their wear is controlled by hysteretic energy dissipation, asperity abrasion against roads, and thermo-oxidative degradation of elastomeric matrices. The nonlinearities of these wear modes make predictive modeling intractable by means of classical Archard-type wear laws alone. Rather, hybrid methodologies blending multiphysics finite-element simulations and data-driven prognostics are increasingly being utilized to capture the coupled surface evolution and sub-surface microstructural transformations. The complexity associated with contemporary lubricants in terms of multiple modalities is due to their multicomponent add-on packages created for specific functions of anti-wear, friction control, detergent-dispersant, and oxidation inhibition. Conventional methods used for cooling systems do not eliminate localized flash temperatures; consequently, novel alternatives need synergistic advancements in materials with elevated thermal conductivity, intelligent lubricants that incorporate viscosity modifiers adjustable for thermal conditions, and predictive thermal models integrated with sensor-feedback systems [18].

The problems with automotive sub-assemblies specific to tribology are, in some cases, magnified. The piston–ring–liner assembly in internal combustion engines is a tribological interface in which boundary lubrication is prevalent during cold starts, whereas mixed or hydrodynamic regimes dominate during steady-state operation. Although advanced coatings, such as diamond-like carbon (DLC) and molybdenum disulfide (MoS_2)-based films, are deployed, their durability remains questionable when faced with synergistic mechanical, chemical, and thermal assaults. Surface engineering and additional lubricants that can sustain elastohydrodynamic films, even under extreme pressure, are the only hope to alleviate the micro-pitting and scuffing challenges in gear transmissions. The brake system involves the additional challenge of low particulate emission to comply with the legal requirements of emission controls. For that, the synthesis of brake materials exhibiting a stable friction coefficient in the low-copper or copper-free range is essential. Tires, as multifunction tribo-structures, must provide traction, energy efficiency, and resistance to wear over a wide range of environmental conditions. Adhesive wear, thermal fatigue, and synergistic degradation with lubricants pose challenges with coatings in lightweight alloy engine blocks and electric drivetrains. Such a case-dependent tribological environment heterogeneity only amplifies the need for tailored solutions instead of the imagined universal ones [19].

Transforming automotive tribology into an intelligent system requires, first and foremost, acquiring data, which proves to be the largest bottleneck. High-speed acoustic and vibration sensors can yield substantial data. However, the tribology community finds it extremely challenging to extract relevant information that advances the understanding of tribology. Data concerning brake wear and tire–road interactions are even more challenging as the information must be collected in the real world, which is dynamic and chaotic, and in the presence of stochastic perturbations such as road surface roughness, environmental contaminants, and driver behaviors [20]. Inadequate conventional methods of data acquisition fail to capture the coupled, nonlinear interactions across disparate scales, from nano-asperity contacts to macro-system performance. Therefore, advanced data fusion, uncertainty quantification methods, and AI-based inference systems must be integrated to manage and synchronize multisource inconsistencies, deal with noise and incompleteness, and enable real-time predictive control. Finally, the construction of open, standard tribological datasets becomes even more difficult because of the ethical and practical issues of proprietary data within automotive companies. Without overcoming such epistemic and infrastructural challenges, the outlook on predictive, adaptive, and self-optimizing auto-tribo-systems continues to be utopian [21].

10.4 CASE STUDIES: AI IN AUTOMOTIVE TRIBOLOGY

The progresses made in employing AI techniques within automotive tribology from preliminary experimental studies to sophisticated application-focused frameworks describing the investigation and optimization of the multifaceted and multiphysics complexities of automotive interfaces are commendable. AI is devoted to various subsystems of automotive engines and parts, transmissions, brakes, lubricants and additives, and surface engineered coatings. As an illustration, automotive tribology AI within these domains has facilitated the predictive and adaptive closure of traditional mechanistic modeling and gone a step further to replace it with predictive data-driven mechanistic modeling. Numerous AI-driven techniques have been incorporated in automotive tribology; the following studies, for instance, depict the multifaceted shifts in knowledge and technology within automotive tribology pivoted on the specific AI techniques adopted and the functional improvements achieved. Within the engine, the highly dissipative and wear intensive friction engines contain tribological interfaces like cam–follower systems, piston rings, and cylinder liners. These interfaces, which are marked by coupled elastohydrodynamic, boundary, and mixed lubrication regimes, are extremely sensitive to microstructural material states, lubrication chemistry, and dynamic loading profiles. Traditional regression-based wear or friction laws, although valuable in controlled environments, do not have a place for the stochastic, nonstationary, and extremely nonlinear characteristics of real-world operational regimes [22].

AI-based approaches have come in the form of supervised and DL frameworks trained on experimental datasets including acoustic emissions, friction force signals, morphology of wear debris, and temperature gradients. For example, multilayer perceptrons (MLPs) and support vector machines (SVMs) have been used to predict piston–liner contact conditions as being in scuffing versus non-scuffing states with predictive accuracy levels in excess of 90%, even under varying operating conditions. Deep networks, specifically convolutional neural networks (CNNs), have shown to be able to handle scanning electron microscopy (SEM) images of worn surfaces without human expert annotation, automatically distinguishing adhesive, abrasive, and fatigue wear patterns previously delineated by an expert. For cam–follower systems, LSTM networks have been trained on sequences of vibrations and temperatures to predict the onset of pitting or spalling failure with tens of operating hours of lead time versus traditional diagnostic methods [23]. In terms of maintenance-initiated downtimes, condition-based maintenance makes predictive maintenance approaches virtually seamless. The predictive capabilities of hybrid approaches are of notable importance in real-time optimization of design parameters in piston-ring sets, utilizing coupled finite-element tribo-dynamic simulations as a primary predictive tool and resolving the antagonistic objectives of low friction and wear. The design and optimization of transmission subsystems such as gears and clutches offer limited scope for the adoption of tribology-focused maintenance approaches. Instead, the design of such subsystems provides a unique challenge of predicting and resolving tribo-mechanical phenomena such as micropitting, scuffing, and rolling contact fatigue and devising maintenance approaches that can resolve highly localized contact stresses and asperity contacts within highly compliant lubricants. ML algorithms, especially random forests and gradient boosting architectures, have been effectively used for large experimental datasets of gear failure modes to probabilistically predict the micropitting susceptibility under different load and lubrication conditions. CNNs have been utilized for automatic classification of images of the gear tooth surface with accuracy levels higher than those of conventional human inspection. In addition, Bayesian inference models, learned on acoustic emission and vibration data, have been used for bearing fault diagnostics and in the context of probabilistic health assessment under noisy, uncertain operating conditions [24].

AI frameworks have played a key role in rational lubricant formulation. Multi-objective evolutionary algorithms (EAs) have been used to optimize additive chemistry, reconciling anti-wear performance, oxidative stability, and eco-toxicity. Rheological and tribochemical data-trained neural networks have made predictive modeling of viscosity–temperature–pressure behavior over broad operation envelopes possible, allowing ultra-low-viscosity lubricants to be designed without affecting the film stability. In lubricant health monitoring, supervised learning algorithms developed using infrared spectroscopy, dielectric constant, and particle count data have achieved high accuracy in forecasting the remaining useful life (RUL) of engine and transmission oils. LSTM neural networks have taken this function to the next level with the ability to identify sequential degradation patterns, providing predictive warnings several months in advance of standard laboratory oil analyses of impending failure [25].

Deep reinforcement learning (RL) has become a cutting-edge methodology for surface engineering, in which AI agents engage iteratively with simulations of coating designs to learn optimal combinations of parameters to reduce wear and friction while increasing durability under thermal cycling. Hybrid AI–simulation systems, in which finite-element models of coating stress distributions are augmented with surrogate neural networks, have accelerated the exploration of large design spaces otherwise computationally infeasible. The value of such strategies is illustrated through particular case studies. Adopting AI applications that associate deposition parameters with tribological outcomes has resolved some of the issues with performance inconsistency of DLC thin films that were previously attributed to hydrogen concentration and other deposition parameters. Consequently, such thin films produced 15%–20% lower wear coatings compared to those manufactured with conventional methods. Likewise, AI-enabled control of porosity and microhardness distribution within thermal spray coatings has allowed for the optimization of brake disks, reducing particulate emissions while friction coefficients remain stable [26].

10.5 COMPARATIVE ANALYSIS OF AI TECHNIQUES

AI, especially in the areas of tribology and predictive modeling within the automotive domain, is gaining significance. The phenomena of surface deterioration, wear, and lubrication deal with multiple, complex, and nonlinear interactions across different scales. As the first applications of bio-inspired learning systems, artificial neural networks (ANNs) deal exceptionally well with the modeling of nonlinear relationships that exist among various inputs—load, sliding speed, surface finish, and lubricant viscosity—and outputs such as friction coefficient, wear rate, and failure rate. The ability of neural networks to approximate any mathematical function, while a core mathematical principle, translates practically into overfitting risks within sparse datasets and requires considerable efforts in hyperparameter tuning. Moreover, in the absence of uncertainty quantification, ANNs are fragile and will extrapolate poorly to situations where the operating conditions fall outside the distribution of the training data. Apart from these limitations, they are still popular for small- to medium-sized datasets because they have relatively low computational costs and generalizability to classification and regression problems [27].

SVMs, on the other hand, follow a conceptually different approach of structural risk minimization, where hyperplanes maximize the margin between classes in high-dimensional spaces of features. Kernel functions, like radial basis or polynomial types, allow for nonlinear mapping without directly augmenting dimensionality, which is beneficial for tribological datasets usually described by small samples but intricate feature interactions. SVMs generally have better generalization compared to shallow ANNs when data sparsity is significant, thus making them extremely well adapted to experimental tribology studies where full-scale testing is not feasible. However, they perform poorly in very high-dimensional contexts, and kernel parameter tuning may be computationally expensive. Also, SVMs do not inherently scale to multiclass classification without decomposition techniques, limiting their immediate scalability. Random forests are exemplary of ensemble learning methods by building an array of decision trees and combining their predictions using bagging. Their main advantage is overfit protection and inherent support for mixed data types and missing values. In tribological diagnostics, RFs have proven to remain robust when dealing with noisy sensor signals, uneven sampling, and disparate feature sets, which commonly occur in condition monitoring of automotive systems. RFs also offer measures of feature importance, enabling interpretability as well as discovering key operating parameters leading to tribological performance. Yet, their forecasting accuracy might be behind that of SVMs or DL models when feature relationships are extremely nonlinear, and their tendency to produce large ensembles can be bottlenecked for computational effectiveness in real-time situations [28]. Figure 10.3 shows the ANN architecture used for predicting lubrication performance.

DL is the most sophisticated echelon of AI, which utilizes multilayered architectures like CNNs, RNNs, and transformer-based models to progressively extract representations from raw data. DL is especially effective in tribology in the processing of high-dimensional modalities like acoustic emissions, thermal images, or *in situ* surface topography maps and the ability to perform end-to-end feature extraction and prediction without handcrafted descriptors. The inherent nature of DL to model spatiotemporal dependencies renders it essential for prognostics in highly complex tribo-systems such as automotive transmissions and engine bearings. However, DL's superior performance is balanced against outrageous data needs, immense compute expenditures, and notorious opaqueness ("black-box" nature), which makes mechanistic interpretability challenging and stifles adoption in safety-critical applications without stringent testing [30]. Figure 10.4 shows a deep neural network (DNN) architecture used for the prediction of lubrication performance from experimental and simulation data. The input layer receives the preprocessed experimental and computational data which include the material, operating, surface properties, and lubricant parameters. Normalized input data are handed over to the first hidden layer, and neurons therein identify the relationships and interdependencies of the variables through weighted connections and activation functions of varying nonlinearity. The augmentation of the hidden layers enhances the model's extraction of higher-order features, enabling the discovery of intricate patterns governing the friction, wear, and lubrication characteristics and

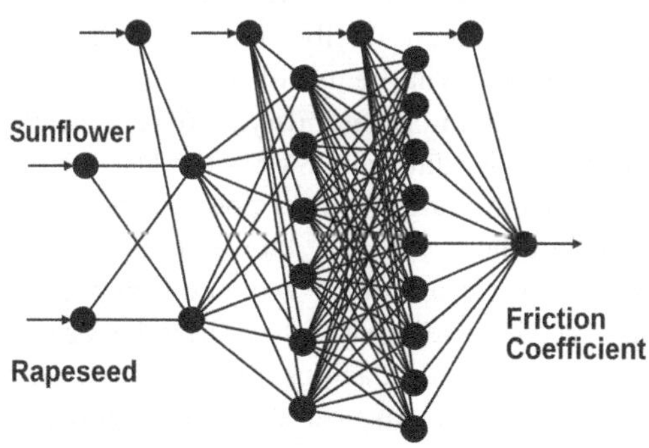

FIGURE 10.3 ANN architecture [29]. (CC 4.0 Attribution.)

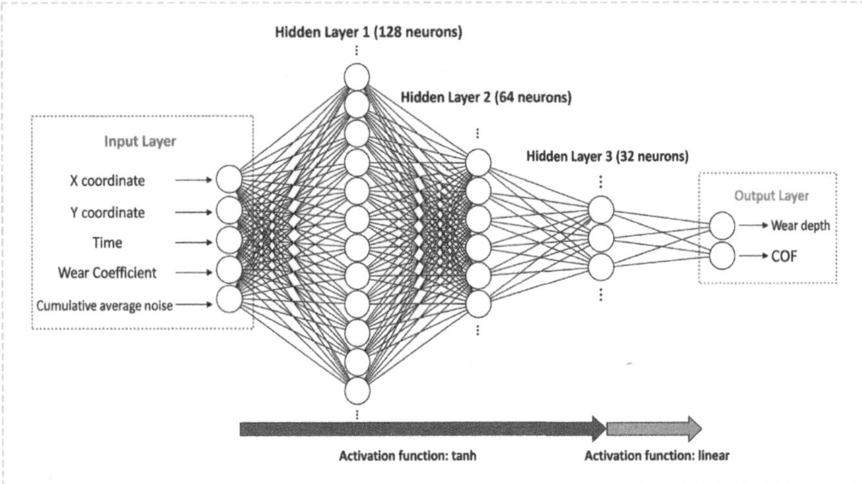

FIGURE 10.4 DNN framework [31]. (CC 4.0 Attribution.)

performance. The optimization of weights and biases in the model is done through the backpropagation algorithm, which aims to reduce the loss function to achieve convergence indicative of a predictor model of acceptable quality.

Apart from specific algorithms, a key divide exists between data-driven and physics-based models. Data-driven models solely depend on the empirical correlation and statistical pattern, permitting them to demonstrate outstanding predictive performance in adequately sampled regimes. But their extrapolation ability is inherently vulnerable as they have no ingrained knowledge of laws of physical governing like Hertzian contact mechanics, Archard's wear law, or elastohydrodynamic lubrication theory. Instead, physics-informed machine learning (PIML) aims to unify empirical data with analytical formulations and conservation principles, thus restricting solution spaces to physically realistic domains. For instance, infusing Reynolds equation constraints into neural structures greatly improves their accuracy in lubrication film thickness prediction for previously untested working conditions. The physics-informed framework also alleviates data sparsity, an ever-present hurdle in experimental tribology, through the minimization of reliance on comprehensive datasets via the use of mechanistic priors. When evaluated on the dimensions of accuracy, interpretability, and generalizability, the landscape becomes apparent. With smaller datasets, SVMs tend to perform better than ANNs since they benefit from strong margin-based optimization. However, ANNs eventually surpass SVMs when the volume of data increases. Of the more transparent frameworks, random forests are optimally balanced on accuracy and clarity, even though they may underperform relative to kernelized SVMs and deep models on capturing extremely intricate nonlinearities. In the presence of large and diverse datasets, DL models consistently surpass shallow learning models on accuracy, particularly with spatiotemporal and imaging tribological data, and especially on all metrics. However, the high cost of computation and lack of interpretability still present considerable barriers to practical use. Physics-informed DL appears to be the next frontier, attempting to integrate the predictive capabilities of DL with the high modeling derived from physical constructs [32,33].

10.6 INTEGRATION WITH IOT AND DIGITAL TWINS

The convergence of AI and Internet of Things (IoT) technologies along with digital twins (DTs) opens up new opportunities for advanced real-time monitoring and predictive maintenance in automotive-engineering tribology for bearings, gears, piston–cylinder interfaces, and lubrication systems. The acquisition of tribological data streams comprising continuously interfaced vibrations, acoustic emission, thermal and cooling lubricant, viscosity, and surface wear metrics can be accomplished through sensor networks and cyberphysical interoperability. Multimodal Pretrained Language Models (MPLM) and AI facilitate the processing of multimodal data streams and advanced predictive analytics directed at nascent degradation phenomena, making them detectable for the first time to modern audits. The synthesis of IoT–AI offers fault detection and the predictive capability to forecast several operationally dynamic health trend scenarios of the systems [34]. Figure 10.5 shows the framework model for prioritizing and systematically evaluating the risks, obstacles, and drivers of DT adoption to facilitate efficient decision-making.

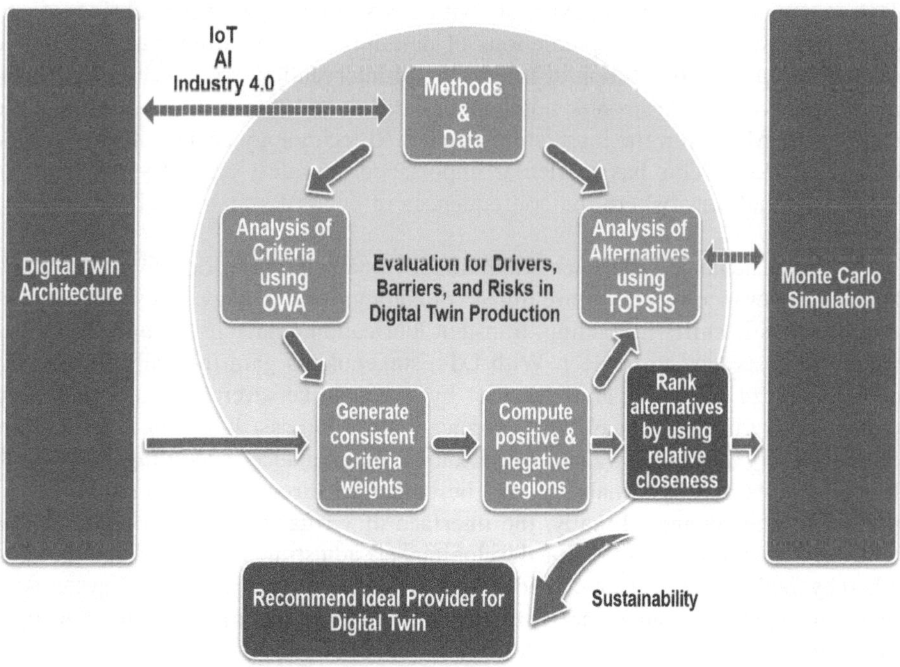

FIGURE 10.5 The model for ranking drivers, barriers, and risks for DT [35]. (CC 4.0 Attribution.)

The DT technology enhances this capability further by creating high-fidelity virtual copies of the physical tribological systems, dynamically synchronized with real-time sensor data via IoT interfaces. In contrast to static computer models, DTs develop simultaneously with the operating state of their physical counterparts, including wear-generated alterations in surface topography, lubricant rheology, or thermal stress fields. Artificially intelligent DTs utilize ML, DL, and hybrid physics-constrained strategies to rescale themselves continuously, diminishing consequently the epistemic uncertainties generally inherent in predictive simulations. In automotive tribology, DTs can, for example, simulate the development of scuffing in transmission gears or the development of rolling contact fatigue in wheel bearings, while at the same time correlating simulated paths with sensor-based evidence. A two-way feedback loop facilitates an adaptive decision-support environment where virtual predictions direct actions taken in the real world and where real world actions are subsequently modified by empirical data acquired from sensors to revise the virtual model [36].

In relation to real-time monitoring, the integration of AI with IoT technologies encourages a pivot away from scheduled, periodic maintenance and toward condition-based and predictive continuous maintenance practices. Smart Microelectromechanical System (MEMS) sensors can detect and record tribological stressors including lubricant nano-vibrations, transient load spikes, and lubricant contamination. Wireless IoT nodes have access to unobstructed data streams to cloud and edge-computing facilities, where advanced processing takes place. These centers use predictive AI tools with CNNs to analyze wear debris imagery and RNNs to assess trend predictions of deterioration over time. Such predictive models facilitate optimization of the timing of interventions to reduce unscheduled downtime, prevent catastrophic component failures, and extend the useful life of the component. Moreover, the integration of edge AI technology enables near real-time decision-making. This has particular importance for safety-critical systems in the automotive industry, where the consequences of delayed system total failures may be disastrous [37].

AI-driven advanced analytics evaluating IoT-enabled real-world driving scenarios has great relevance for the future impact of automotive tribology. These IoT devices track stochastic shifts in loading, transient lubrication, and start–stop cycles, all of which lab tests tend to neglect. With DTs, stakeholders gain the ability to predict cycles, analyze scenarios, and "what-if" hypotheses like severe climates and accelerated fatigue usage in heavy-duty applications to forecast lubricant deterioration. Supply chain logistics, spare part provisioning, maintenance scheduling, and overall lifecycle costs and sustainability will be enhanced through the automation of AI predictive maintenance. Finally, the interface of virtual models and physical systems will open opportunities for developing self-adjusting systems in automobiles, whereby real-time prognostic feedback will allow systems to control systems of lubricant injection, contact pressure control, and reset automated systems to optimized feedback control systems [38].

AI with DTs and IoT incorporates a new paradigm in autonomous tribological system prescribing. While AI anomaly detection systems focus on procedure violations, a major leap in system control is the coupling of extensive sensing with AI- and

IoT-oriented predictive control. DT technology facilitates the development of predictive dynamic simulations constrained by relevant physics. The integration of these technologies developed predictive maintenance mechanisms that are fundamentally sound and both data-driven and reliable. This, in essence, satisfies the Fourth Industrial Revolution principles of predictability, functionality, and environmental responsibility. Although there are still issues with data incoherence and heterogeneity, computing bleed over, and security, the innovative development of edge computing, federated learning, and safe IoT promises to reduce the negative impacts [39].

10.7 CHALLENGES, GAPS, AND FUTURE PERSPECTIVES

While AI has the potential to transform tribology and the automotive industry, certain issues still need to be solved before adoption becomes seamless. One of the main challenges to the adoption of AI is data scarcity. This is particularly bad in the field of tribology because high-quality data are expensive and time-consuming to develop as well as complicated to obtain. The acquisition of representative datasets is particularly challenging because tribological systems have stochastic wear mechanisms, nonlinear material responses, and coupled thermo-mechanical-chemical processes. Moreover, AI models trained on limited or unbalanced datasets become brittle and lose the ability to make reliable predictions in unexamined scenarios. Additionally, the problem of overfitting—where the model captures noise in the small dataset, leading to illusory training accuracy and poor practical outcome—further compromises the model. Although the application of regularization techniques, cross-validation, and transfer learning can mitigate the problem to some extent, the lack of robust data remains a significant barrier to the use of AI in tribological diagnostics [40].

One potential use of RL is in adaptive designs and optimization of elements in tribology such as lubrication control, surface texturing, and load balancing, where agents learn from and navigate simulated tribological systems. In contrast to supervised learning, which works with fixed datasets, RL benefits from ongoing iterative cycles of exploration and exploitation. This characteristic makes RL particularly suitable for online adaptive control applications in automotive tribo-systems. Generative AI represents another frontier and offers tools for generating synthetic data, which can to some extent remove the shortage of data by generating realistic wear traces, surface topologies, or lubricant degradation conditions for training. Generative adversarial networks (GANs) and diffusion models, if used with sparse experimental datasets, can significantly increase the input space for model training, which in turn will increase generalization and resilience [41]. Integrating hybrid models incorporating contact mechanics, elastohydrodynamic lubrication, or wear kinetics equations serves as a great advancement toward incorporating data-driven predictive techniques with dependable data-based predictability within the confines of physical theory. Fusing AI technologies with predictive models helps hybrid models avoid data-driven restriction, aids extrapolative construction, enhances ascertainability, and decreases data dependence. Such models, receiving triable data as input, can predict theoretical data as output and constitute the next stage of advancement in intelligent tribology [42].

10.8 CONCLUSION

The integration of AI into automobile tribology is a clear sign of a standard that incorporates changes from deterministic modeling and empiricism into adaptive data-driven and physics-constrained inference. For nonlinear tribo-dynamics, wear trajectory prediction, lubrication regime optimization, and predictive maintenance, frameworks based on AI such as ANNs, SVMs, random forests, DL, and hybrid physics have great potential. The convergence of IoT-enabled sensors and DT frameworks makes cyberphysical tribo-systems with self-adaptive control and real-time prognostics feasible. Nevertheless, widespread industrial adoption is still limited by persistent challenges, such as interpretability issues, overfitting, sparse data, and high implementation costs. Emerging domains like generative AI, RL, and hybrid physics-informed approaches offer new ways to address these practical and epistemic issues. The current AI-enabled tribology offers the most promise in the design of intelligent, self-optimizing, and environmentally sustainable automobile systems, especially in the electrification and autonomous vehicle integration markets, where operational efficiency and reliability are critical.

REFERENCES

1. Ji Y, Bao J, Yin Y, Ma C. Applications of artificial intelligence in tribology. *Recent Patents on Mechanical Engineering*. 2016 Aug 1;9(3):193–205.
2. Hoque MA, Yao CW. Nanoscale characterization of nanomaterial-based systems: Mechanisms, experimental methods, and challenges in probing corrosion, mechanical, and tribological properties. *Nanomaterials*. 2025 Dec 2;15(23):1824.
3. Senatore A. Special issue "automotive tribology". *Lubricants*. 2020 Apr 17;8(4):48.
4. Marian M, Tremmel S. Current trends and applications of machine learning in tribology—A review. *Lubricants*. 2021 Sep 1;9(9):86.
5. Korkmaz ME, Gupta MK. Nano lubricants in machining and tribology applications: A state of the art review on challenges and future trend. *Journal of Molecular Liquids*. 2024 Aug 1;407:125261.
6. Shah R, Jaramillo R, Thomas G, Rayhan T, Hossain N, Kchaou M, Profito FJ, Rosenkranz A. Artificial intelligence and machine learning in tribology: Selected case studies and overall potential. *Advanced Engineering Materials*. 2025 Jan 15;27(23):2401944.
7. Humelnicu C, Ciortan S, Amortila V. Artificial neural network-based analysis of the tribological behavior of vegetable oil–diesel fuel mixtures. *Lubricants*. 2019 Apr 9;7(4):32.
8. Sose AT, Joshi SY, Kunche LK, Wang F, Deshmukh SA. A review of recent advances and applications of machine learning in tribology. *Physical Chemistry Chemical Physics*. 2023;25(6):4408–43.
9. Tung SC, McMillan ML. Automotive tribology overview of current advances and challenges for the future. *Tribology International*. 2004 Jul 1;37(7):517–36.
10. Tomanik E, Tomanik V, Morais P. Use of tribological and AI models on vehicle emission tests to predict fuel savings through lower oil viscosity. SAE Technical Paper 2021-36-0038; 2022 Feb 4.
11. Rosenkranz A, Marian M, Profito FJ, Aragon N, Shah R. The use of artificial intelligence in tribology—A perspective. *Lubricants*. 2020 Dec 26;9(1):2.
12. Khakbiz M, Akhlaghi F. A hybrid artificial intelligence and tribological approaches for surface characterization of nanocomposites. *Tribology International*. 2025 Aug 18;214:111105.

13. Ghorpade U, Suryawanshi S, Bhosale D, Babar G, Kshirsagar V, Singh M. Artificial intelligence in high-temperature tribology. In Prashar G, Vasudev H, Thakur L (Eds) *Thermal Spray Coatings for High-Temperature Conditions*; 2025 June 23 (pp. 185–200). Boca Raton, FL: CRC Press.
14. Motamedi N. *Towards the prediction and understanding of tribological effects on system performance through artificial intelligence* (Doctoral dissertation, Université de Lille).
15. Balitskii O, Kolesnikov V. Identification of wear products in the automotive tribotechnical system using computer vision methods, artificial intelligence and big data. In *2019 XIth International Scientific and Practical Conference on Electronics and Information Technologies (ELIT)*. 2019 Sep 16 (pp. 24–27). New York: IEEE.
16. Rouf S, Raina A, Ul Haq MI, Naveed N. Sensors and tribological systems: Applications for industry 4.0. *Industrial Robot: The International Journal of Robotics Research and Application*. 2022 Apr 21;49(3):442–60.
17. Mahadeshwara MR, Kumar S, Dastidar AG. Artificial intelligence in the tribology. In *Emerging Research in Computing, Information, Communication and Applications: Proceedings of ERCICA 2022*. 2022 Dec 13 (pp. 351–67). Singapore: Springer Nature.
18. Kano S. A short review: Tribology in machining to understand conventional and latest modeling methods with machine learning. *Machines*. 2025 Jan 23;13(2):81.
19. Priest M, Taylor CM. Automobile engine tribology—Approaching the surface. *Wear*. 2000 Jul 31;241(2):193–203.
20. Taylor CM. Automobile engine tribology—Design considerations for efficiency and durability. *Wear*. 1998 Oct 1;221(1):1–8.
21. Campillo NE, Talavante P, Ponzoni I, Soto AJ, Martínez MJ, Naveiro R, Gómez-Arrayas R, Franco M, Lee SH, Mauleón P. Artificial intelligence in tribology: Design of new dispersants using artificial intelligence tools. *Proceedings of the 23rd International Colloquium Tribology: Industrial and Automotive Lubrication*. 2022 Feb 14;23:423–4.
22. Drees D, Lopes L, Baião P, Koutsomichalis A, Georgiou E. Solving real-life wear problems: Streamlined lab tests for diverse industrial applications. Can AI help?. *Wear*. 2025 Feb 3;571:205853.
23. Selvaraj SK, Raj A, Dharnidharka M, Chadha U, Sachdeva I, Kapruan C, Paramasivam V. A cutting-edge survey of tribological behavior evaluation using artificial and computational intelligence models. *Advances in Materials Science and Engineering*. 2021;2021(1):9529199.
24. Ajuka LO, Ogedengbe TS, Adeyi T, Ikumapayi OM, Akinlabi ET. Wear characteristics, reduction techniques and its application in automotive parts–A review. *Cogent Engineering*. 2023 Dec 31;10(1):2170741.
25. Subramani R, Leon RR, Nageswaren R, Rusho MA, Shankar KV. Tribological performance enhancement in FDM and SLA additive manufacturing: materials, mechanisms, surface engineering, and hybrid strategies—a holistic review. *Lubricants*. 2025 Jul 7;13(7):298.
26. Bhaumik S, Pathak SD, Dey S, Datta S. Artificial intelligence based design of multiple friction modifiers dispersed castor oil and evaluating its tribological properties. *Tribology International*. 2019 Dec 1;140:105813.
27. Karthick NK, Bharathiraja M. Ai-optimized Al6061-T6 hybrid metal matrix composites reinforced with Al_2O_3, graphite, and bauxite residue (SiO_2) for advanced brake hub applications: Tribological and machining performance. *Journal of Polymer & Composites*. 2025;13(6):182–98.
28. Ball AD. *A tribological study of the design and performance of automotive cams* (Doctoral dissertation, University of Leeds).
29. Chimeno-Trinchet C, Murru C, Díaz-García ME, Fernández-González A, Badía-Laíño R. Artificial Intelligence and fourier-transform infrared spectroscopy for evaluating water-mediated degradation of lubricant oils. *Talanta*. 2020 Nov 1;219:121312.

30. Fouly A, Taha M, Albahkali T, Shar MA, Abdo HS, Nabhan A. Developing artificial intelligence models for predicting the tribo-mechanical properties of HDPE nanocomposite used in artificial hip joints. *IEEE Access*. 2024 Jan 10;12:14787–99.
31. Tian Y, Zheng B, Khan M, Yang Y. Real-time prediction of wear morphology and coefficient of friction using acoustic signals and deep neural networks in a tribological system. *Processes*. 2025 Jun 3;13(6):1762.
32. Hasan MS, Kordijazi A, Rohatgi PK, Nosonovsky M. Triboinformatic modeling of dry friction and wear of aluminum base alloys using machine learning algorithms. *Tribology International*. 2021 Sep 1;161:107065.
33. Franek F, Dörr N, Badisch E, Pauschitz A. Rethinking tribology–tracking trends, their presence at the ECOTRIB 2019 conference, and their impact on tribology research in Austria. *Lubricants*. 2020 Jul 31;8(8):80.
34. Johns-Rahnejat PM, Rahmani R, Rahnejat H. Current and future trends in tribological research. *Lubricants*. 2023;11:391.
35. Nabeeh NA, Abdel-Basset M, Gamal A, Chang V. Evaluation of production of digital twins based on blockchain technology. *Electronics*. 2022 Apr 17;11(8):1268.
36. Pasha MB, Rao RN, Ismail S, Gupta M, Prasad PS. Tribo-informatics approach to predict wear and friction coefficient of Mg/Si_3N_4 composites using machine learning techniques. *Tribology International*. 2024 Aug 1;196:109696.
37. Ibrahim MA, Çamur H, Savaş MA, Sabo AK, Mustapha M, Abba SI. Hybrid artificial intelligence models with multi objective optimization for prediction of tribological behavior of polytetrafluoroethylene matrix composites. *Applied Sciences*. 2022 Aug 30;12(17):8671.
38. Golla CB, Rao RN, Ismail S, Özcan M, Prasad PS. Development of hybrid aluminum nanocomposites for automotive applications: An in-depth analysis using experimental approaches and predictive machine learning techniques. *Journal of Materials Research and Technology*. 2025 May 1;36:4383–99.
39. Zahoor A, Rahman A, Mursaleen M. Accelerating tribological material discovery: A machine learning framework with predictive insights. *Tribology International*. 2025 Jun 6;211:110862.
40. Prasad SV, Asthana R. Aluminum metal-matrix composites for automotive applications: Tribological considerations. *Tribology Letters*. 2004 Oct;17(3):445–53.
41. Herwig N, Peng Z, Borghesani P. Bridging the trust gap: Evaluating feature relevance in neural network-based gear wear mechanism analysis with explainable AI. *Tribology International*. 2023 Sep 1;187:108670.
42. Rohatgi PK, Liu Y, Barr TL. Tribological behavior and surface analysis of tribodeformed AI Alloy-50 Pet graphite particle composites. *Metallurgical Transactions A*. 1991 Jun;22(6):1435–41.

11 Case Study on Artificial Intelligence in Biomedical Tribology

Jibrin Muhammad Yelwa

11.1 INTRODUCTION

Biomedical tribology, which is known as biotribology, is the discipline that is focused on understanding the interactions of friction, lubrication, and wear in both living organisms and medical devices (Kumar et al., 2024a). It is a fundamental factor in guaranteeing the safety and stability of the above-mentioned systems in addition to others like cardiovascular systems and medical interfacial areas where mechanical contact can't be avoided. Such systems require the best tribological properties since really bad conditions can lead to intensive wear, creation of dangerous particles, and so forth, which cause the device to break down prematurely, and also sometimes the device cannot be used anymore due to its adverse effects on the patient, for instance, causing infections (Lu et al., 2020). Such issues impair not only the mechanical function of implants but also trigger adverse biological reactions, including inflammation, osteolysis, and tissue degeneration, which may necessitate costly and invasive revision procedures (Liu et al., 2021).

A schematic diagram of a hip joint simulator is illustrated in Figure 11.1, which is applicable in the tribological *in vitro* test case. The device models the physiological

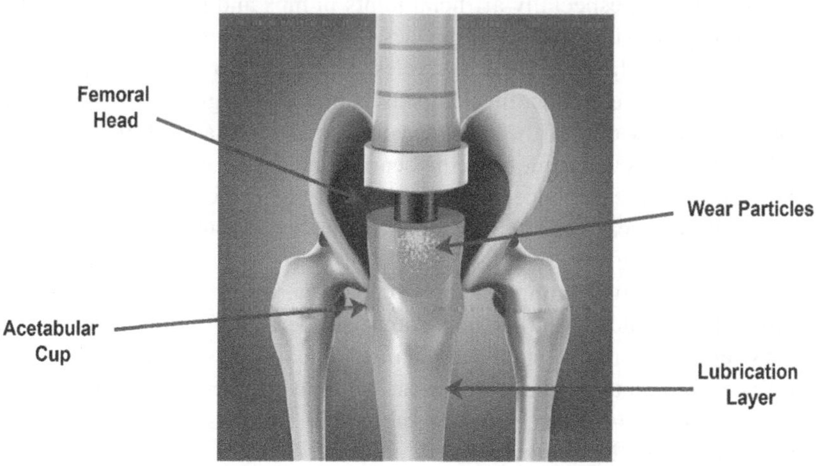

FIGURE 11.1 Schematic of tribological phenomena in biomedical implants.

DOI: 10.1201/9781003624165-11

gait cycles by loading arms and motion axes under controlled conditions, which allow the measurement of wear, friction, and lubrication behavior under clinically relevant conditions. Artificial intelligence (AI) has become a disruptive device to solve these tribological issues. Conventional methods of wear and friction research in biomedical contexts are based on both empirical experimentation, which could involve pin-on-disk experiments and experiments on hip joint simulators, and on computational simulation based on physics, including finite-element analysis (FEA) or the lubrication theory (Barber et al., 2021). Although these techniques were invaluable, they are constrained by simplification of assumptions, high costs, and time-consuming calculations (Toh et al., 2021). Another paradigm presented by AI is the use of machine learning (ML) and deep learning (DL) to create predictive models that may learn directly using data (Kumar et al., 2024b; Zhang et al., 2020). Such models are demonstrated to be more effective to describe complex nonlinear relations in tribological systems than more traditional regression or analytical methods (Marian & Tremmel, 2021; Shah et al., 2025). Recent developments include neural networks that use acoustic or vibration signals to estimate the wear morphology and friction coefficient in real time and surrogate models that simulate computationally expensive lubrication simulations with high fidelity at low computational costs (König et al., 2024; Tian et al., 2025). This chapter discusses the use of AI in the field of biomedical tribology and especially in hip joint prostheses. The extent of the analysis will include the definition of important tribological issues in biomedical implants, the comparison between traditional and AI-based solutions, and the combination of different types of data, i.e., lab experiments, clinical results, and models. This chapter demonstrates the role of AI in the enhancement of predictive modeling and optimization in biotribology and analyzes this context to draw conclusions about future research and clinical practice implications.

11.2 BACKGROUND

Biomedical implants, especially artificial joints in hips and knees, function in complicated tribological situations. The devices in question must endure till the end of their service life millions of loading cycles, and their performance is very much affected by the surface contact mechanisms, lubrication regimes, and material degradation. Abrasion is one of the most serious issues (Ackland et al., 2015; Milone et al., 2022). Particles worn out from the bearing materials of polyethylene, ceramic, or metallic implants can trigger biological reactions like osteolysis and loosening of the implant in the case of hip and knee replacements (Van De Steen et al., 2021). Wear may be adhesive, abrasive, fatigue-related, or corrosive, and it is influenced by load magnitude, relative sliding velocity, material hardness, and environmental factors such as body fluids. Another important parameter is friction as it affects both the energy required for movement and the mechanical forces exerted on surrounding tissues. Excessive friction accelerates wear, while insufficient lubrication may cause a transition from fluid-film to boundary lubrication, increasing contact stresses and material degradation (Nitish Prasad & Ramkumar, 2024).

Figure 11.2 presents a schematic model of a hip joint simulator to be used in *in vitro* tribological testing. The device models the physiological gait cycles by means

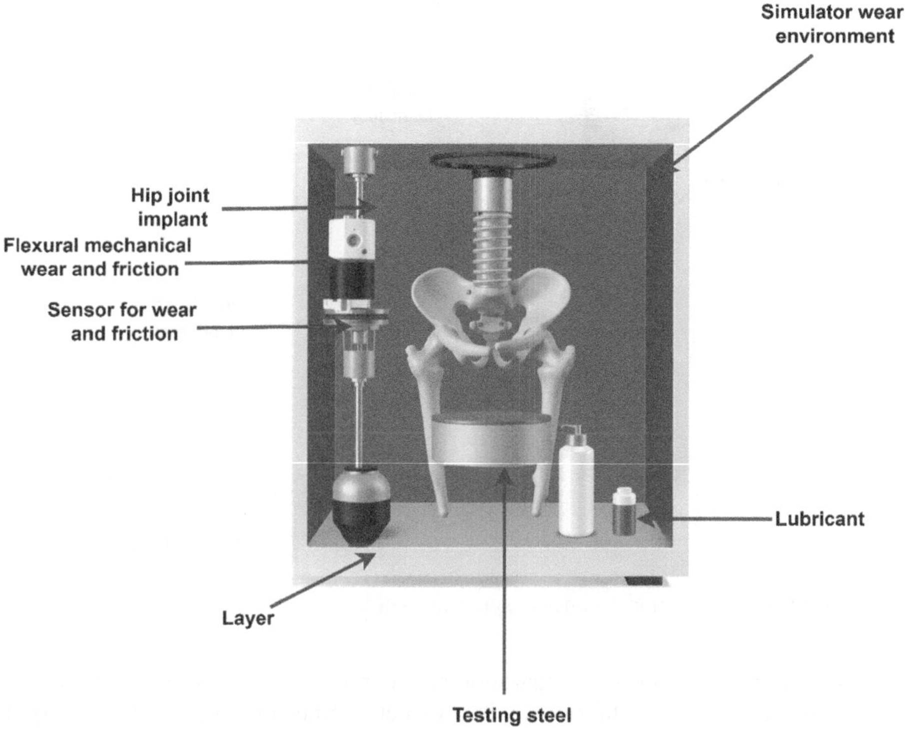

FIGURE 11.2 Hip joint tribological testing apparatus.

of controlled loading arms and motion axes, which allows the accurate measurement of wear, friction, and lubrication behavior under clinically relevant conditions. Experimental results of this nature form the basis of training AI models capable of forecasting the performance of the implants more effectively than traditional long-term testing. There is a tendency to consider synovial fluid in the biomedical devices as a control of lubrication on the boundaries in the joints and as the source of mixed lubrication due to its protein and lipid contents. The capability of an implant to run in a desirable lubrication regime is reliant on the size of the femoral head, choice of materials, roughness of the surfaces, and clearance of a joint. It has been numerically and experimentally demonstrated that minute changes in the geometry of implants or loading conditions can switch lubrication regimes and have direct effects on frictional performance and wear (Nitish Prasad & Ramkumar, 2024; Stanković et al., 2019; Van De Steen et al., 2021). In addition to wear and lubrication, the corrosion and fatigue degradation of materials may impair the stability of implants especially in metallic systems where surface oxidation or delamination takes place. In order to overcome these mechanisms, novel coatings like zirconia were examined due to the two-fold advantages of wear reduction and antibacterial protection (Kedia et al., 2022).

Figure 11.3 shows the key lubrication regimes that control the mechanics of synovial joints. Boundary lubrication occurs when there is a direct surface contact with no separating fluid. Mixed lubrication is a partial separation between fluids, and fluid

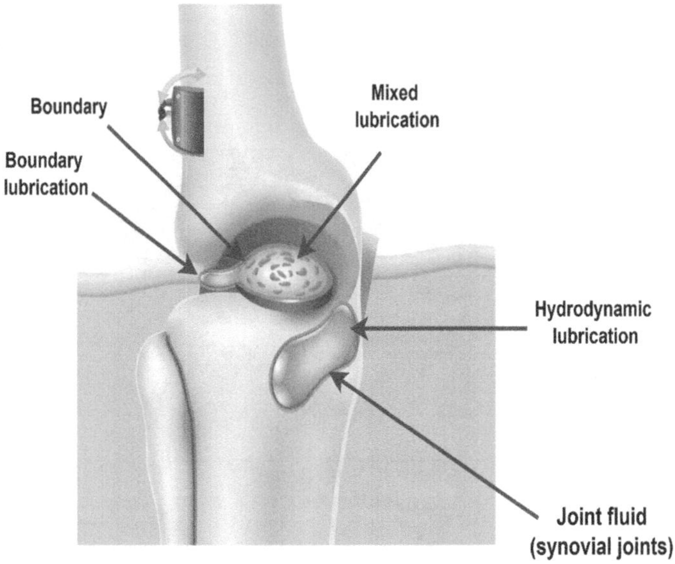

FIGURE 11.3 Lubrication regimes in synovial joints.

film lubrication is an ultimate separation between fluids by a continuous film. Both AI-based optimization of lubrication in prosthetic joints and implant design require the accurate modeling of such regimes. Traditional methods involving the study of these tribological phenomena are based on experimental tests and computer simulations. Physiological loading cycles can be simulated on hip joint simulators, such as, and volumetric wear can be measured during millions of simulated steps (Ackland et al., 2015; Ma et al., 2020). Pin-on-disk tests are a controlled way to gain knowledge of frictional and wear behavior of material pairs (Stanković et al., 2019). The computational models (including FEA and lubrication analysis) complement these techniques and offer the scheme to estimate the wear volume and the lubrication film thickness under a specific loading and geometrical condition (Liu et al., 2021; Lu et al., 2020). However, these methods have apparent limitations. Experiments are both resource-intensive and time-consuming, and they can take months to yield information that is of significance in terms of wear. Computational methods, being more relaxed, involve simplification of assumptions, which need not be representative of variation and complexity in the actual biological situation (Toh et al., 2021). These problems can be resolved very powerfully using AI-based strategies. ML models, in particular, artificial neural networks, support vectors machines, random forests, and gradient boosting algorithms, have achieved increasingly successful approaches to tribology problems (Akkad et al., 2023; Eldesouky & Hofy, 2020; Marian & Tremmel, 2021). By training on experimental or simulation datasets, these models are able to make rapid predictions of wear rate, coefficient of friction, or lubrication regime (Chanda et al., 2016a). Sensor signals (e.g., acoustic emissions) have also been analyzed with DL structures (e.g., convolutional neural networks) and subsequently tracked in real time to monitor tribological states (Tian et al., 2025; Zhang et al.,

2020). Surrogate models that can be used to address large-scale simulations give the possibility to replace computationally expensive finite-element or lubrication models by predictions that can be essentially computed by the user in real time and remain accurate (König et al., 2024).

It is important to note that AI solutions are more efficient not only in the field of prediction but also in optimization, which enables scientists to find out what material combinations, geometries, and operating conditions make it possible to reduce wear and friction (Barber et al., 2021; Chanda et al., 2016b). The recent researches have pointed toward the transformative characteristic of AI in tribology (Zhang & Yin, 2024). Shah et al. (2025) observed that AI-based methods are not restrictive to prediction, but they may support a mechanistic perception and condition-based observation as well as the intelligent design of tribological systems. Marian and Tremmel (2021) confirm that AI is already effectively applied in the field of wear prediction, formulation of lubricants, and fault detection in real time. With biomedical uses, adoption is still immature, but AI shows significant potential to better characterize complex interactions in implants than physics-based methods. The newly created subdiscipline of triboinformatics that is a blend of AI, big data, and traditional tribology is the manifestation of this paradigm shift (Arun et al., 2023; Wu et al., 2017).

11.3 CASE CONTEXT

This case study focuses on an artificial hip joint, which is one of the widely used biomedical implants across the globe. The main purpose of hip replacement design is to enhance patients' movement and to remove pain produced by either degenerative joint disease or trauma. The tribological behavior of hip implants is the main point of their triumph. On the articulating surfaces, which usually consist of material combinations like metal-on-polyethylene, ceramic-on-ceramic, or metal-on-metal, cyclic loading is applied (Ackland et al., 2015). The bad tribological optimization of these interfaces can cause inflammation and implant loosening due to the creation of wear debris over time; thus, tribological optimization is of utmost importance in clinical practice (Chevalier et al., 2011). One of the problems that need to be solved is estimation of the wear rate and the coefficient of friction during the realistic gait cycle, thus controlling the lubrication methods for implant longevity. The wear predictions for hip implants are complicated by the presence of very different parameters, like the weight of the person, the speed of walking, the diameter of the femoral head, the space between the femur and the acetabulum, the quality of the surface, and the type of material used. Classical computational models have attempted to explain these factors using frictional work or contact pressure as predictive parameters; however, their accuracy is limited (Liu et al., 2021). AI offers the ability to integrate diverse data streams into powerful predictive models capable of processing nonlinearities and cross-relationships between variables.

The artificial hip replacement illustrated in Figure 11.4 shows principal geometric characteristics. In the figure, the femur, the acetabular cup, and the acetabular liner and the stem have representative clearances at the articulating interface, which play a big role in tribological performance and implant longevity. The introduction of these geometric variables into AI predictive models facilitates more accurate wear

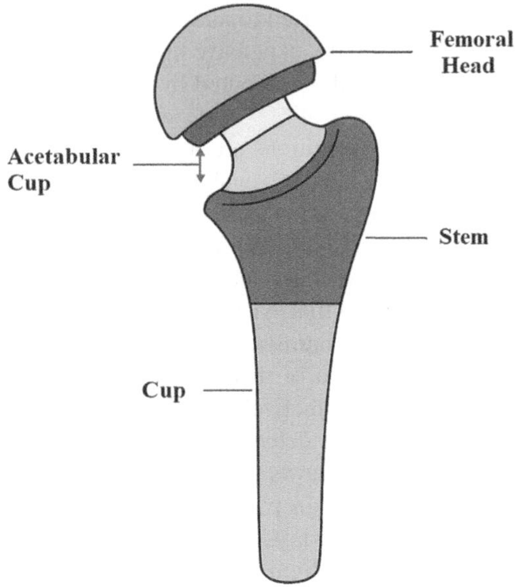

FIGURE 11.4 Artificial hip implant geometry.

forecasting and allows optimization of the implants to be patient-specific. Another major area of study is the optimization of lubrication. It has been demonstrated that the regimes of lubrication in hip joints are dichotomous to the geometry of implants and are patient-specific (Nitish Prasad & Ramkumar, 2024). It is possible to train AI models to consider which combination of geometric and operational parameters may result in maximum fluid film thickness and minimum friction and wear (Barber et al., 2021; Shinmori et al., 2020). This and especially the design of new-generation implants that are specific to the characteristics of the patient are new ground, which coincides with personalized medicine.

The data sources for this case study are wide-ranging. Experimental tribological tests such as hip joint simulators and pin-on-disk tests are available to provide empirical measurements of wear and friction. Clinical results from retrieval analyses and radiograph studies provide real-world data of implant wear patterns and failure rates. The use of computational simulation datasets such as FEA and lubrication modeling has practically shown the ways of wear and lubrication behavior (Lu et al., 2020; Van De Steen et al., 2021). Among the different studies to surface modifications, researchers working with zirconia coatings have been able to obtain insights from a study about material optimization (Kedia et al., 2022). It is the mesh of these datasets that constitutes the basis of AI-informed modeling, which is capable of creating predictive models that are exact and relatable to those produced by our ML systems (Sose et al., 2023). This case study puts the artificial hip joint at the center of relevant biomedical systems and demonstrates how AI can resolve critical tribological problems. This indicates that the integration of experimental, clinical, and computational data with ML models is a very effective way to predict wear, optimize lubrication and, eventually, implant design and patient outcomes (Kumar et al., 2024b).

11.4 METHODOLOGY

This research followed a systematic approach that can be described in quite a few organized steps: it started with data collection and preprocessing; then the researchers applied AI techniques and other computational methods; and finally, they performed a thorough evaluation and validation. The overall objective was to create a strong basis for forecasting the wear and friction behavior of artificial hip implants and detecting the best tribological conditions. Data collection was mainly from the *in vitro*, imaging, clinical, and computational sources that were related to the tribological behavior. On hip joint simulators, which imitate the gait loading cycles of human walking, *in vitro* tribological experiments were done. These simulators measure wear, frictional torque, and wear-particle characterization directly and over a long time. Besides, high-resolution imaging methods, such as radiographs, CT scans, and MRI, were applied to map implant geometry, bone structure, and wear patterns in the area of the extracted components. Then, the tribological outcomes were linked to the patient-specific factors drawn from clinical records, which included age, gender, body mass index, activity level, and time since implantation, as well as demographic, clinical, and imaging data. The use of mature and validated computational models (specifically FEA) supplied the researchers with the ability to generate continuous distributions of contact pressure, sliding velocity, lubrication film thickness, and stress during dynamic gait cycles as additional data to support the experimental work (Ruggiero & Sicilia, 2020; Wang et al., 2019). Through simulations, the researchers were able to cover a wide range of parameter combinations that would otherwise be restricted to either ethical or practical experimentation only.

A representation of multimodal data acquisition concerning hip implantation tribology is given in Figure 11.5. The data sources being dealt with are *in vitro* simulator testing, radiological studies, and computational modeling, which are all processed further before being gathered into structured datasets. This setup works excellently for neural network training since they will be able to comprehend nonlinear relations as well as to predict the performance of the studies regarding the friction and lubrication of the implanted parts. ML pipelines were used for the various sources of raw data which were preprocessed to ensure their compatibility. Imaging data

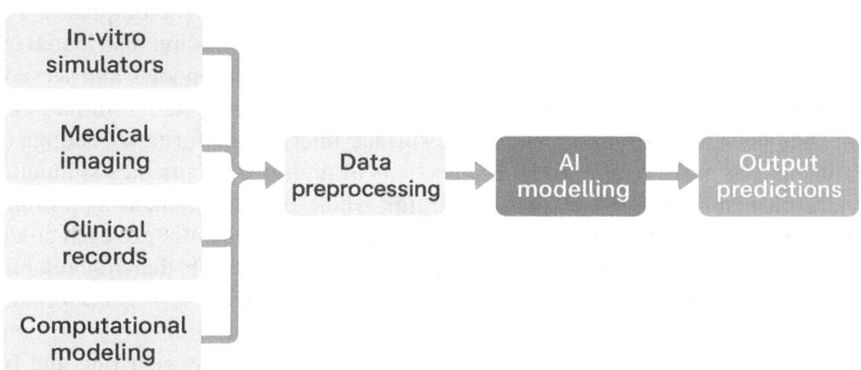

FIGURE 11.5 Data collection pipeline.

were segmented and normalized so that the extraction of geometrical parameters, like head and cup clearance, surface curvature, and volumetric wear, could be done. Derived wear metrics and simulator outputs were first processed through calibration, denoising, and grouping into structured data tables. The clinical variables were then standardized for consistency, regression models were used to impute missing values, and one-hot encoding was applied to categorical variables. The simulation results obtained were then interpolated by means of resampling and standardized gait cycles. Feature selection was done using variables that are known to have physical effects like contact pressure, sliding speed, material pairing, surface roughness/lubrication state, patient weight, and implant geometry, thus making sure that the models reflected both intuitively understood factors from physics and data-driven patterns. The dataset partitioning was done according to the best practices, with the data split into training, validation, and testing sets, while a temporal gap was created in the clinical cases to replicate real-world deployment. Three different levels were used for applying AI methods. In the first level, decision tree-based approaches such as random forests, support vector machines (SVMs), and gradient boosting machines (Sose et al., 2023) were included in the list. All of these models created an interpretive basis for the prediction of wear volume, friction coefficient, and lubrication regime; random forests, for instance, produced feature importance metrics that pointed out the importance of contact pressure and surface roughness as key variables. During the second stage, DL methods like neural networks designed with multiple layer perceptrons (MLPs) and convolutional neural networks (CNNs) were implemented, especially for the features obtained via imaging and gait cycle data taken in space and time. CNNs not only forecasted the locations of wear-scar patches but also the chance of local wear; on the other hand, recurrent neural networks (RNNs) predicted the changes over time in the pressure and velocity fields that were tagged by the gait phases. By the end, hybrid approaches were paralleled by the outputs coming from the simulations based on physics and the AI-ready models which were using neural networks as surrogates to approximate the finite element-based simulations. With the use of AI, the prediction of lubrication film thickness or wear distribution according to different loads and shapes was practically instantaneous. AI along with FEA has always been an important method for predicting contact mechanics and wear development. One of the outputs of FEA simulations included the maps of stress, strain, and relative motion distributed over the contact surfaces. These outputs both trained AI surrogate models and served as ground-truth labels for comparison. Molecular dynamics, despite its limitations in scale, has been used to model nanoscale surface interactions for new coatings and treatments (Gröning et al., 2013). Biomechanical multibody dynamics simulations provided kinematics and load patterns to link whole-body movement with implant contact behavior. Employing the techniques that are complementary to each other in some ways, AI trained with these methods would produce results that are both physically realistic and computationally efficient (Gröning et al., 2013). Performance of the model was measured through different metrics, determined by the output type. In evaluating the performance of continuous targets like wear rate and friction coefficient, mean absolute error (MAE), root mean square error (RMSE), coefficient of determination (R^2), and, where appropriate, Bland–Altman analysis to ascertain the agreement with experimental results were applied to evaluate the performance. Accuracy, precision, recall, F1-score, and area under the receiver operating characteristic curve (AUC-ROC) were used for the evaluation of classification tasks like lubrication regime identification or wear mechanism categorization. To guarantee the internal performance estimation was solid, cross-validation methods, such as k-fold and leave-one-case-out, were utilized. For external validation, the model's generalization to practical situations was determined by using clinical and simulator data that were not previously observed. The understanding of the model was measured through the use of permutation feature importance, Shapley additive explanations (SHAP), and layer-wise relevance propagation (LRP), especially for DL models. Sensitivity analysis was done to examine the stability of predictions when inputs were altered, thereby providing support for the robustness of the method used.

11.5 RESULTS AND DISCUSSION

Predictions were made by AI-trained models, and they were amazing. The MAE per million cycles for volumetric wear, which was under 0.05 mm^3, was reached by random forests, a decision tree ensemble, alongside an R^2 value of 0.92 on the testing data. SVM regressors were as accurate but not quite interpretable. DL models, on the other hand, to analyze wear-scar images, employed CNNs, and these reached near 90% accuracy in the classification of high-wear areas. On the contrary, RNNs were applied to the study of gait phases, and they made predictions of friction fluctuations with less than 10% relative error over cycles. The results of AI predictions and the data from experiments and clinics were compared, and they were very consistent. Regression analysis confirmed that predicted wear volumes of the simulators were lying within the confidence intervals of actual data for more than 85% of experimental test cases. Additionally, radiographic measurements of clinical wear were highly correlated with model predictions, especially in patients who had undergone extensive preoperative imaging. Hybrid surrogate models represented finite-element wear distributions with a mean spatial deviation of less than 0.1 mm, which made it possible to quickly and efficiently map out the design parameter space without the need for heavy computation. Data that pertained to specific cases were acquired. The models indicated that elevating the diameter of the femoral head from the standard values resulted in a decrease in the friction coefficient owing to the establishment of a more preferable lubrication regime; however, the wear rate increased unless the roughness of the polymer surface was controlled. The levels of surface finish that were optimized led to a roughly 20% reduction in wear volume without a corresponding increase in friction. The material combinations such as ceramic-on-polyethylene were always superior to metal-on-polyethylene as far as wear and friction are concerned, which is in agreement with previous physics-based predictions but now quickly predicted by using AI (Figure 11.6). The predictions that were specific to the patient made it possible to optimize the geometry of the implant in accordance with the weight and activity profile of the individual; hence, the personalized tribological design was supported. The results clearly point out that AI has played a major role in the field of biomedical tribology. The accuracy of the models was very high, and they consumed less computational

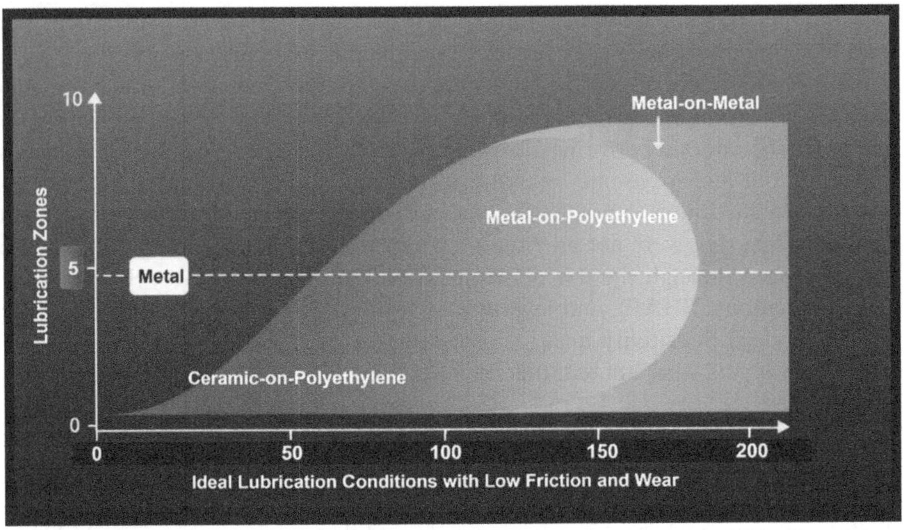

FIGURE 11.6 Optimized lubrication regimes and material pairings.

resources than the traditional simulations based on FEA. Real-time prediction by AI models will make interactive design and surgical planning possible. Moreover, the models applied interpretability techniques that pointed out the important features, such as contact pressure and surface roughness, that are in line with physical understanding; hence, the trust of clinicians is built, and transparency through interpretability techniques is provided.

Nevertheless, data scarcity has been one of the constraints. Attempts were made to integrate simulator, imaging, and clinical data, but it was not fully possible to cover all combinations of parameters. The models demonstrated reduced performance in the extrapolation to large loads or of new material pairs that are not found in the training data. While SHAP and LRP can be used to improve interpretability, it is still limited in deep architectures. The models made use of spurious correlations in imaging artifacts rather than actual wear patterns. Generalization was problematic: models trained on a single simulator/clinical population did not consistently generalize to other populations without retraining or fine-tuning, implying the presence of domain shift problems. Clinical, ethical, and regulatory considerations are critical. Deployment in surgical planning or implant design must take medical device regulatory pathways into account. The opacity of DL models may make approval difficult to obtain, unless accompanied by stringent validation and interpretability. Patient data present a number of ethical concerns that basically include managing imaging and clinical data with strict privacy and consent rules; unintentional biases in models could, for example, make patients with unusual anatomy or activity profiles suffer. In hospitals and clinics, AI predictions must be the case if and when an expert decides so, with the models being decision support tools that are not only explainable and transparent but also hard to ignore.

11.6 FUTURE DIRECTIONS

The field of biomedical tribology is continuously being transformed by the combined use of real-time monitoring technologies and the development of digital twin, federated learning, and engineered biological interfaces (Kumar et al., 2024a). AI-powered implantable sensors, which are among these technologies, have the greatest potential to have an impact on the *in vivo* monitoring of tribological parameters (Johns-Rahnejat et al., 2023). The latest advances in implantable autonomous bioelectronics, which are powered by triboelectric nanogenerators, allow the uninterrupted collection of physiological data like pressure, motion, and friction in prosthetic joints without using any external power sources (Wang et al., 2023). AI-equipped sensors have the ability to observe wear or lubrication deterioration in real time, and this capability will be a huge step forward for application of preventive interventions as well as the design of adaptable implants. Along with the advancements in sensor technologies, digital twin methods are developing into the main technological infrastructures for user-specific tribological simulation. These digital models of actual implants integrate consistently biological and mechanical data uninterrupted, thus making it possible not only to retrospectively estimate wear but also to proactively model various use cases. The utilization of digital twins in orthopedic trauma evidences a great deal of potential already, virtual recreation of individual cases, and implant stress distributions being predicted under patient-specific biomechanics (Andres et al., 2025). This application for joint implants allows a digital twin to continually adjust the tribological performance by generating scenarios of walking, varying loads, and slickness switches, which are determined by the physical characteristics and living patterns of every patient. The confidentiality of data-rich environments is of utmost importance. Federated learning proves to be a powerful response by permitting the model training of cooperation among devices or organizations in such a way that sensitive raw data are not being moved. In addition, the medical and health care community is reaping the benefits of federated learning with datasets coming from devices like wearables and Internet of Things (IoT) for privacy persuasion and better model generalizability (Ali et al., 2023). The use of federated learning can be further strengthened through the employment of cryptographic techniques like homomorphic encryption and secure multiparty computation that will not only ensure the highest level of protection to sensitive tribological and patient data but also facilitate decentralized AI development (Moulahi et al., 2023).

Now tribology is being applied in new ways through the development of tissue-engineered implants as well as drug–device interfaces. Biohybrid methods are a very interesting field because it integrates living tissues and synthetic materials. Implants with cells that produce cartilage or skilled gels that alter their surface properties depending on the amount of mechanical stress are a couple of examples of this direction (Menciassi et al., 2020). The other direction is ‘bioinstructive’ surface design that modifies the implants with growth factors or proteins, which are then less prone to cause inflammation due to friction and more likely to have a stronger interface with the host tissue (Gonzalez-Fernandez et al., 2019). All these techniques indirectly point that currently implants are not just plastic devices anymore, but rather they are highly advanced, adaptable data-driven systems. AI-equipped sensors will send real-time tribological information to custom-made digital twins that apply predictive models to extend the life of the implant and its compatibility with the body. The combination of these innovations brings to the market the possibility of truly smart, strong, and patient-specific biomedical implants.

11.7 CONCLUSION

The application of tribology in medicine integrated with AI is a revolutionary approach in the field of medical implantation itself. AI is capable of handling complex data from various sources such as biomechanics, molecular dynamics, and real-world tests to predict the wearing out of materials, the selection of materials, and the design process with a high degree of accuracy. This method narrows the dissimilarity between clinical and laboratory research and so, implants become safer and more reliable to be used by specific patients. In the near future, the emergence of AI-powered sensors and digital twins will lead to metamorphosis of implants from being a one-size-fits-all approach to being patient-oriented solutions. In the end, AI-infected tribology would result in longer implant life, a decrease in the number of repeat surgeries, and provide patients with better living standards.

REFERENCES

Ackland, D. C., Moskaljuk, A., Hart, C., Vee Sin Lee, P., & Dimitroulis, G. (2015). Prosthesis loading after temporomandibular joint replacement surgery: A musculoskeletal modeling study. *Journal of Biomechanical Engineering*, *137*(4), 041001. https://doi.org/10.1115/1.4029503

Akkad, K., Mehboob, H., Alyamani, R., & Tarlochan, F. (2023). A machine-learning-based approach for predicting mechanical performance of semi-porous hip stems. *Journal of Functional Biomaterials*, *14*(3), 156. https://doi.org/10.3390/jfb14030156

Ali, M., Naeem, F., Tariq, M., & Kaddoum, G. (2023). Federated learning for privacy preservation in smart healthcare systems: A comprehensive survey. *IEEE Journal of Biomedical and Health Informatics*, *27*(2), 778–789. https://doi.org/10.1109/JBHI.2022.3181823

Andres, A., Roland, M., Wickert, K., Diebels, S., Stöckl, J., Herrmann, S., Reinauer, F., Leibinger, R., Pavlov, A., Schuppener, L., Schäfer, D., Histing, T., & Braun, B. J. (2025). Advantages of digital twin technology in orthopedic trauma surgery – exploring different clinical use cases. *Scientific Reports*, *15*(1), 19987. https://doi.org/10.1038/s41598-025-04792-w

Arun, K. K., Jasmin, N. M., Kamesh, V. V., Pramod, V. R., Krishnaraj, S., Suresh, V., & Subbiah, R. (2023). Applications of artificial neural network simulation for prediction of wear rate and coefficient of friction titanium matrix composites. *Materials Research*, *26*, e20220306. https://doi.org/10.1590/1980-5373-mr-2022-0306

Barber, H., Kelly, C. N., Abar, B., Allen, N., Adams, S. B., & Gall, K. (2021). Rotational wear and friction of Ti-6Al-4V and CoCrMo against polyethylene and polycarbonate urethane. *Biotribology*, *26*, 100167. https://doi.org/10.1016/j.biotri.2021.100167

Chanda, S., Gupta, S., & Pratihar, D. K. (2016a). A combined neural network and genetic algorithm based approach for optimally designed femoral implant having improved primary stability. *Applied Soft Computing*, *38*, 296–307. https://doi.org/10.1016/j.asoc.2015.10.020

Chanda, S., Gupta, S., & Pratihar, D. K. (2016b). Effects of interfacial conditions on shape optimization of cementless hip stem: An investigation based on a hybrid framework. *Structural and Multidisciplinary Optimization*, *53*(5), 1143–1155. https://doi.org/10.1007/s00158-015-1382-1

Chevalier, J., Taddei, P., Gremillard, L., Deville, S., Fantozzi, G., Bartolomé, J. F., Pecharroman, C., Moya, J. S., Diaz, L. A., & Torrecillas, R. (2011). Reliability assessment in advanced nanocomposite materials for orthopaedic applications. *Journal of the Mechanical Behavior of Biomedical Materials*, *4*(3), 303–314. https://doi.org/10.1016/j.jmbbm.2010.10.010

Eldesouky, I., & Hofy, H. E. (2020). Design and prototyping of a novel low stiffness cementless hip stem. *International Journal of Biomedical Engineering and Technology*, *32*(3), 229. https://doi.org/10.1504/IJBET.2020.106033

Gonzalez-Fernandez, T., Sikorski, P., & Leach, J. K. (2019). Bio-instructive materials for musculoskeletal regeneration. *Acta Biomaterialia*, *96*, 20–34. https://doi.org/10.1016/j.actbio.2019.07.014

Gröning, F., Jones, M. E. H., Curtis, N., Herrel, A., O'Higgins, P., Evans, S. E., & Fagan, M. J. (2013). The importance of accurate muscle modelling for biomechanical analyses: A case study with a lizard skull. *Journal of The Royal Society Interface*, *10*(84), 20130216. https://doi.org/10.1098/rsif.2013.0216

Johns-Rahnejat, P. M., Rahmani, R., & Rahnejat, H. (2023). Current and future trends in tribological research. *Lubricants*, *11*(9), 391. https://doi.org/10.3390/lubricants11090391

Kedia, S., Das, A., Patro, B. S., & Nilaya, J. P. (2022). Potential tribological and antibacterial benefits of pulsed laser-deposited zirconia thin film on Ti6Al4V bio-alloy. *Applied Physics A*, *128*(8), 670. https://doi.org/10.1007/s00339-022-05835-z

König, F., Wirsing, F., Singh, A., & Jacobs, G. (2024). Machine-learning-based wear prediction in journal bearings under start–stop conditions. *Lubricants*, *12*(8), 290. https://doi.org/10.3390/lubricants12080290

Kumar, A., Kumar, A., & Kumar, A. (Eds.). (2024a). *Applications of Biotribology in Biomedical Systems*. Cham, Switzerland: Springer Nature Switzerland. https://doi.org/10.1007/978-3-031-58327-8

Kumar, V., Tewari, R. P., & Rawat, A. (2024b). Machine learning-based investigations of the effect of surface texture geometry on the wear behaviour of UHMWPE bearings in hip joint implants. *Biosurface and Biotribology*, *10*(4), 143–158. https://doi.org/10.1049/bsb2.12085

Liu, F., He, Y., Gao, Z., & Jiao, D. (2021). Enhanced computational modelling of UHMWPE wear in total hip joint replacements: The role of frictional work and contact pressure. *Wear*, *482–483*, 203985. https://doi.org/10.1016/j.wear.2021.203985

Lu, X., Nečas, D., Meng, Q., Rebenda, D., Vrbka, M., Hartl, M., & Jin, Z. (2020). Towards the direct validation of computational lubrication modelling of hip replacements. *Tribology International*, *146*, 106240. https://doi.org/10.1016/j.triboint.2020.106240

Ma, H.-D., Wang, Q.-Y., Teng, H.-D., Zheng, T.-H., & Liu, Z. (2020). Evaluation of the therapeutic effect of Bi-maxillary osteotomy using the stress distribution on the temporomandibular joint when doing anterior teeth occlusion. *Journal of Biomechanical Engineering*, *142*(12), 121010. https://doi.org/10.1115/1.4047425

Marian, M., & Tremmel, S. (2021). Current trends and applications of machine learning in tribology—A review. *Lubricants*, *9*(9), 86. https://doi.org/10.3390/lubricants9090086

Menciassi, A., Takeuchi, S., & Kamm, R. D. (2020). Biohybrid systems: Borrowing from nature to make better machines. *APL Bioengineering*, *4*(2), 020401. https://doi.org/10.1063/5.0014918

Milone, D., Risitano, G., Pistone, A., Crisafulli, D., & Alberti, F. (2022). A new approach for the tribological and mechanical characterization of a hip prosthesis trough a numerical model based on artificial intelligence algorithms and humanoid multibody model. *Lubricants*, *10*(7), 160. https://doi.org/10.3390/lubricants10070160

Moulahi, W., Jdey, I., Moulahi, T., Alawida, M., & Alabdulatif, A. (2023). A blockchain-based federated learning mechanism for privacy preservation of healthcare IoT data. *Computers in Biology and Medicine*, *167*, 107630. https://doi.org/10.1016/j.compbiomed.2023.107630

Nitish Prasad, K., & Ramkumar, P. (2024). Effect of geometrical, operational and material parameters in the lubrication regime of hard-on-hard hip implants. *Materials Chemistry and Physics*, *317*, 129171. https://doi.org/10.1016/j.matchemphys.2024.129171

Ruggiero, A., & Sicilia, A. (2020). Lubrication modeling and wear calculation in artificial hip joint during the gait. *Tribology International*, *142*, 105993. https://doi.org/10.1016/j.triboint.2019.105993

Shah, R., Jaramillo, R., Thomas, G., Rayhan, T., Hossain, N., Kchaou, M., Profito, F. J., & Rosenkranz, A. (2025). Artificial intelligence and machine learning in tribology: Selected case studies and overall potential. *Advanced Engineering Materials*, *27*, 2401944. https://doi.org/10.1002/adem.202401944

Shinmori, H., Kubota, M., Morita, T., Yamaguchi, T., & Sawae, Y. (2020). Effects of synovial fluid constituents on friction between UHMWPE and CoCrMo. *Tribology Online*, *15*(4), 283–292. https://doi.org/10.2474/trol.15.283

Sose, A. T., Joshi, S. Y., Kunche, L. K., Wang, F., & Deshmukh, S. A. (2023). A review of recent advances and applications of machine learning in tribology. *Physical Chemistry Chemical Physics*, *25*(6), 4408–4443. https://doi.org/10.1039/D2CP03692D

Stanković, M., Marinković, A., Grbović, A., Mišković, Ž., Rosić, B., & Mitrović, R. (2019). Determination of Archard's wear coefficient and wear simulation of sliding bearings. *Industrial Lubrication and Tribology*, *71*(1), 119–125. https://doi.org/10.1108/ILT-08-2018-0302

Tian, Y., Zheng, B., Khan, M., & Yang, Y. (2025). Real-time prediction of wear morphology and coefficient of friction using acoustic signals and deep neural networks in a tribological system. *Processes*, *13*(6), 1762. https://doi.org/10.3390/pr13061762

Toh, S. M. S., Ashkanfar, A., English, R., & Rothwell, G. (2021). Computational method for bearing surface wear prediction in total hip replacements. *Journal of the Mechanical Behavior of Biomedical Materials*, *119*, 104507. https://doi.org/10.1016/j.jmbbm.2021.104507

Van De Steen, A., Khalife, R., Colant, N., Mustafa Khan, H., Deveikis, M., Charalambous, S., Robinson, C. M., Dabas, R., Esteban Serna, S., Catana, D. A., Pildish, K., Kalinovskiy, V., Gustafsson, K., & Frank, S. (2021). Bioengineering bacterial encapsulin nanocompartments as targeted drug delivery system. *Synthetic and Systems Biotechnology*, *6*(3), 231–241. https://doi.org/10.1016/j.synbio.2021.09.001

Wang, C., He, T., Zhou, H., Zhang, Z., & Lee, C. (2023). Artificial intelligence enhanced sensors—Enabling technologies to next-generation healthcare and biomedical platform. *Bioelectronic Medicine*, *9*(1), 17. https://doi.org/10.1186/s42234-023-00118-1

Wang, L., Isaac, G., Wilcox, R., Jones, A., & Thompson, J. (2019). Finite element analysis of polyethylene wear in total hip replacement: A literature review. *Proceedings of the Institution of Mechanical Engineers, Part H: Journal of Engineering in Medicine*, *233*(11), 1067–1088. https://doi.org/10.1177/0954411919872630

Wu, D., Jennings, C., Terpenny, J., Gao, R., & Kumara, S. (2017). Data-driven prognostics using random forests: Prediction of tool wear. *Manufacturing Equipment and Systems*, *3*, V003T04A048. https://doi.org/10.1115/MSEC2017-2679

Zhang, S., Zhang, S., Wang, B., & Habetler, T. G. (2020). Deep learning algorithms for bearing fault diagnostics—A comprehensive review. *IEEE Access*, *8*, 29857–29881. https://doi.org/10.1109/ACCESS.2020.2972859

Zhang, Z., & Yin, N. (2024). *Tribo-Informatics: The Systematic Fusion of AI and Tribology* (1st ed.). Boca Raton, FL: CRC Press. https://doi.org/10.1201/9781003467991

12 AI-Driven Predictive Maintenance and Wear Prediction in Tribology

Industrial Case Studies and Applications

Virinder Kumar

12.1 INTRODUCTION

The implementation of artificial intelligence (AI) to predict wear in rotating machines represents a significant innovation in maintenance methods for industries, with the application of this concept subjectively altering the way tribology and predictive maintenance (PdM) are executed [1,2]. It is a very important area due to the heavy application of rotating machinery in various industries, such as power generation, the transport industry, manufacturing, and mining, where constant processes under extreme conditions often lead to wear and possible system failures [3–6]. The tribological science of the interaction between surfaces in relative motion also encompasses the study of friction, wear, and lubrication [7–9]. In rotating machines, tribological relationships are the key points of operational activity and part life. Wear, which is quite an issue, is the process of removing or deforming material as a result of surface contact and motion [9]. Friction, lubrication, and wear dynamics directly affect the reliability and efficiency of mechanical-based systems, and tribological considerations usually reduce life cycles [10]. A case in point is that tooth duration is impacted by gear wear, reducing the remaining useful life (RUL) of transmission systems, which can cause severe failures [11]. In highly developed polymer gears, the constraints of load-bearing, operating temperature, and life are closely linked to tribological characteristics [12]. The issues are, hence, essential to optimise the design and performance of industrial and transport machinery.

12.2 CHALLENGES IN TRADITIONAL WEAR PREDICTION

The conventional methods of maintenance have several limitations, including high maintenance costs, unplanned downtime, and poor prediction accuracy [13–15]. Whether it is through corrective maintenance that attempts to rectify failures after they have occurred or through preventive maintenance that tries to address failures

DOI: 10.1201/9781003624165-12

at specific time intervals, both approaches most likely would result in costly breakdowns and loss of production, as well as premature servicing or unscheduled early failures [16]. Wear is also complicated by nature and depends on a variety of factors such as material composition, contact stress, lubrication, and environmental conditions, making it difficult to model with reasonable accuracy using conventional statistical or physics-based procedures [17]. Although it is useful, laboratory-scale tribological experiments often employ accelerating or high-pressure conditions, which are not analogous to real-world operation and may lead to discrepancies in long-term wear prediction [18]. In addition, the old systems do not support real-time monitoring and analysis, which can slow down fault detection before it significantly affects the system [19]. Not only is the machine's behaviour dynamic in the face of various faults, but it is also impossible to avoid the use of expert opinion, which complicates the diagnostics further [20,21].

12.3 ROLE OF AI IN PdM

AI and machine learning (ML) solutions to these problems are indeed radical when it comes to implementation in PdM models [22]. AI-driven predictive maintenance (AI-PdM) utilises information from large-scale sensor data, including vibration, temperature, acoustic emissions, and electrical variances, to identify common wear or failure radiographs [23,24]. This is a more accurate solution compared to the traditional methods because it identifies the minor anomalies that otherwise could not have been identified [25,26]. Convolutional neural networks and Long Short-Term Memory (LSTM) networks are deep learning models that can be applied to sensor signals to identify faults in rotating machines and detect anomalies [27,28]. AI algorithms can categorise the health of the machine, anticipate the degradation of its components, and predict the RUL accurately, enabling the scheduling of maintenance at the most opportune time [13,24]. Combining the Internet of Things (IoT) further allows real-time monitoring and automated decision-making [29]. Although these generative models are typically expensive, relying on large, labelled datasets, recent methods, including generative adversarial networks, can alleviate the issue of data scarcity by synthesising samples [21]. Prediction is not the only AI application used in multifault diagnosis based on multisensor data fusion and explainable AI (XAI) systems [30,31]. XAI tools, such as LIME, SHAP, PDP, and ICE, make AI models more interpretable, which in turn makes them more reliable in high-risk areas like component health monitoring. The combination of AI, IoT, and XAI aligns with the principles of Industry 4.0, facilitating the creation of smart manufacturing environments that support intelligent and adaptive maintenance systems [32]. Figure 12.1 represents the overall procedure of a system based on AI-PdM. Sensors record and store data, including temperature, vibration, and pressure. This information is analysed with the help of ML algorithms that find the operational anomalies and forecast possible failures or the RUL of the components. This is a proactive structure that enables the maintenance staff to implement appropriate remedies before any severe failures are experienced [31].

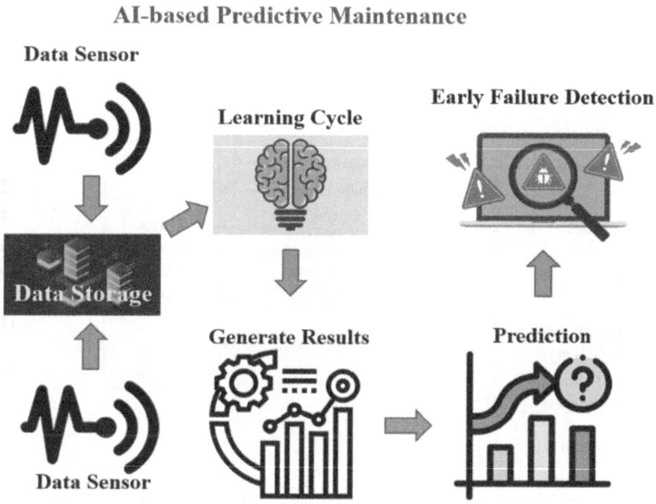

FIGURE 12.1 The framework of an AI-PdM [31]. (CC by 4.0.)

Based on the overall structure of AI-PdM systems, it is necessary to understand the types of typical failures that occur in rotating equipment and are the primary focus of AI-PdM. Such faults can be summarised as bearing faults, gear faults, and rotor faults, all of which influence system reliability and operational stability [33]. Rotor faults may be further categorised as shaft faults (alteration, a bent shaft, shaft crack, rub impact, corrosion, or wear) or structural rotor faults (misalignment (parallel, angular, or combined), unbalance (static, coupled, or dynamic), and looseness (component or structural)). The broken rotor bars and the presence of compound faults, along with various simultaneous problems, are also highlighted in the diagram. Both these problems pose significant challenges to AI-based diagnostic systems [34].

The AI-based tool wear prediction research methodology is illustrated in Figure 12.2. Information from sources such as spindle current, spindle vibration, and cutting forces is gathered, processed, and scaled through feature extraction and scaling [35]. The predictive modelling and performance evaluation are then carried out by implementing ML algorithms, such as AdaBoost Regression.

This case study on AI wear prediction in rotating machinery aims to determine the efficacy, advantages, and drawbacks of incorporating AI and ML technologies into PdM models of vital industrial equipment. It is characterised by the shift from reactive or schedule-oriented approaches to condition-oriented models that utilise advanced analytics to anticipate wear and prevent failures early.

12.4 FUNDAMENTALS OF WEAR IN ROTATING MACHINERY

Wear of rotating machines can be considered a significant challenge in the field of mechanical engineering, which has tribology as its principal underpinning. Sciences and technologies such as friction, wear, and lubrication are concerned with the study

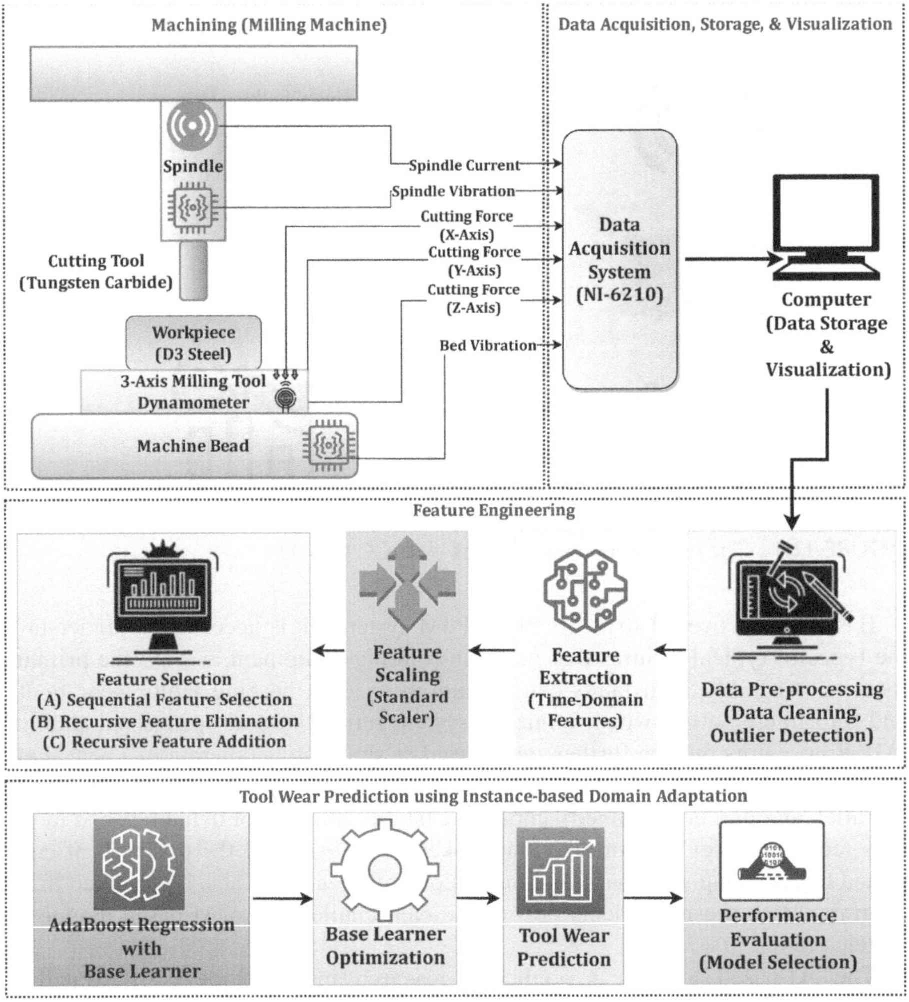

FIGURE 12.2 Overall research methodology for tool wear prediction using domain adaptation approach [35]. (Reprint Permission by Elsevier.)

of the relative motion between surfaces that interact with each other, known as tribology [36–38]. The principles play a significant role in the reliability, safety, and service life of mechanical systems, including those used in power generation, manufacturing, and transportation facilities [39].

12.4.1 Tribological Principles: Friction, Wear, and Lubrication

Friction is the factor that acts against motion between two surfaces that come into contact with one another. Although friction is a desirable attribute of mechanical processes, when applied excessively, it has undesirable effects, including the generation of heat, energy consumption, and rapid deterioration of materials [40]. Lubrication, a

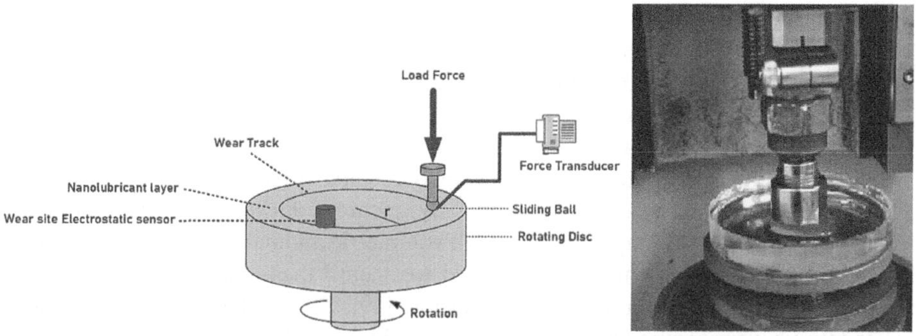

FIGURE 12.3 Schematic representation of the rotating disc tribotester [44]. (Reprint Permission by Elsevier.)

crucial component of tribological systems, reduces friction and wear of interacting surfaces by creating a lubricant film [41–43]. It is a gaseous, liquid, or solid film that isolates the surfaces of contact, reducing direct contact between metals and allowing for the reduction of heat generation, as well as inhibiting enduring wear of the material. The properties and selection of lubricants are fundamental in dictating tribological performance. An example is nano-lubricants, such as acidic functionalised graphene suspended in hydraulic oil, which have shown reduced friction and wear resistance, as well as improved thermal conductivity and viscosity in rotating machines [44]. Figure 12.3 presents a schematic diagram of the rotating disc tribotester used to assess such friction and wear features.

Wear, which is a direct result of friction and lack of lubrication, is a gradual loss or deformation of a material on a solid surface as a result of mechanical interaction [45]. The concept of wearing has been researched since Leonardo da Vinci's initial article, which also explored the connection between friction and lubrication on surfaces [46]. The findings on modern tribology are further developed with the help of materials science, surface engineering, and computational modelling. The mechanism of wear is complicated, and the interplay of mechanical, chemical, and environmental conditions controls it. Wear behaviour is affected by surface roughness, pressure at the contact point, sliding velocity, temperature, and the availability of lubricants. Using inadequate or incompatible materials as lubricants can lead to early wear and ultimately reduce the life of the parts, as well as the serviceability of the system. Thus, an effective tribological design of rotating machinery, such as material selection, surface finish, mode of lubrication, or loading conditions, is instrumental in improving the working conditions of the rotating equipment and increasing its life cycle [10].

12.4.2 Common Wear Mechanisms in Turbines, Pumps, and Bearings

Rotating machinery components, such as turbines, pumps, and bearings, are highly susceptible to various mechanisms of wear due to their constant exposure to changing and often hostile industrial environments. Abrasive wear is the most widespread occurrence type, where a rough surface is in contact with a much softer one or a hard particle is sandwiched between two surfaces, particularly when it is removed by the softer surface [47]. Abrasive wear is a significant concern in wear systems that interact with slurries, such as slurry pumps and pipes, where the solid portions continually collide with internal system components, like impellers, casings, and seals, leading to erosion, wear, and decreased efficiency over time [48]. The other common mechanism is referred to as adhesive wear, which is shaped as a consequence of localised joining or micro-welding among linking surfaces and ensuing fragmentation and transfer of material between surfaces. This is especially apparent during the initial running-in phases of the machine component, in which volumetric anomalies in the form of asperities come into contact with each other and wear until these asperities are smoothed out, achieving a steady and consistent wear rate [49]. Contact fatigue is the term used when fatigue wear occurs under cyclic stress conditions in both surface and subsurface layers, whereby a stress causes the formation and growth of microcracks, eventually resulting in the dislocation of the material. Contact fatigue may affect gears, especially in situations where changes in loads, high temperatures, and variations in lubricant consistency contribute to fatigue [50]. The other major issue affecting rotating equipment is erosive wear, which is caused by the continuous collisions of solid or liquid particles against a surface that wears away the material through micro-cutting and deformation [48]. The mechanism is commonly found in hydraulic turbines and other fluid-moving systems that require high flow velocities for operation. Moreover, corrosion wear is caused by chemical or electro-chemical reactions between a metallic surface and its surroundings, frequently enhanced by mechanical wear at the same time or improper lubrication [47]. The combined effects of corrosion and abrasion often result in severe damage to parts that are subjected to moisture, oxygen, or reactive chemicals. Sliding and rotating parts of machinery (pistons and cylinder blocks) in axial piston-type pumps are highly efficient in transforming rotational energy into hydraulic energy. Still, they severely restrict the total lifetime of the machinery [51]. Equally, wear in wind turbine gearbox planetary bearing systems can significantly influence turbine reliability and energy efficiency when the load and temperature change over time [52]. The bearings used in journaled wind turbines are particularly susceptible to accelerated abrasive wear in dynamic and low-speed environments, and the condition is exacerbated by the failure of lubrication or the application of maximum mechanical loads [53]. Although materials, such as ultra-high molecular weight polyethene used in artificial joints, are known to have excellent wear resistance, they can change their tribological wear behaviour and processes during thermo-oxidative ageing, which modifies the surface morphology and hardness of the material [54]. Figure 12.4 presents a schematic diagram of the testing machine used in these tribological tests.

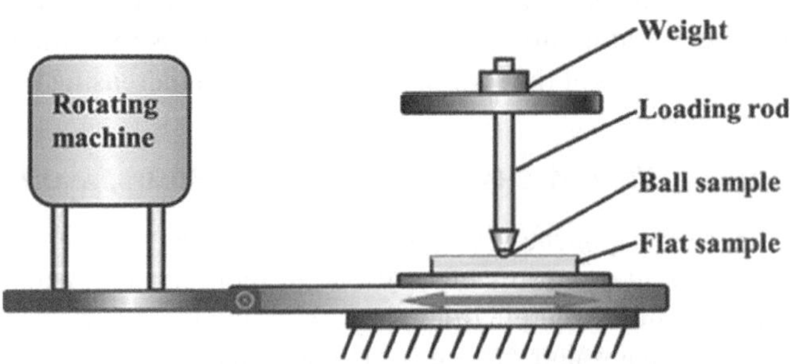

FIGURE 12.4 Schematic diagram of the testing machine [54]. (CC BY 4.0.)

12.5 MEASUREMENT PARAMETERS: VIBRATION, TEMPERATURE, AND LUBRICANT DEGRADATION

Continuous monitoring of critical operating parameters, including vibration, temperature, and the condition of the lubricant, is crucial in predicting and maintaining the health of rotating machinery. Of these, vibration analysis is regarded as the foundation of condition-based monitoring for rotating systems [55]. The differences in vibration patterns, particularly the appearance of high-frequency components, tend to indicate faults in bearings, imbalance, misalignment, and lubrication problems at an earlier stage [56,57]. ML algorithms are now more active in terms of utilising vibration and orbital signal elements to recognise these types of anomalies, allowing for the diagnosis of mechanical wear and tear as early as possible. Additionally, mechanical seals typically employed in rotating shafts can be evaluated for wear and performance loss by estimating vibration acceleration measurements taken across the seal faces [58]. By using such methods, micro-vibrations can be detected in relation to wear on surfaces or the collapse of lubrication films, which are valid indicators of impending failure.

Temperature monitoring is vital in identifying wear and determining the condition of machinery. High temperatures can be one of the signs of increased friction levels, a lack of lubrication, or an impending component failure [52]. High temperatures result in thin layers of lubricant and oxidation of gear and bearing mechanisms, which promote pitting and scuffing of the surfaces, thereby increasing wear and fatigue damage [50]. Temperature stabilisation values will give necessary feedback on real-time health parameters. In combination with the AI analysis, it helps identify temporary heating of operations and permanent temperature fluctuations, which can be regarded as indicative of mechanical problems. The rate at which the lubricant is degraded is yet another critical parameter to consider in analysing tribological performance and the integrity of any given machinery [43]. Lubricants can be damaged due to oxidation, heat, or contamination, which directly affect the viscosity, the film that prevents metal-to-metal contact, and contact time. Indicators of lubricant condition, such as wear debris analysis, viscosity examinations, and oil oxidation analysis, may offer useful information on the wear and well-being of components and the system as a whole [59]. As an illustration, the presence of abrasive particles or high amounts of moisture in lubricants can significantly contribute to surface fatigue wear in journal bearings and gears. This is enabled by sophisticated oil analysis, which provides evidence for the formation of PdM structures through the identification of wear metals and additives used to inform the condition of vital components. Quality assessment of lubricants used in the aviation industry has expanded to include not only physical and chemical indicators but also tribological and performance parameters as the extreme working temperatures in the industry significantly impact safety and reliability [60].

12.6 DATA CHARACTERISTICS AND COMPLEXITY IN TRIBOLOGICAL SYSTEMS

The substantial use of physical, chemical, and mechanical interacting factors in tribology systems makes them complex and multifaceted systems [61]. The objective values gathered by the rotating machinery through various sensors share common undesirable characteristics that complicate the analysis and modelling process. High dimensionality is one of those difficulties because sensor networks can record numerous simultaneous measurements, including vibration, temperature, pressure, acoustic emissions, and motor current [62]. The method is multidimensional and requires more sophisticated processing methods to extract valuable features and distinguish between patterns in terms of wear development. The other distinguishing feature is nonlinearity. There is seldom a direct linear relationship between tribological behaviours and operational parameters (speed, load, and lubrication conditions); therefore, the traditional methods of analysis or the application of statistical tools for modelling system behaviour are not always straightforward [63]. Even a slight variation in the operating conditions of a tribological interface can lead to a sudden shift in response, and the transition between friction regimes is erratic. This complexity is also facilitated by the variability, where rotating machinery is a variable-rate machine with varying loads and environmental conditions, and wear mechanisms and wear rates vary. [64]. As a result, it becomes difficult to set specific limits or general wear indicators that characterise wear use because similar components may exhibit different wear marks under changed working conditions [65].

The interpretation of data is also complicated by the fact that parameters are interdependent. An example is that friction is greater, resulting in high temperatures that accelerate the oxidation of the lubricant and decrease its viscosity, thus increasing wear rate [9]. Such cyclical dependence on having parameters in support of analytical frameworks that can capture dynamics coupled together is required. Moreover, early fault signatures tend to be overwhelmed by the presence of noise and uncertainty in sensor data due to variations in environmental conditions, signal distortion, and equipment-related variability. Consequently, the use of sound diagnostics and wear prediction depends on solid data preprocessing and filtering as well as fusion approaches. Triboinformatics, a new branch of AI and ML that integrates tribology, is one possible solution for revealing unseen data patterns and trends in significant, multidimensional friction and wear data [66]. With the help of triboinformatics, it is possible to model sufficient surface interactions with data, allowing for the prediction of wear in different working conditions. The tribological and material parameters can be incorporated into the ML algorithm, and wear rates are related to material properties, including hardness and yield strength, as well as acquisition methods. For example, data-driven models have been able to predict the tribological behaviour and wear mechanisms of PTFE composites at various loads and sliding velocities [67]. Similarly, surrogate models based on neural networks have been developed to predict the wear volume of journal bearings operating under start–stop conditions, using data collected from mixed electrohydrodynamic lubrication and wear simulations [68]. These achievements demonstrate the capabilities of AI and ML methods to handle, match, and analyse sophisticated tribological data, making PdM an active and exact decision-making instrument [69]. With the help of intelligent data analytics, engineers will be able to learn more about wear changes, optimise lubrication plans, and predict the RUL of components more accurately.

12.7 LUBRICATION, WEAR, AND MATERIAL OPTIMISATION WITH AI

Lubrication systems play a crucial role in maintaining the performance and service life of mechanical equipment by reducing friction and wear [70]. AI is an effective tool in predicting and optimising AI parameters, such as oil consumption rate, viscosity variability, and film stability. Zhou et al. [71] trained a hybrid AI system based on the Kernel Principal Component Analysis, LSTM networks, and ensemble learners (LightGBM and CatBoost) using a meta-learners framework to predict the consumption rate of aero-engine lubricating oil. Such dynamic and static combinations of features used for extraction make this integration useful in both the capture of complex lubrication behaviour and effective PdM. AI can also be used in the design and choice of lubricants. Experiments with silicone rubber at various lubrication conditions, including water, engine oil, dimethyl silicone oil, and dimethyl silicone oil–graphene oxide, demonstrate the applicability of data-driven modelling in optimising lubricant formulations [72]. AI has the potential to predict the lubricant–surface interactions and the compositions that increase performance and minimise friction at different loads and temperatures. In the field of materials research, AI can be utilised to design alloys, coatings, and composite structures, predicting their material properties based on their chemical composition and microstructure [73]. In additive fabrication, AI models are used to optimise mechanical strength, toughness, and weight efficiency over complex contours [74,75]. Miret et al. [76] presented a closed-loop AI-guided materials development cycle that entails design, automated synthesis, and characterisation, with the feedback information improving future developments. Figure 12.5 illustrates how AI can be utilised to integrate design, synthesis, and analysis, thereby accelerating the process of identifying materials with customised tribological characteristics.

They also use AI in the selection of materials and their sustainability. The ML algorithms consider strength, cost, and environmental impact to suggest the most suitable materials for use in reconstruction and manufacturing industries [77].

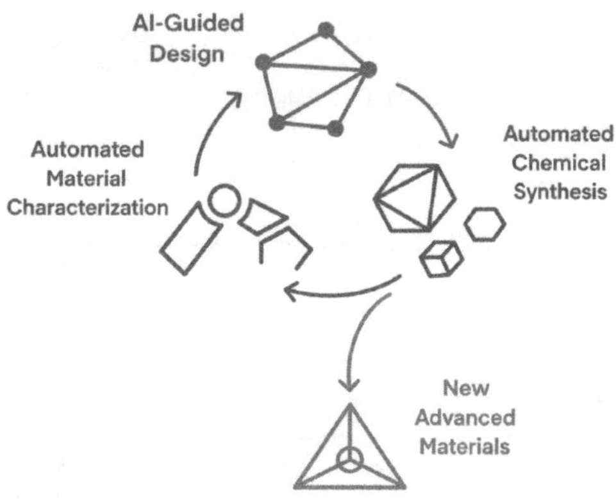

FIGURE 12.5 AI-guided cyclical process for materials development [76]. (CC by 4.0.)

Neural networks are also applied to the design of friction materials in the automotive industry to create brake system friction materials that correlate composition and performance, thereby enhancing wear resistance and stability [78]. AI-based methods enhance lubrication, wear prediction, and material design by combining sensor information, predictive modelling, and feedback learning. These approaches lead to more innovative maintenance planning, lubrication optimisation, and material discovery, resulting in a higher level of performance, reduced energy loss, and sustainable development of industries.

12.8 CASE STUDIES: AI-PDM AND WEAR PREDICTION IN TRIBOLOGY

The AI-PdM has revolutionised tribological systems in the industrial sector through the improvement of accuracy, minimisation of expenditure, and extension of the life of vital parts. Several representative studies demonstrate the effectiveness of AI applications in real-world maintenance contexts.

Zhuang et al. [79] presented an advanced digital twin-driven tool wear monitoring and prediction framework for the CNC turning of Inconel 718, aimed at integrating real-time data analytics and AI for PdM in machining tribology. The system configuration, shown in Figure 12.6, illustrates the overall architecture of the digital twin, which establishes a two-way data flow between the physical machining setup and its virtual model. The digital twin continuously receives and processes signals from a dynamometer operating at 30 kHz to record cutting forces and from a long-wave infrared thermal camera with a 406×240-pixel resolution to capture the temperature distribution at the tool–chip interface. These high-frequency, multimodal data streams formed the sensory foundation for characterising tool wear progression in real time.

The detailed tribological behaviour during machining was analysed using temperature-field evolution, as shown in Figure 12.7. The LWIR thermal images

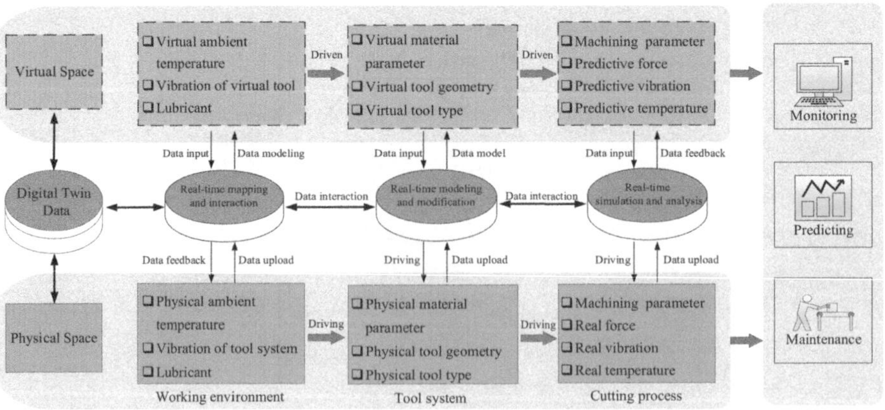

FIGURE 12.6 Digital twin-driven architecture for tool wear monitoring and prediction in CNC turning [79]. (CC by 4.0.)

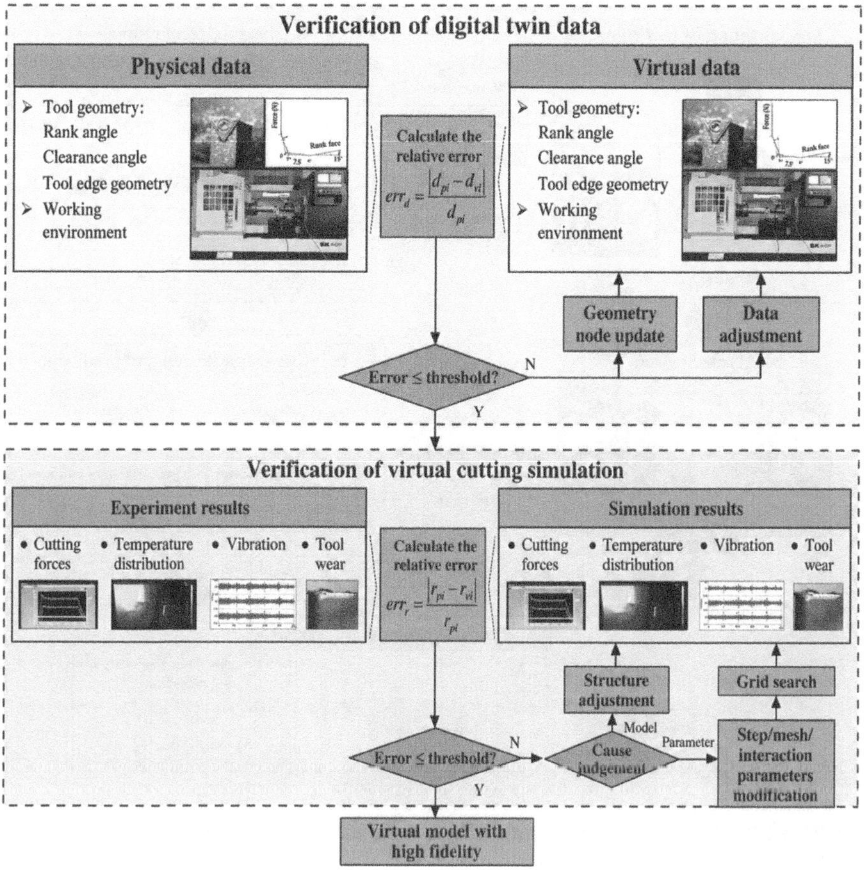

FIGURE 12.7 Thermal field evolution and tribological behaviour during the turning of Inconel 718 [79]. (CC by 4.0.)

clearly captured the temperature rise along the tool rake face and the tool–chip contact zone as wear developed. The increasing thermal gradient and localised hot spots revealed strong correlations between friction-induced heat generation and acceleration of tool degradation. This demonstrated that temperature data, when integrated with force signals, serve as critical input for accurate wear prediction within the digital twin framework.

To enable intelligent prediction, Zhuang et al. [79] developed a hybrid AI model combining k-nearest neighbours for discrete wear-state classification and support vector machine regression for continuous wear-rate estimation. The model was optimised using particle swarm optimisation, allowing the system to adjust model parameters and improve prediction robustness dynamically. Validation results, presented in Figure 12.8, showed excellent agreement between the predicted and experimentally measured wear curves. The wear prediction accuracy of the AI-enhanced digital twin was 40 times higher than that of standard empirical models, and the dimensional

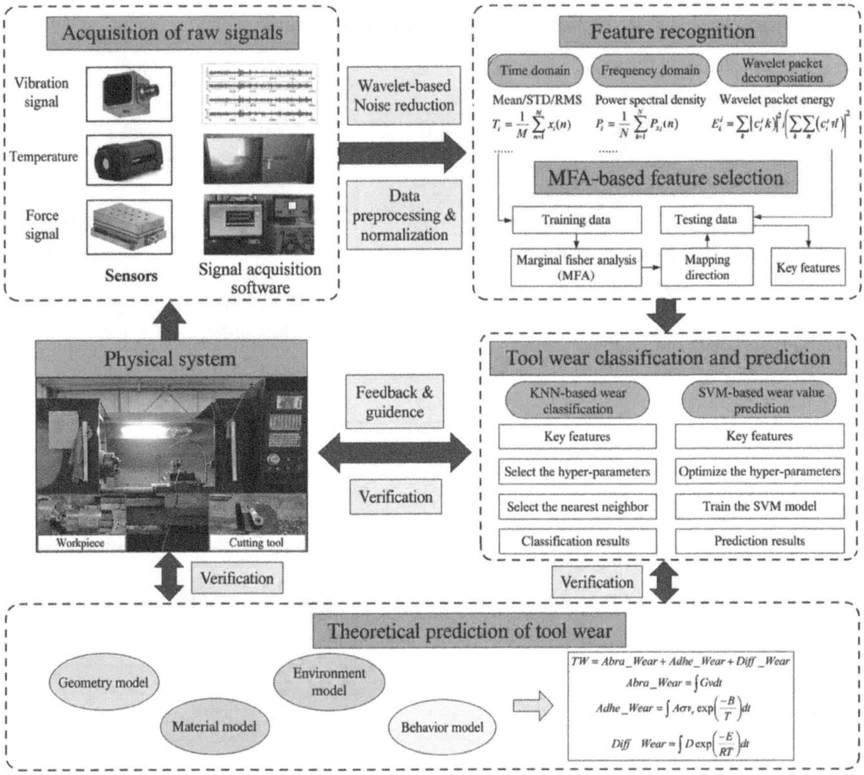

FIGURE 12.8 Comparison between predicted and measured tool wear using the AI-based digital twin model [79]. (CC by 4.0.)

accuracy of machining was kept at ±5 μm. Additionally, providing real-time feedback enabled the CNC controller to automatically correct tool wear, ensuring consistent machining accuracy through a closed-loop cyber-physical interaction.

The study demonstrates how digital twins, based on data, can operate as self-adaptive systems, thereby enabling real-time tribological measurements, predictive patterns, and autonomous decision-making. This study has identified the possibilities of integrating sensor fusion, AI-driven modelling, and steady digital feedback within the context of bridging the divide between conventional wear monitoring and intelligent maintenance in precision manufacturing.

Deshmukh [80] explored the application of AI-PdM in rotating machinery, specifically in rural industrial and agro-processing industries of Osmanabad district, Maharashtra, India. The research addressed the long-standing issues of regular technical breakdowns, unplanned downtime, and high maintenance expenditures that plague small industries and irrigation systems in resource-rich regions. To overcome these challenges, Deshmukh [80] designed a low-cost sensor-based architecture for PdM that combines vibration, temperature, and acoustic sensing with high-level ML algorithms, enabling real-time fault detection. Data from a centrifugal pump, a belt-driven compressor, and a small turbine were recorded using ADXL335 acceleration sensors, DS18B20 temperature sensors, and piezoelectric microphones, which operated under various conditions in the field. These models were coded in Python, utilising Python frameworks such as TensorFlow and Scikit-learn, and were tested and validated using five-fold cross-validation.

The relative usefulness of the implemented models is summarised in Figure 12.9, which shows the accuracy of fault detection in machines of various types. A CNN achieved the highest accuracy of 99.89% in detecting defects on impeller blades in centrifugal pumps, proving to be more effective with vibration-signal characteristics. The LSTM network performed well, achieving an accuracy of 98.5% in detecting turbine shaft anomalies based on vibration, as it effectively modelled the time correlation of sequential data. The Isolation Forest and hybrid Temporal Convolutional Network Autoencoder models also showed promising results, with accuracies of 95.2% and 98.7%, respectively, in identifying compressor pressure outliers and multisensor faults.

These results have established that AI-PdM systems do not simply work in a hostile field of operation, which includes noisy and rural conditions; they have proven to be not only technically achievable but also highly efficient in identifying mechanical faults at the early stages of their occurrence. The application of the models resulted in a 71% decrease in machine downtime and a 30% decline in monthly maintenance, highlighting the practical operational role of AI-based maintenance plans in low-infrastructure environments. This chapter has also highlighted the significance of XAI tools, such as fault-heat map visualisations and interpretable dashboards, in building trust between technicians and plant managers.

Overall, this study suggests that AI-PdM can have a significant impact on the tribological stability, productivity, and dependability of rotating equipment used in the rural industrial sector when adapted locally and designed in accordance with local

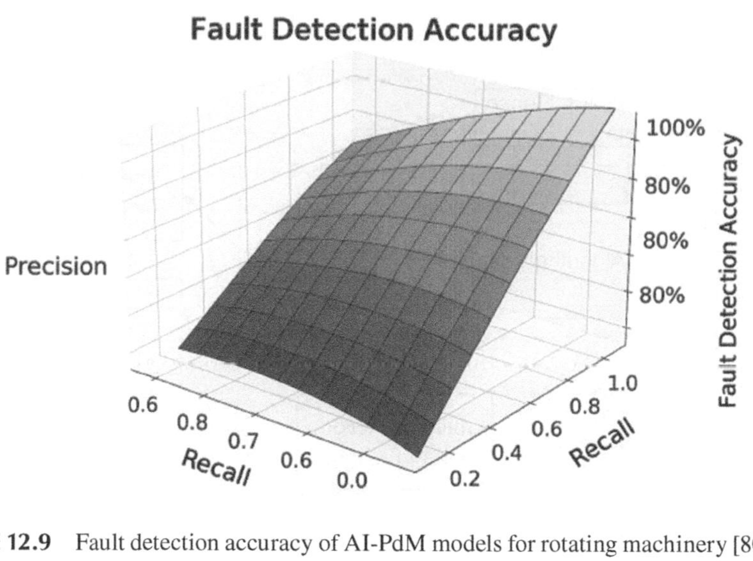

FIGURE 12.9 Fault detection accuracy of AI-PdM models for rotating machinery [80]. (CC by 4.0.)

requirements. The framework will close the technological divide between urban smart factories and rural industrial clusters, driving the participation of the poor in Industry 4.0 by utilising low-cost sensors and open-source AI platforms, which can serve as a foundational base in resource-intensive areas.

Almuraia et al. [81] introduced an AI-based maintenance optimisation approach for natural gas liquid (NGL) pumps used in the oil and gas industry, in response to the constraints of traditional time-based maintenance scheduling and reactive repair methodologies. Their strategy was to minimise operational downtimes and maintenance expenses and enhance the dependability and durability of vital rotating machines that were experiencing high tribological and thermal loads. The general layout of the suggested system, shown in Figure 12.10, represents the AI-supported optimisation model of NGL pump maintenance. The system combines real-time sensor data (vibration, leakage pressure, and temperature) with prior reliability data, as well as diagnostic algorithms based on AI technology. Through constant observation, the system determines the fitness of significant subsystems related to tribological behaviour, such as bearings, mechanical seals, and couplings. The mean time between failure and RUL of each component is determined, and predictive and adaptive maintenance are, therefore, scheduled by calculating them using a reliability analysis module. The digital tool, with the assistance of AI, is connected to the physical state of the pump, providing proactive maintenance options and serving as a functional domain-specific digital twin structure.

The dynamic optimisation process is represented as a step-wise process, as indicated in Figure 12.11. The dynamic optimisation flow diagram illustrates how reliability testing, sensor-based condition assessment, and maintenance cost analysis can be integrated into a single decision-making cycle. The model will apply fuzzy logic inference and multiobjective optimisation algorithms to determine the most appropriate time and the grouping rule for maintenance interventions. This balance of cost, equipment availability, and component wear severity is ensured by

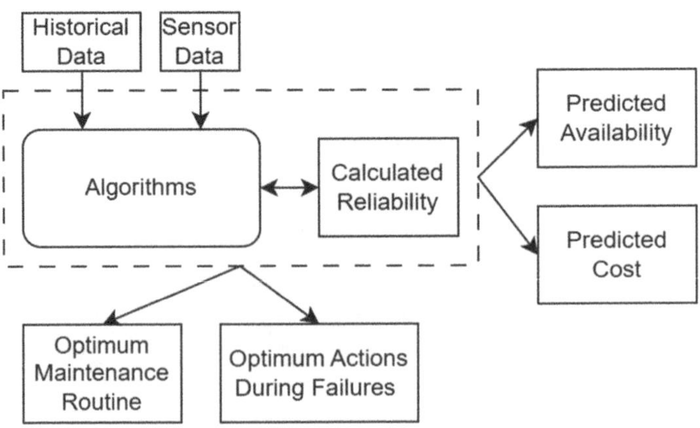

FIGURE 12.10 AI-supported maintenance optimisation model for NGL pumps [81]. (CC by 4.0.)

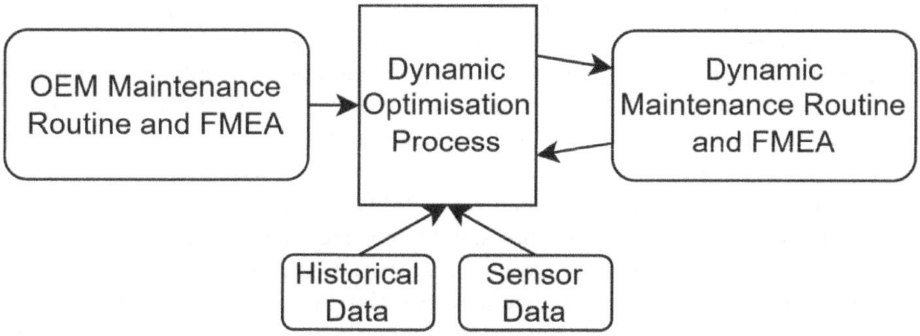

FIGURE 12.11 Dynamic optimisation flow diagram for AI-based maintenance scheduling [81]. (CC by 4.0.)

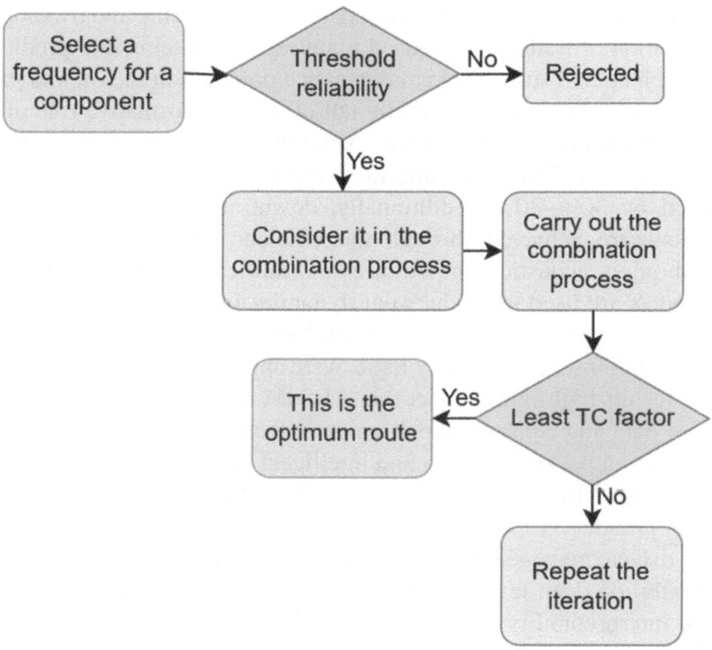

FIGURE 12.12 Optimisation logic flow diagram for intelligent maintenance decision-making [81]. (CC by 4.0.)

the methodology, leading to data-driven scheduling that is more effective than traditional maintenance cycles.

The sequence of optimisation decision-making as shown in Figure 12.12 exemplifies the optimisation logic flow diagram through which AI modules continuously process condition indicators and take maintenance measures when the degradation thresholds are attained. Data on pump G221 of one of the NGL processing plants were used to validate the predictive ability of the system, with the optimised maintenance schedule resulting in an 80% reduction in total maintenance expenditure. This led to a considerable reduction in unscheduled downtime compared to other, more traditional methods. Additionally, the successful prediction of wear-induced degradation of seals and bearings by the framework also prevented disastrous events of failure, which underscores its usefulness in tribological maintenance.

A survey of 15 industry professionals assessing the operational reliability and transparency of the system confirmed the validity of the tool's recommendations, with more than 90% of respondents stating that the tool improved maintenance decision-making. In tribological terms, the AI-based optimisation scheme provides a connection between wear sensing, reliability estimation, and cost-effective maintenance schemes (innovative and adaptive maintenance ecosystems) of high-resistance rotating gears.

This study demonstrates how a system architecture is developed into dynamic optimisation and decision logic. It offers an overall insight into the ways AI can make condition-based, reliability-centred maintenance. This chapter demonstrates that combining AI with the principles of reliability engineering and tribology enables the transformation of traditionally based maintenance systems into predictive, economically efficient, and sustainable maintenance practices in the oil and gas sector.

In these studies, the themes generally tell of the transformative role of AI-PdM. The results of all systems showed a significant increase in accuracy, reliability, and operational efficiency. Fault detection rates exceeded 95%, and maintenance costs were reduced by 30%–80%. Additionally, downtime was significantly reduced. These results were achieved through the efficient combination of sensor data, including vibration, acoustic, temperature, and lubrication data, within AI models. Complex models are used to model wear dynamics in data-rich conditions, such as those encountered during machining with CNC. Reliability and ease of use: The interpretable solutions, such as fuzzy logic, were applied in resource-constrained settings. These results highlight the fact that AI-PdM can be successful in cases where the quality of data and context adaptation are effective, as well as in cases of a successful cooperation between human and intelligent systems.

Together, these case studies indicate that AI-PdM represents a paradigm shift in tribology practice, replacing reactive or periodic interventions with more information-driven management of assets. The use of digital twins offers accuracy and flexibility, deep learning provides practical predictive features, and fuzzy logic offers interpretability and reliability. These technologies, when combined, help develop sustainable, cost-effective, and reliable maintenance systems for both high-tech industry establishments and rural infrastructure. A comparative summary of the case studies on industrial applications, the AI methods used, and their characterised performance enhancements is presented in Table 12.1.

12.9 CONCLUSION

This chapter discussed the use of AI in wear prediction and PdM of rotating machines, combining the concepts of tribology with multisensor data and ML. The nonlinear and time-varying behaviour of wear in turbines, pumps, and bearings is beyond the scope of conventional empirical techniques; however, AI techniques are suitable for simulating

TABLE 12.1
Comparative Summary of AI-PdM Case Studies in Tribology

Study	Application Area	AI Methods Used	Sensor Modalities	Key Performance Outcomes
Zhuang et al. [79]	CNC turning process (Inconel 718)	K-nearest neighbours, support vector machine regression, and particle swarm optimisation	Dynamometer (30 kHz) and thermal camera (406 × 240 LWIR)	40% improvement in wear prediction accuracy; maintained machining precision within ±5 μm
Deshmukh [80]	Rotating machinery in rural industries (pumps, compressors, and turbines)	CNN, LSTM, isolation forest, and hybrid TCN-Autoencoder	Vibration (ADXL335), temperature (DS18B20), and acoustic sensors	99.9% fault detection accuracy; 71% downtime reduction; 30% cost reduction
Almuraia et al. [81]	Natural gas liquid (NGL) pumps in the oil and gas sector	Fuzzy logic, reliability modelling, and multiobjective AI optimisation	Vibration, pressure, temperature, and leakage sensors	80% maintenance cost reduction; major downtime reduction; 90% expert approval rating

the multifactorial relationships between friction, temperature, and lubrication events. The case studies have shown that AI-based solutions, such as digital twins, hybrid neural networks, and fuzzy logic systems, become more precise with maintenance, minimise downtime, and are more reliable at the same time. By combining vibration, acoustic, and lubricant data, AI models can predict RUL with high sensitivity and flexibility in response to actual mechanical situations in the industry. Outside of maintenance, AI also helps optimise lubrication, surface engineering, and material design by predicting performance and supporting sustainable engineering practices. All these developments together herald a change of paradigm, where proactive and data-driven maintenance is undertaken, which reduces costs, environmental footprint, and extends equipment life. The further development of the hybrid AI framework and digital twin systems will offer the newest generation of intelligent tribological design and operation.

REFERENCES

1. Shah, R., Jaramillo, R., Thomas, G., Rayhan, T., Hossain, N., Kchaou, M., Profito, F. J., & Rosenkranz, A. (2025). Artificial Intelligence and Machine Learning in Tribology: Selected Case Studies and Overall Potential. *Advanced Engineering Materials*, *27*, 2401944.
2. Gawde, S., Patil, S., Kumar, S., Kamat, P., Kotecha, K., & Alfarhood, S. (2024). Explainable Predictive Maintenance of Rotating Machines Using LIME, SHAP, PDP, ICE. *IEEE Access*, *12*, 29345–29361.

3. Xu, Z., Guo, Y., & Saleh, J. H. (2021). *Remaining Useful Life Prediction with Uncertainty Quantification: Development of a Highly Accurate Model for Rotating Machinery* (Version 1). arXiv:2109.11579.
4. Apeiranthitis, S., Zacharia, P., Chatzopoulos, A., & Papoutsidakis, M. (2024). Predictive Maintenance of Machinery with Rotating Parts Using Convolutional Neural Networks. *Electronics, 13*(2), 460.
5. Erhueh, O. V., Nwakile, C., Akano, O. A., Aderamo, A. T., & Hanson, E. (2024). Advanced Maintenance Strategies for Energy Infrastructure: Lessons for Optimising Rotating Machinery. *Global Journal of Research in Science and Technology, 2*(2), 065–093.
6. Li, G., Yu, L., He, J., Zhang, W., Liu, S., & Wei, J. (2024). A Remaining Useful Life Prediction Method for Rotating Machinery Based on Interactive BiLSTM. *Measurement Science and Technology, 36*(1), 016128.
7. Bayer, R. G. (2012). Tribological and Wear Testing. In: Ahmed, S., & Darke, V. S. (Eds) *Characterisation of Materials* (pp. 1–16). Hoboken, NJ: Wiley.
8. Kayode, J. F., Lawal, S. L., & Afolalu, S. A. (2023). An Overview of Tribology and Its Industrial Applications. *2023 International Conference on Science, Engineering and Business for Sustainable Development Goals.* (SEB-SDG), Omu-Aran, Nigeria, pp. 1–15.
9. Jeyaprakash, N., & Yang, C.-H. (2021). Friction, Lubrication, and Wear. In: Kukshal, V., Singh, T., & Patnaik, A. (Eds) *Tribology in Materials and Manufacturing - Wear, Friction and Lubrication.* London: IntechOpen.
10. Zhang, G. X., Li, M., & Cao, J. W. (2014). The Tribology Design in Transport Machinery Design. *Applied Mechanics and Materials, 602–605*, 148–150.
11. Feng, K., Ji, J. C., Ni, Q., & Beer, M. (2023). A Review of Vibration-based Gear Wear Monitoring and Prediction Techniques. *Mechanical Systems and Signal Processing, 182*, 109605.
12. Xu, S., Zhang, Y., Zhou, J., Chen, Z., Fu, H., Liu, B., & Ji, J. (2025). Tribology of Polymer Gears: Friction Coefficient and Wear. *Polymer Gears, 4*, 495–514.
13. Kamel, H. (2022). Artificial Intelligence for Predictive Maintenance. *Journal of Physics: Conference Series, 2299*(1), 012001.
14. Fabić, M. (2025). Application of AI-based Predictive Maintenance for Industrial Processes. *Tehnički Glasnik, 19*(Si1), 154–158.
15. Shah, J. P. (2024). Integration of AI-based Predictive Maintenance for Energy-Efficient Mechanical Systems. *World Journal of Advanced Engineering Technology and Sciences, 11*(2), 664–673.
16. Maurya, M., Panigrahi, I., Dash, D., & Malla, C. (2023). Intelligent Fault Diagnostic System for Rotating Machinery based on IoT with Cloud Computing and Artificial Intelligence Techniques: A Review. *Soft Computing, 28*(1), 477–494.
17. Kalifa, M., Starr, A., & Khan, M. (2025). Current Research and Challenges in Modelling Wear, Friction, and Noise in Mechanical Contacts. *Proceedings of the Institution of Mechanical Engineers, Part J: Journal of Engineering Tribology, 239*(8), 985–1005.
18. Georgiou, E. P., Drees, D., Van der Donck, T., Economou, S., & Celis, J.-P. (2021). Parallel Wear Tests: The Need for Statistical Analysis in Tribology. In: Ponthiaux, P., Celis, J.-P. (Eds) *Mechanical and Electro-Chemical Interactions Under Tribocorrosion* (pp. 83–109). Amsterdam, Netherlands: Elsevier.
19. Saini, K., Dhami, S. S., & Vanraj. (2022). Predictive Monitoring of Incipient Faults in Rotating Machinery: A Systematic Review from Data Acquisition to Artificial Intelligence. *Archives of Computational Methods in Engineering, 29*(6), 4005–4026.
20. Gawde, S., Patil, S., Kumar, S., Kamat, P., & Kotecha, K. (2024). An Explainable Predictive Maintenance Strategy for Multi-Fault Diagnosis of Rotating Machines Using Multi-Sensor Data Fusion. *Decision Analytics Journal, 10*, 100425.

21. Ding, Y., Ma, L., Ma, J., Wang, C., & Lu, C. (2019). A Generative Adversarial Network-Based Intelligent Fault Diagnosis Method for Rotating Machinery Under Small Sample Size Conditions. *IEEE Access, 7*, 149736–149749.
22. Zeb, S., & Lodhi, S. K. (2025). AI for Predictive Maintenance: Reducing Downtime and Enhancing Efficiency. *Enrichment: Journal of Multidisciplinary Research and Development, 3*(1), 135–150.
23. Zhou, J., Yang, J., Xiang, S., & Qin, Y. (2025). Remaining Useful Life Prediction Methodologies With Health Indicator Dependence for Rotating Machinery: A Comprehensive Review. *IEEE Transactions on Instrumentation and Measurement, 74*, 1–19.
24. Gawde, S., Patil, S., Kumar, S., Kamat, P., Kotecha, K., & Abraham, A. (2023). Multi-Fault Diagnosis of Industrial Rotating Machines using Data-Driven Approach : A Review of Two Decades of Research. *Engineering Applications of Artificial Intelligence, 123*, 106139.
25. Chambi, N., Sanga, C., Ortiz, J., Sanga, A., Sanga, P., Manrique, R., & Lu-Chang-Say, J. (2025). Predictive Maintenance in Underground Mining Equipment Using Artificial Intelligence. *Eng, 6*(10), 261.
26. Mian, T., Choudhary, A., Fatima, S., & Panigrahi, B. K. (2023). Artificial Intelligence of Things based Approach for Anomaly Detection in Rotating Machines. *Computers and Electrical Engineering, 109*, 108760.
27. Azim, M., Afzal, M., Ariz, M., & Samad, A. (2023). Predictive Maintenance in Rotating Machinery Using Deep Learning Techniques. *Journal of Fluid Mechanics and Mechanical Design, 5*(3), 1–9.
28. Kim, D.-G., & Choi, J.-Y. (2021). Optimisation of Design Parameters in LSTM Model for Predictive Maintenance. *Applied Sciences, 11*(14), 6450.
29. Boughardini, I. E., Mechkouri, M. H., & Reklaoui, K. (2023). A Predictive Maintenance System Based on Vibration Analysis for Rotating Machinery Using Wireless Sensor Network (WSN). In: Kacprzyk, J., Ezziyyani, M., & Balas, V.E. (Eds) *International Conference on Advanced Intelligent Systems for Sustainable Development. AI2SD 2022. Lecture Notes in Networks and Systems*. Cham: Springer.
30. Krishnamurthy, V., Nezafati, K., Stayton, E., & Singh, V. (2020). Explainable AI Framework for Imaging-Based Predictive Maintenance for Automotive Applications and Beyond. *Data-Enabled Discovery and Applications, 4*(1), 7.
31. Hosseinzadeh, A., Frank Chen, F., Shahin, M., & Bouzary, H. (2023). A Predictive Maintenance Approach in Manufacturing Systems via AI-based Early Failure Detection. *Manufacturing Letters, 35*, 1179–1186.
32. Shetty, R., & Jadhav, P. (2023). Predictive Maintenance in Industrial Automation and Smart Manufacturing Applications using Artificial Intelligence. *International Academic Journal of Science and Engineering, 10*(3), 30–34.
33. Nath, A. G., Udmale, S. S., & Singh, S. K. (2020). Role of Artificial Intelligence in Rotor Fault Diagnosis: A Comprehensive Review. *Artificial Intelligence Review, 54*(4), 2609–2668.
34. Zhang, X., Ma, R., Zhao, Y., Wang, C., Zhao, Z., & Chen, X. (2024). Differentiable Sampling based Efficient Architecture Search for Automatic Fault Diagnosis. *Engineering Applications of Artificial Intelligence, 127*, 107214.
35. Warke, V., Kumar, S., Bongale, A., & Kotecha, K. (2024). Robust Tool Wear Prediction using Multi-Sensor Fusion and Time-Domain Features for the Milling Process using Instance-based Domain Adaptation. *Knowledge-Based Systems, 288*, 111454.
36. Yust, C. S. (1985). Tribology and wear. *International Metals Reviews, 30*(1), 141–154.
37. Uchiyama, Y. (2016). Tribology (The Science and Technology of Friction, Wear and Lubrication) - The Elementary Knowledge to Application, and the Latest Progress. *Nippon Gomu Kyokaishi, 89*(9), 281–286.

38. Jin, C., Wei, W., & Narayan, R. (2020). 10.2 Friction, Lubrication, and Wear. *Biomedical Materials, 365*, 1–17.
39. Sharma, S. C. (2013). Tribology in Machine Components. In: Menezes, P. (Eds) *Tribology for Scientists and Engineers* (pp. 821–879). New York: Springer.
40. Czichos, H. (1983). Tribology: Scope and Future Directions of Friction and Wear Research. *JOM, 35*(9), 18–20.
41. Sahoo, P. (2015). Tribology Measurements. In: Myer, K. (Eds) *Mechanical Engineers' Handbook* (pp. 1–24). New York: Wiley.
42. Vellanki, C., Choudhury, S., Kumar, S., Vimson, G., & Paul, G. (2022). Influence of Lubrication on the Friction and Wear Characteristics of Low Carbon Steel under Sliding Reciprocation Conditions. *IOP Conference Series: Materials Science and Engineering, 1248*(1), 012033.
43. Sukhodolya, A., Akulov, A., Podgorny, A., & Taldykin, D. (2025). Relationship between Lubricant Characteristics and Tribological Properties. *AIP Conference Proceedings, 3347*, 020052.
44. Kader, A., Selvaraj, V., Ramasamy, P., & Senthilkumar, K. (2023). Experimental Investigation on the Thermo-Physical Properties and Tribological Performance of Acidic Functionalised Graphene Dispersed VG-68 Hydraulic Oil-based Nanolubricant. *Diamond and Related Materials, 133*, 109740.
45. Verhoef, P. N. W. (2017). Concepts of Wear Processes from Tribology. In: Mahmut, C. B., Osman, K. S., Michael, R., & Richard, G. (Eds) *Wear of Rock Cutting Tools* (pp. 29–48). London: Routledge.
46. Betancourt-Parra, S. (2021). Leonardo da Vinci's Tribological Intuitions. *Tribology International, 153*, 106664.
47. Abdullayev, H., Huseynzade, E., & Sable, H. (2025). A Comprehensive Review of Wear Mechanisms and Mitigation Strategies for Tribological Systems. *Tribology in Industry, 47*(2), 294–312.
48. Singh, J. (2021). A Review on Mechanisms and Testing of Wear in Slurry Pumps, Pipeline Circuits, and Hydraulic Turbines. *Journal of Tribology, 143*(9), 090801.
49. Lijesh, K. P., Khonsari, M. M., & Kailas, S. V. (2018). On the Integrated Degradation Coefficient for Adhesive Wear: A Thermodynamic Approach. *Wear*, 408–409, 138–150.
50. Amarnath, M., & Lee, S.-K. (2015). Assessment of Surface Contact Fatigue Failure in a Spur Geared System Based on the Tribological and Vibration Parameter Analysis. *Measurement, 76*, 32–44.
51. Schuhler, G., Jourani, A., Bouvier, S., & Perrochat, J.-M. (2017). Multi Technical Analysis of Wear Mechanisms in Axial Piston Pumps. *Journal of Physics: Conference Series, 843*, 012077.
52. Yin, J., Meng, X., & Cheng, S. (2025). Enhancing Wind Turbine Energy Efficiency: Tribo-Dynamics Modeling and Shape Modification. *Renewable and Sustainable Energy Reviews, 208*, 115071.
53. Lehmann, B., Trompetter, P., Guzmán, F. G., & Jacobs, G. (2023). Evaluation of Wear Models for the Wear Calculation of Journal Bearings for Planetary Gears in Wind Turbines. *Lubricants, 11*(9), 364.
54. Zhang, X., Tan, D., Tang, Q., Hou, B., Tian, J., & Wei, M. (2024). Impact of Thermo-Oxidative Aging on the Dry Tribological Performance and Wear Mechanisms of UHMWPE/ZrO_2 Friction Pairs. *AIP Advances, 14*(3), 035143.
55. Jablon, L. S., Avila, S. L., Borba, B., Mourão, G. L., Freitas, F. L., & Penz, C. A. (2020). Diagnosis of Rotating Machine Unbalance Using Machine Learning Algorithms on Vibration Orbital Features. *Journal of Vibration and Control, 27*(3–4), 468–476.

56. Sateesh Kumar, P., Amarnath, M., Kamarapu, S. K., Varatharajan, G., Chelladurai, H., Vardhaman, B. A., & Ramkumar, J. (2024). A Study of Surface Fatigue Wear and Influence of Contaminated Lubricant in Journal Bearing System Using Tribological and Vibration Analyses. *Proceedings of the Institution of Mechanical Engineers, Part E: Journal of Process Mechanical Engineering*, *239*(6), 3622–3636.
57. Hemati, A. (2025). Bearing Failure Prevention Through Predictive Lubrication Monitoring Using Vibration Signals and Deep Learning. *Journal of Failure Analysis and Prevention*, *25*(5), 2165–2175.
58. Wang, Q., Song, Y., Li, H., Shu, Y., & Xiao, Y. (2023). Tribological Behavior Characterisation and Fault Detection of Mechanical Seals Based on Face Vibration Acceleration Measurements. *Lubricants*, *11*(10), 430.
59. Radulescu, A. V., Radulescu, I., & Petrescu, F. (2022). Influence of Lubricants Degradation Level over Tribological Properties. *INCAS Bulletin*, *14*(1), 125–136.
60. Mikosianchyk, O., & Ilina, O. (2023). Research of Tribological Characteristics of Modern Aviation Oils. In: Karakoc, T. H., Atipan, S., Dalkiran, A., Ercan, A. H., Kongsamutr, N., & Sripawadkul, V. (Eds) *Sustainable Aviation* (pp. 371–377). Cham: Springer International Publishing.
61. Hasan, M. S., Kordijazi, A., Rohatgi, P. K., & Nosonovsky, M. (2021). Triboinformatic Modeling of Dry Friction and Wear of Aluminum base Alloys Using Machine Learning Algorithms. *Tribology International*, *161*, 107065.
62. Chen, S., Cheng, G., Guo, F., Jia, X., & Wen, X. (2024). Integrating Friction Noise for In Situ Monitoring of Polymer Wear Performance: A Machine Learning Approach in Tribology. *Journal of Tribology*, *147*(6), 061701.
63. Cai, H., Hao, L., & Su, Y. (2018). ISAR Target Recognition Based on Two-dimensional Locality Preserving Projection. *Journal of Physics: Conference Series*, *1060*, 012006.
64. Jeon, H.- G., Cho, D.- H., Yoo, J.- H., & Lee, Y.- Z. (2018). Wear Prediction of Earth-Moving Machinery Joint Bearing via Correlation between Wear Coefficient and Film Parameter: Experimental Study. *Tribology Transactions*, *61*(5), 808–815.
65. Stembalski, M., Jankowski, T., Roszkowski, A., & Skoczynski, W. (2025). Gear Wear Mechanisms, Monitoring Techniques and Their Potential Use in Gear Predictive Maintenance. *Journal of Machine Engineering*, *25*(2), 43–63.
66. Nosonovsky, M., & Aglikov, A. S. (2024). Triboinformatics: Machine Learning Methods for Frictional Instabilities. *Facta Universitatis, Series: Mechanical Engineering*, *22*(3), 423.
67. Wang, Q., Wang, X., Zhang, X., Li, S., & Wang, T. (2023). Tribological Properties Study and Prediction of PTFE Composites based on Experiments and Machine Learning. *Tribology International*, *188*, 108815.
68. König, F., Wirsing, F., Singh, A., & Jacobs, G. (2024). Machine-Learning-Based Wear Prediction in Journal Bearings under Start–Stop Conditions. *Lubricants*, *12*(8), 290.
69. Chattoraj, C. (2021). Tribology and Maintenance of Machinery - A Review. *Indian Journal of Production and Thermal Engineering*, *1*(2), 16–20.
70. Ametov, V. A., Sarkisov, Y. S., & Spirin, E. N. (2004). Effect of Combined Effects on Processes in Tribocouples. *Chemistry and Technology of Fuels and Oils*, *40*(5), 333–341.
71. Zhou, Q., Guo, Y., Xu, K., Chai, B., Li, G., Wang, K., & Dong, Y. (2024). Research on the Prediction Algorithm of Aero Engine Lubricating Oil Consumption based on Multi-Feature Information Fusion. *Applied Intelligence*, *54*(22), 11845–11875.
72. Liu, R., Su, J., Zhang, T., & Ke, L. (2024). Fretting Wear Behaviors of Silicone Rubber under Dry Friction and Different Lubrication Conditions. *Materials*, *17*(11), 2598.
73. Nimal, R. J. G. R., Sangamaeswaran, R., Hariharan, R., Divahar, S. R., Selvi, S. S., & Ebinraj, S. (2025). AI-Driven Material Design and Discovery: Revolutionising the Development of Novel Alloys and Composites. *2025 IEEE 6th International Conference in Robotics and Manufacturing Automation (ROMA)*, pp. 113–118. Bangi, Malaysia.

74. Prakash, J., Nahar, R., Raj, A., Silvester, J., Pandey, S. K., & Patil, P. D. (2025). Advancements in Additive Manufacturing: Optimising Mechanical Properties Through AI-Based Design. *2025 First International Conference on Advances in Computer Science, Electrical, Electronics, and Communication Technologies (CE2CT)*, pp. 899–904. Bhimtal, Nainital, India.
75. Hassan, M., Misra, M., Taylor, G. W., & Mohanty, A. K. (2024). A Review of AI for Optimisation of 3D Printing of Sustainable Polymers and Composites. *Composites Part C: Open Access*, *15*, 100513.
76. Miret, S., Krishnan, N. M. A., Sanchez-Lengeling, B., Skreta, M., Venugopal, V., & Wei, J. N. (2024). Perspective on AI for Accelerated Materials Design at the AI4Mat-2023 Workshop at NeurIPS 2023. *Digital Discovery*, *3*(6), 1081–1085.
77. Xu, S. (2025). AI-Assisted Sustainability Assessment of Building Materials and Its Application in Green Architectural Design. *Journal of Industrial Engineering and Applied Science*, *3*(4), 1–13.
78. Lin, Y. X., Gao, C. H., & Chen, Z. H. (2011). CAD of Friction Material for Automobile Brake Using Artificial Neural Networks. *Applied Mechanics and Materials*, *121–126*, 3832–3836.
79. Zhuang, K., Shi, Z., Sun, Y., Gao, Z., & Wang, L. (2021). Digital Twin-Driven Tool Wear Monitoring and Predicting Method for the Turning Process. *Symmetry*, *13*(8), 1438.
80. Deshmukh, S. (2023). Artificial Intelligence in Predictive Maintenance of Rotating Machinery: A Case Study from Rural India. *Saudi Journal of Engineering and Technology*, *8*(12), 316–322.
81. Almuraia, A., He, F., & Khan, M. (2025). AI-Driven Maintenance Optimisation for Natural Gas Liquid Pumps in the Oil and Gas Industry: A Digital Tool Approach. *Processes*, *13*(5), 1611.

13 Artificial Intelligence in Marine Tribology

Applications, Challenges, and Future Directions

Vikas Sharma

13.1 INTRODUCTION

Marine tribology is the science and practice of the way surfaces are moving against each other in moving relative to each other. This is particularly so in marine equipment and parts of ships which include ship engines, propellers, bearings, shafting, and other mechanical systems on board ships [1–3]. It deals with the wear of moving components, those processes that maintain them in a lubricated state, and their interactions during extreme conditions of the sea like saltwater exposure, heavy loads, change of temperature, and biofouling. The study of marine tribology is concentrated on friction and wear of important moving components of ship engines such as cylinder liners and piston rings, crankshafts and connecting rods, and bearings [4].

This is to ensure that engines are good and have a long duration even when subjected to high stress levels. Propeller and rudder tribology involves the study of the means to decrease the friction between the hull of the ship and the surrounding water [5]. This not only increases the efficiency of propulsion but also reduces the wear and corrosion, as well as energy losses. Bearing and shafting system tribology considers how lubrication and materials can be used to reduce wear, prevent seizing, and lengthen the life of parts even when they are in salt water and are under mechanical stress. The objectives of marine tribology include improved energy conservation, reduced material degradation by wear, increased lifespan of component parts, improved operational efficiency, and improved adverse environmental effects of tribological components over their life time [6,7]. This is extremely significant to the safe, effective, and sustainable operation of marine vessels and machinery. Figure 13.1 highlights sustainability benefits, energy efficiency, material durability, performance optimization, environmental impact reduction, corrosion control and safety management.

DOI: 10.1201/9781003624165-13

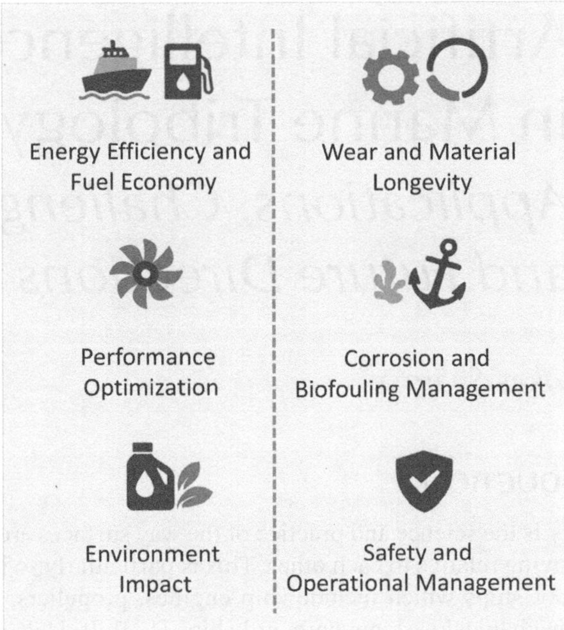

FIGURE 13.1 Importance of tribology in marine machinery.

13.2 MARINE COMPONENTS AND THEIR TRIBOLOGICAL INTERFACES

There are a number of components that constitute marine tribological systems that are in motion relative to each other, experience extreme mechanical loads, and are in a corrosive environment [8–10]. From the unique interfaces of both systems, friction, wear, and lubrication influence performance, efficiency, and dependability. To reduce the amount of energy wasted, for maintenance expenses, and to ensure that valuable maritime equipment can serve as long as possible, these relationships should be understood. Propeller shafts, gears, thrust bearings, and couplings are propulsion systems, which have varying loads and speeds. The lubrication is highly essential at the points of contact of shaft-bushings, gear teeth, and bearing surfaces. It prevents friction and worsening of surfaces. Good tribological management ensures that the transmission of torque is smooth, vibration is minimized, and efficiency of propulsion is enhanced. This reduces fuel consumption and equipment damage and wear.

Components of an engine that are at elevated levels of pressure, temperature, and speed of movement include piston rings, cylinder liners, crankshaft journals, and cam-followers. These components are highly critical in the generation of power, and the performance of these components determines the length of the engine life, the quantity of gasoline consumed, as well as the quantity of pollution. The right lubricants, advanced coating, and non-easy-wearing materials can stabilize the engine performance of a vehicle and reduce mechanical losses.

Journal and rolling-element bearings are also very essential in maintaining radial and axial loads in engines, pumps, turbines, and other auxiliary equipment. They should be lubricated with a good lubricating coating to work well. This reduces friction and heat. Early failure can occur due to bad lubrication or contamination, and this can render the entire maritime system less reliable and safe [11]. Sliding and rotating seals and valves are made to ensure the lubricants, fuels, and other fluids do not leak out under constant mechanical pressures and friction. These components can be damaged by changes in temperature, become disfigured, and disintegrated, thereby rendering the system less stable. Tribological studies can be used to create more durable and less demanding materials and surface finishes. Deck machinery and mooring gear (winches, wire ropes, pulleys, and anchor chains) operate in varied and often deplorable sea conditions. When exposed to saltwater, salt, and fluctuating mechanical stress, things corrode and wear off faster. Use of proper lubricants, coating, and materials that do not wear off easily can facilitate the equipment to be in a good functional state, last long, and be safe when anchoring and hauling cargo. The tribological interface of the performance of every one of these parts influences the reliability, efficiency, and durability of the system. Tribological solutions have to be application-specific due to the variations in geometry, material mixes, and operating regimes [12].

13.3 ADVANTAGES OF TRIBOLOGY IN MARINE MACHINERY

Marine engines, propellers, and gear systems incur a significant amount of fuel in the form of frictional losses, and so energy efficiency and fuel economy are extremely significant to the operation of a ship. A lot of energy can be saved through good lubrication and reduced wear, which reduce the emissions and operating costs. The wear-out of marine parts, such as bearings, piston rings, and propeller shafts, is caused by cavitation, salt water exposure, and high-load cycles. The implementation of materials that do not wear out and improved lubrication strategies also increases the life of parts, reduces maintenance, and saves money, in general, through tribology. Correct tribological management will also enhance the performance of engine, turbine, and propulsion systems [13]. This enables them to operate more efficiently with reduced vibrating and noise, and the ships will be more stable on long journeys or engagements at sea.

Severe maritime environments cause more rapid corrosion and biofouling, accelerating wear and surface degradation. Oil protective tribo-coatings and special lubricants are required to ensure smooth operations in the long run. The reduction of tribological wear reduces the quantity of lubricant applied and prevents leakage of oil through the reduction of friction and wear [14]. This is environmental-friendly and reduces greenhouse gas emissions. In addition, the tribological performance of the naval equipment must be robust so that it is safe and dependable. Failure of such vital components as bearings or gearboxes can lead to disastrous failures. It allows identifying wear at an early stage with the help of advanced tribological experience and predictive maintenance based on AI, ensuring the absence of accidents and continued and safe operation of vessels.

13.4 ENVIRONMENTAL INFLUENCES ON MARINE TRIBOLOGY

Tribological performance is affected by harsh environments that are posed by the maritime environment. Marine tribological systems operate in harsh and varying conditions, and wear and lubrication behavior is significantly influenced by factors such as sea water exposure, pressure, and fluctuations in temperature [15–17]. Saltwater contains a lot of chlorides; hence, it accelerates corrosion, pitting, and tribocorrosion at contact surfaces. This destroys the protective oxide layers and compromises the integrity of the coating as well as the metal. This degradation does not only render structural components weak but also alters the surface chemistry, which augments friction and dissipation. Meanwhile, parts submerged below the waterline like propeller shafts, submersible bearings, and tidal turbine components are highly pressurized hydrostatically. Such pressure affects the formation of lubricant films, increases the strength of squeeze-film effects, and leads to a direct contact with the surface in case of lubrication failure [18,19]. In addition, varying temperatures, such as the cooling capacity of salt and the high intensity heat generated by engines, influence the viscosity of lubricants, accelerate thermal fatigue, and cause wear of materials in shorter durations [20]. All these reduce the efficiency of operation and component lifetime.

Besides environmental stress, marine gear must be able to cope with biofouling, pollution, and dynamic mechanical stresses. Abrasion of surface and alteration of coarseness are brought about by the growth of marine organisms (barnacles, algae, etc.) and fine saltwater particles that accumulate on the surface. This increases friction and consumption of energy. These deposits can also prevent the flow of fluids, hence leading to unequal pressure distributions in the lubrication systems. Besides, the constructions erected on the water and ships are never at constant weights and dynamic forces due to the waves, alteration in the current, and movements in the operation [21,22]. These unpredictable forces lead to unequal contact patterns and stress distribution and, hence, complicate the prediction of wear and manipulation of lubrication. In order to come up with tribological systems to be used in the sea that are durable, reliable, and efficient, the synergistic effects of the environment and operations should be understood and minimized.

13.4.1 Materials and Lubricants Used in Marine Applications

It is necessary to select the appropriate materials and lubricants in order to ensure that marine tribosystems are durable and effective. These systems are highly loaded, temperature varies, and are exposed to saltwater often. The selection of the appropriate materials can help find the optimal wear resistance, corrosion resistance, and mechanical strength [23]. Marine components are frequently made of high-strength steels, while those required in corrosion resistance are typically made of bronze and nickel-based alloys, and those required for light weight and reduced friction are typically made of polymer composites. In order to increase their resistance to cavitation, pitting, and tribocorrosion, other techniques such as advanced ceramics and surface-engineered layers such as hardfacing layer or anticorrosive films are also used. The selection process must not only consider the pressures and environmental exposure that will exist during operation but also ensure that the materials which will be coming into contact with one another are compatible such that galvanic corrosion or adhesive wear does not occur.

The correct selection of lubricants is also of paramount importance because it is the initial line of defense against wear and corrosion and friction. Marine lubricants, such as mineral-based lubricants, greases, or synthetic blends, are made to resist washout by saltwater, emulsification, and heat deterioration. Additives are incorporated to enhance the performance by ensuring that the equipment lasts longer and performs better by stopping rust, wear, and oxidation. Increased interest in environmental sustainability has resulted in application of lubricated bearings using water and lubricants that are ecologically acceptable (EALs) [24,25]. They protect wear and tear and result in fewer damages to the environment in case of a leakage of these products. The selection and maintenance of these lubricants are highly significant to ensure that machines perform efficiently in a wide range of maritime conditions.

A new production process using surface engineering techniques has been increasingly significant in enhancing tribological behavior in marine environments as well as optimization of materials and lubricants. Surface modification by altering the surface properties of a material to render it harder, less slippery, and more resistant to corrosion and wear may be achieved by a variety of methods: including nitriding, physical and chemical vapor deposition (PVD/CVD), and thermal spraying. High-quality tribo-coatings, e.g., diamond-like-carbon (DLC), ceramic or composite, and tribo-coatings, provide customized protection to parts experiencing high levels of contact stress or other harsh marine applications [26]. Such engineered surfaces cause things to wear longer, require less maintenance, and wear in a more predictable way. The modern design of a marine tribosystem is based on the combination of material science, lubrication technology, and surface engineering. This enables the maritime operations to be efficient, reliable, and durable.

13.4.2 Limitations of Conventional Monitoring and Maintenance

Although they are long-lasting, the traditional methods of monitoring and maintenance of marine tribological systems have several shortcomings that limit their effectiveness, reliability, as well as efficiency. Indicatively, time-based maintenance schedules are reliant on set servicing schedules which do not consider the actual wear condition of the components or the speed at which the lubricant is worn out. This creates excessive or insufficient maintenance that wastes resources and inflates the cost of operation or results in insufficient maintenance that causes the risk of component failure between periods [27]. Likewise, eye inspection and operator skills are also essential elements of manual inspections; however, they may be subjective and erratic. Lubricant contamination, micro-cracking, or other signs of wear are not something that are always detected early enough before massive damage is caused. Moreover, inspections typically require disassembling equipment of some form, which can result in a prolonged lack of availability as well as costly and labor-intensive service procedures.

Another significant issue with traditional approaches is the impossibility to track the situation in real time. Conventional maritime systems are often lacking sensors and automated diagnostic tools that can record continuous performance data, including vibration, temperature, and lubricant quality [28]. With this limitation, there is no way of identifying when things are not operating according to how it should and

TABLE 13.1
Common Wear Mechanisms in Marine Systems

Wear Mechanism	Description	Typical Occurrence
Abrasion	Material removal caused by hard particles or asperities sliding across surfaces	Piston rings, cylinder liners, and shaft bearings
Scuffing/Adhesion	Severe localized welding and tearing due to insufficient lubrication	Gear teeth, bearings, and piston assemblies
Corrosion-Wear	Combined action of mechanical wear and chemical/corrosive attack due to saltwater	Propeller shafts, rudders, and exposed metal surfaces
Fatigue	Progressive surface/subsurface cracking from cyclic loading	Gears, crankshafts, and bearings
Cavitation Erosion	Damage from vapor bubble collapse in lubricated or water-exposed regions	Wet cylinder liners and pump components
Fretting	Damage from small amplitude, repeated relative motion at contact surfaces	Bearing mounts and coupling interfaces

correcting them in due time. This makes fault identification to be far too long, and thus the relatively minor problems such as anomalous friction, misalignment, or lubricant deterioration are further complicated, such as disastrous breakdowns or other serious mechanical breakdowns. The operational reliability and safety in the sea could be enhanced through predictive maintenance techniques that would not be utilized continuously because of the absence of constant data flow.

13.5 MARINE TRIBOLOGY AND WEAR MECHANISMS IN MARINE SYSTEMS

Marine tribology is a scientific discipline that evaluates the impact of wear, friction, and lubrication on moving components that are implemented in maritime engineering, such as bearings, engines, and propellers. The tribological concerns of marine equipment are more challenging than those of land-based systems due to environmental considerations, saltwater corrosion, changing loads, and mixed lubrication conditions. The tribological phenomena, internal (piston rings, cylinder liners, and crankshaft-bearings) and external (contact between the propeller and the hull), have a great influence on the efficiency and reliability of the equipment of marine engines. Friction in offshore structures and maritime vessels is a significant cause of fuel and power losses, and wear is the leading cause of failure of materials and maintenance costs [29,30]. Table 13.1 lists the common wear mechanisms in marine systems.

13.6 ENVIRONMENTAL INFLUENCES ON MARINE TRIBOLOGY

Corrosion in Saltwater: Saltwater is known to promote the speed of corrosion and accelerate the rate of corrosion on the surface. Two of the ways through which saltwater influences the efficacy of lubricants are emulsification and insufficiency of lubrication. Marine gears such as diesel engines and propellers experience frequent and unpredictable changes in load due to changes in sea conditions, operating profiles, and propulsion requirements [31]. This is referred to as variable load cycles. These changes in load cause fluctuations in stress and wear processes. Mixed lubrication: It is common to find that the regimes of operation of marine components involve a combination of boundary, hydrodynamic, and elastohydrodynamic lubrication. The risk of disastrous wear and failure is high when a lack of lubrication occurs to support metal-to-metal contact. Environmental changes that affect the tribological interactions and lubricant properties in the maritime environments include temperature, pressure, and humidity. Consequently, there is the need to have materials and lubricants that can withstand such conditions.

13.7 USE OF ML FOR ANALYZING CONDITION MONITORING DATA

ML algorithms are used to analyze data on vibration measured through time by sensors on the shafts or engine. The vibration patterns usually change before mechanical failures such as bearing degeneration and misalignment occur [32]. Information in the time, amplitude, and frequency domain can be extracted on the ML models to differentiate between normal and malfunctioning states. There are sensors at vital interfaces that sense the temperatures and feedback on the changes that might indicate loss of lubrication or excessive frictional heating. ML algorithms analyze these heat patterns to detect possible wear or lubrication failure signals, and preventive maintenance is done. Emissions in the Audible Range: The audible data records are high-frequency signals that are generated by fracture formation or frictional interactions in substances. To detect some types of faults, ML methods permit the filtering and finding of patterns in such noisy data [33].

13.7.1 Common Machine Learning Models Used

Using support vector machines (SVMs), it is possible to effectively divide fault occurrences into binary and multiclass categories by finding the most effective separating hyperplanes in feature space [34,35]. Common applications are marine engine sensor datasets consisting of classified faults. Random Forests: These are ensemble learning algorithms that are capable of operating on high-dimensional noise data of condition monitoring. They give robust categorization. They are less likely to overfit and better at identifying important traits. Automatic learning of complicated hierarchical characteristics from raw sensor data is possible using deep neural networks (DNNs), such as convolutional neural networks (CNNs) and recurrent neural networks (RNNs) [36]. This allows for better accuracy in defect detection, prediction, and classification tasks [37,38]. Continuous real-time monitoring is a good fit for them.

In case of Marine Diesel Engine Fault Detection, preventing disastrous breakdowns in marine diesel engines requires prompt defect diagnosis. Datasets used to train ML models [39] often include vibration, temperature, and audio inputs collected from sensors in addition to tagged fault categories such bearing wear, piston slap, or misfire. These ML systems are able to train models and extract features in a way that allows for condition-based maintenance routines to be implemented with low false alarm rates and high detection rates [40]. These models may be deployed in real time to improve operational safety, efficiency, and downtime reduction.

13.7.2 Deep Learning for Wear Mechanism Identification

Deep learning, especially CNNs, provides robust image-based analytical methods for automated detection of wear processes in marine tribology. This method is similar to research on monitoring the status of gearboxes, except that it looks at marine wear particles discovered in trash, filters, and oil samples taken from marine engines and equipment.

13.7.2.1 CNNs for Image-Based Wear Mechanism Identification

To automate the examination of worn debris in marine tribological systems, CNNs can train hierarchical features directly from raw photos, eliminating the need for human feature engineering. When analyzing wear particle pictures, this skill is vital since various wear processes are indicated by form, size, texture, and edge definition, among other attributes.

CNNs can evaluate pixel patterns to differentiate between various types of particles and outperform traditional visual inspections with experts. Besides data classification, CNNs can evaluate the level of wear on oil or filter samples by considering its size, the distributions of particles, and morphological variations. This may help with predictive maintenance by giving early indications of when wear is going to develop. Training these models allows for near-real-time condition monitoring without human involvement by swiftly processing or filtering pictures from enormous volumes of data [41]. We build trust in AI-driven diagnostics for critical marine applications by using explainability methods like Grad-CAM and Layer-wise Relevance Propagation to visualize which image regions influence CNN decisions. This ensures that classifications are based on physically meaningful features related to wear phenomena.

The input (I/P) layer as shown in Figure 13.2 is in RGB and grayscale and is made up of pixels that need to be normalized to a range between 0 and 1 before being

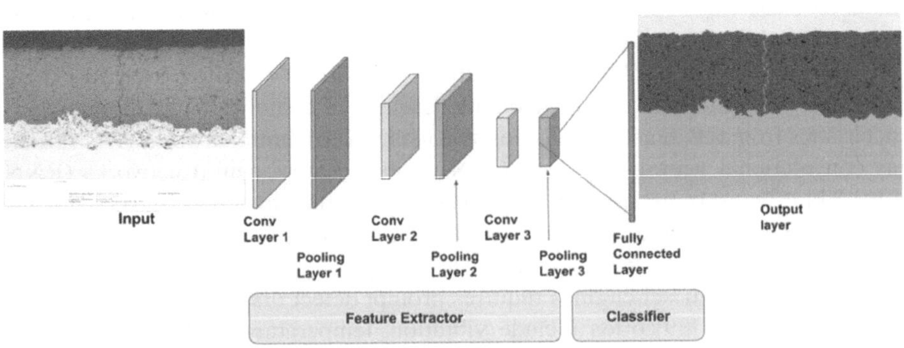

FIGURE 13.2 Generalized CNN model [42].

delivered to the model. In the convolutional layer, the input picture is filtered several times to find or extract its most relevant characteristics. The pooling layer downsamples the feature map after convolution. This helps save important information or characteristics of the input picture, while also making the computer work faster.

13.7.2.2 AI-Driven Optimization of Lubrication and Maintenance Intervals

Qualitative analysis of lubricating oil through AI implies the prediction of its degradation, frequency of contamination, and additive depletion. These are simple measures that should be taken to ensure that the maritime gear is running at its optimum. ML models can be used to predict the breakdown of oil by using sensor data, such as the number of particles, viscosity changes, and spectroscopic research. Such models include SVMs, Random Forests, and deep neural networks (DNNs), which analyze the prior data to determine when an oil has deteriorated to a point where an oil change and a filter are needed [43]. It is possible that dirt, water, metal particles, and other foreign objects can damage the efficacy of lubricants and accelerate the wear. In a way, AI algorithms examine sensor data such as size distribution of particles, acoustic emissions, or spectroscopic fingerprints to detect contamination. An example is the neural networks being trained on images of oil particles and chemical analysis data that are capable of identifying the pollutant type.

Lubricants containing additives are protective against wear, prevent corrosion, and maintain viscosity. These components fade away over time, and this renders the oil ineffective. AI models take sensor data about the chemistry, viscosity, and film strength of oil to compute the quantity of additives in the oil. This assists in establishing most appropriate maintenance and oil refill schemes. Integrated Oil Condition Diagnostics: AI-based models that examine a large number of data streams, including particle counts, spectroscopic data, and acoustical data, give a complete image of lubricant health. This enables you to come up with specific maintenance plans depending on the situation of the operation. Such solutions reduce unplanned downtimes and increase the lifespan of parts, reducing the operating cost and environmental impact.

Significance in Marine Operations Data-driven oil health monitoring complements with condition-based maintenance (CBM), which reduces oil waste, prevents equipment failure, and enhances fuel efficiency. AI-driven diagnostics assist companies in ensuring compliance with the rules by ensuring that lubricants meet the contamination and RUL standards. The devices allow warning signs of problems that are likely to occur beforehand. This is especially crucial in far or offshore sea scenarios where access to maintenance is difficult.

13.8 FUTURE DIRECTIONS IN AI-DRIVEN MARINE TRIBOLOGY

There is a rapid increase in the use of AI in tribological studies. The interest of the researchers is on hybrid models that rely on physics-based simulations and data-driven algorithms to achieve more accurate predictions. An increased push to make AI more explainable (XAI) is undertaken to make predictions of models more transparent and credible, especially to use them in safety-related applications, such as marine propulsion and energy systems [44]. Researchers have used deep reinforcement learning to optimize autonomous systems, including dynamic lubrication control and wear mitigation, through the input supplied by the environment. Transfer learning allows models that are trained in a particular marine condition to function in different marine conditions with minimal retraining, and this is a solution to the issue of inadequate data.

13.8.1 Composite Materials in Marine Engineering

Composite materials are essential in promoting marine engineering as it provides the strength and corrosion resistance to harsh marine environments combined with strength and durability. Fiber-reinforced polymers (FRPs) of glass, carbon, or aramid fiber embedded within an epoxy or polyester matrix and metal and polymer-based composites are common materials. Aluminum alloys that are characterized by their lightness and good strength-to-weight ratio are commonly used in hulls, superstructures, and offshore components where corrosion resistance and structural integrity are essential. Moreover, the polymer composites such as acrylonitrile butadiene styrene (ABS) and polylactic acid (PLA), are easy to mold, are low-density products with environmental advantages [45,46], and are used in interior components, buoyancy structures, and additive-manufactured marine parts. Developed methods of fabrication like resin transfer molding, winding filament, and 3D printing make it possible to make custom designs that have enhanced mechanical and tribology properties. The new hybrid and nano-reinforced composites make use of metals, polymers, and fibers to multifunctional composites that have impact resistance, self-healing properties, and flame retardancy. Composite materials, by and large, are revolutionizing the marine engineering industry through giving marine vessels and structures the opportunity to be lighter, more efficient, and durable.

13.8.2 Integration of Digital Twins for Marine Tribosystems

Digital twins are virtual replicas of physical maritime systems that utilize sensor information to execute, predict, and enhance the functionality of engines, propellers, as well as bearings. AI enhances such models by considering data of systems in real time to forecast the wear, corrosion, and lubrication trends with time. Such an integration allows operators to perform some virtual tests on maintenance plans or upgrade and then implement them. This saves time, reduces the downtime, and streamlines the operations. Future improvement will be focused on multi-physics digital twins with fluid–structure interaction, thermal effects, and tribological influences to establish complete system supervision.

13.9 CORROSION FATIGUE LIFE PREDICTION OF 10CRNICU SHIPBUILDING STEEL IN MARINE ENVIRONMENTS

This case study demonstrates the corrosion fatigue life prediction of 10CrNiCu shipbuilding steel operating under authentic marine conditions by applying a continuum damage mechanics-based approach. The model integrates three sub-models: a pitting corrosion evolution model, an improved fatigue damage evolution model, and a nonlinear damage accumulation model, supported by a fatigue life database derived from experimental data.

The pitting evolution model assumes conical pit geometry, validated through marine exposure tests in the South China Sea under non-wettable, semi-wettable, and wettable conditions. The mean radius-to-depth ratios (3:1, 4:1, and 6:1, respectively) were used to describe corrosion morphology. Pit growth was expressed using Faraday's law and an Arrhenius-type relation incorporating cyclic stress amplitude and time, yielding predictive constants for each wettability condition. The schematic of the proposed corrosion fatigue life prediction model is presented in Figure 13.3. It is used to identify the rate of corrosion of pits in different wettability conditions, and the results of corrosion tests in the marine environment are used initially. The tests considered elements like stress field, properties of the corrosive medium, and properties of external alternating loads. An evolution model of pitting corrosion is then created at varying wettability conditions to explain pit morphology and radius-to-depth ratio. The second step is the fatigue damage accumulation model, whereby the life of fatigue is analyzed at different damage mechanisms in order to create a fatigue life database of different damage cases. Lastly, the physical evolution process of corrosion fatigue is also introduced, and the fatigue life database is applied in accordance with the nonlinear damage accumulation model.

The organism growth rate was determined at 0.1 mm/day, with random locations as well as maximum sizes of the biological attachment points. Figure 13.4 shows

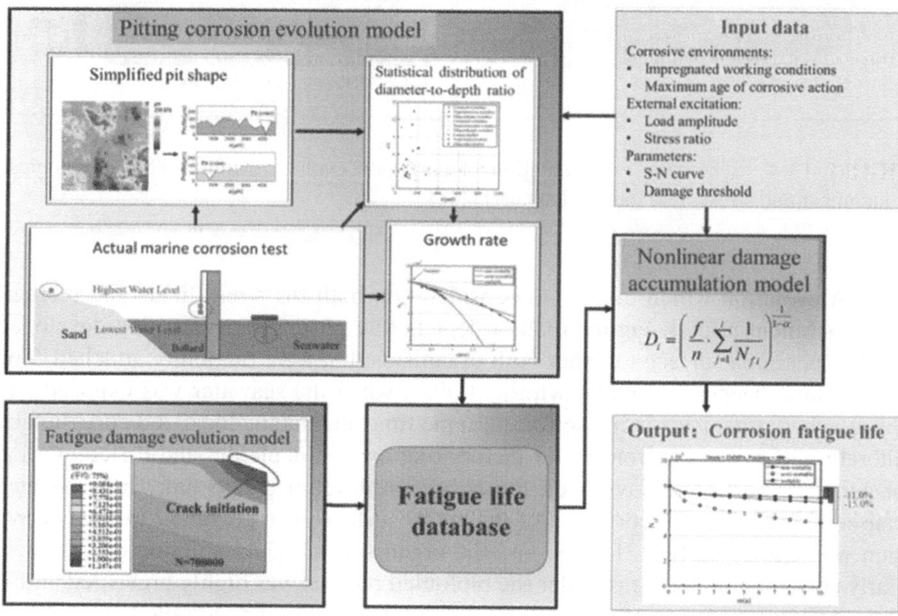

FIGURE 13.3 Schematic representation of the corrosion fatigue life prediction model derived from real marine corrosion experiments [47].

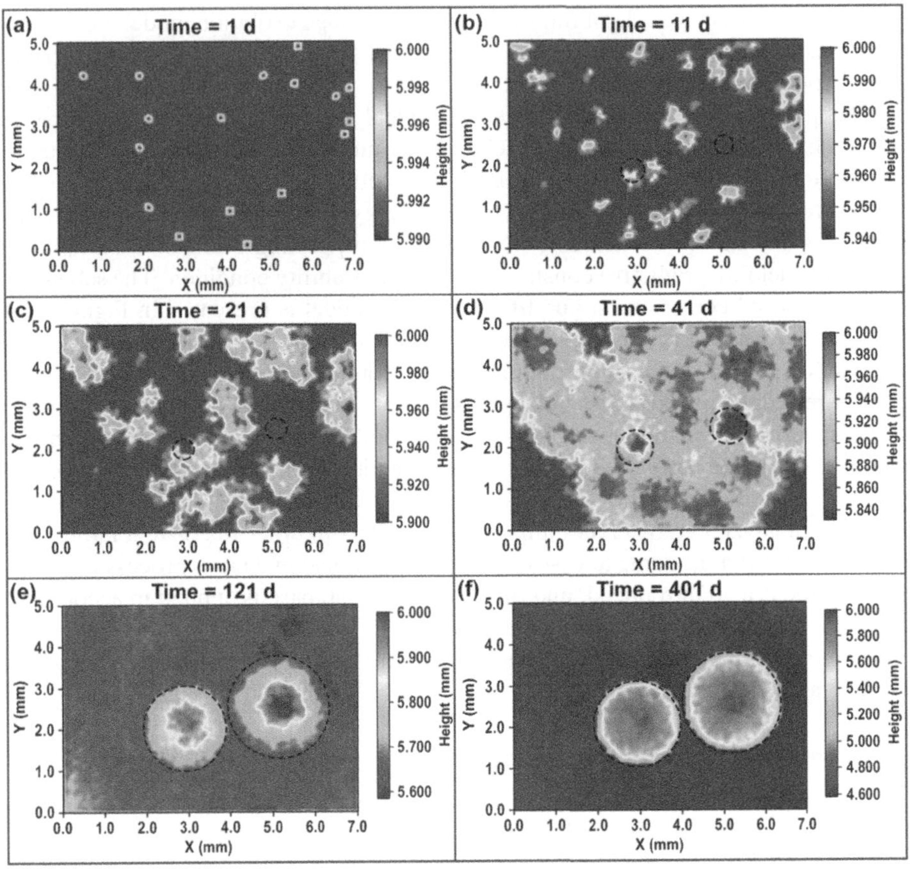

FIGURE 13.4 Schematic representation of corrosion evolution influenced by biological attachment and stress field interactions (a–f) [48].

the CA evolution which takes into consideration both the stress fields and biological attachment effects. Figure 13.4a–f depicts that pitting corrosion first developed on the specimen surface together with organisms that were randomly attached. The pits became deeper and wider within 10 days when the seawater was exposed and the attached organisms increased at the same time, enhancing localized corrosion by alteration of micro-environmental factors like oxygen depletion and differentiation of aeration. Forty-one days later, the corrosion developed into flaky areas, which caused partial uniform corrosion of the entire surface. The prevalent surface corrosion was witnessed by 121 days, and the organisms attached attained critical size. Early corrosion in the areas under the biofouled regions was highly prevented at this stage because diffusion of oxygen was highly hindered in the biofouled areas due to the existence of the biological layers that served to act as a physical barrier to corrosion, which changed the local corrosion relationships.

13.10 CONCLUSION

The marine tribology is being transformed using the AI to regulate wear, friction, and lubrication through intelligent, predictive, and data-driven approaches. Its integration allows locating the problems at an early stage, planning the lubrication more economically, and consuming fewer energy resources, saving money and having a less significant impact on nature. AI helps in the maritime equipment to be more reliable and safe as it continues to be in harsh environments through continuous learning of the operational data. The combination of AI-based automation, IoT-based real-time monitoring, and digital twin technology is the next stage in marine tribology. This will render maritime operations smarter, efficient, and environmentally friendly. Tribologists, AI engineers, and marine technologists will be forced to collaborate in order to achieve optimal results with intelligent tribological management. These new technologies are enhanced by the marine industry, which drives toward longer-living equipment, low emissions, and safe navigation in the open ocean.

REFERENCES

1. Carlton J. *Marine Propellers and Propulsion.* Oxford, UK: Butterworth-Heinemann; 2018.
2. Litwin W. *Water-Lubricated Journal Bearings: Marine Applications, Design, and Operational Problems and Solutions.* Amsterdam, Netherlands: Elsevier; 2023.
3. Guo X, Zhang D. Research on marine shafting transmission in china and global. *Eng Solut Mech Mar Struct Infrastruct.* 2024;1:4.
4. Hong S-H. Tribology in marine diesel engines. *Tribol Mach Elem Appl.* 2022;3:547.
5. Sun Y, Yan X, Yuan C, et al. Insight into tribological problems of green ship and corresponding research progresses. *Friction.* 2018;6:472–483.
6. Woydt M. Material efficiency through wear protection–The contribution of tribology for reducing CO_2 emissions. *Wear.* 2022;488:204134.
7. Ciulli E. Tribology and sustainable development goals. *Int Work IFToMM Sustain Dev Goals.* 2021;48:438–447.
8. Johnson R, Von Der Ohe CB. *Tribocorrosion in marine environments: Tribocorrosion of Passive Metals and Coatings.* Amsterdam, Netherlands: Elsevier; 2011, pp. 441–474.
9. Yimin LUO, Junhua Y, Xia LI, et al. Self-lubricating and wear-resistant and anti-corrosive multi-functional coating technologies applied in tropical marine atmosphere. *Tribology.* 2022;43:104–120.
10. Totten GE. *Friction, Lubrication, and Wear Technology.* Novelty, OH: ASM international; 2017.
11. Mishra CS, Ali F, Adam S, et al. Study of the causes of engine failure in ship and maintenance required to prevent engine failure. *Int J Contemp Res Rev.* 2017;8:20218–20241.
12. Tung SC, McMillan ML. Automotive tribology overview of current advances and challenges for the future. *Tribol Int.* 2004;37:517–536.
13. Adetunla A, Afolalu S, Jen T-C, et al. The advances of tribology in materials and energy conservation and engineering innovation. *E3S Web Conf. EDP Sci.* 2023;391:1014.
14. Erdemir A. Review of engineered tribological interfaces for improved boundary lubrication. *Tribol Int.* 2005;38:249–256.
15. Chengpeng W, Shenghui W, Haizeng W, et al. Research status and prospects of tribological behaviors of key friction pairs of materials in marine equipment. *Mater Sci.* 2021;27:148–154.

16. Wood RJK. Marine wear and tribocorrosion. *Wear.* 2017;376:893–910.
17. Kumar P, Yadav J, Kunwer R. Advancements and challenges in water-lubricated bearings for marine applications: A comprehensive review. *Tribol Online.* 2025;20:1–25.
18. Wordu CC-RP. Effects of good lubricants on engine performance to eliminate frictional force and prolonging engine life span. *DELSU J Educ Res Devel.* 2021;22:364–379.
19. Raghuvanshi NS, Johari A, Saxena M, et al. Thermally stable lubricants. In: Kumar A (Eds) *Performance Characterization of Lubricants.* Boca Raton, FL: CRC Press; 2024, pp. 262–288.
20. Uanekee A. *The Performance, Wear, and Environmental Impact of Lubrication Systems Used in Marine Engines.* Bachelor's Thesis; 2024.
21. Sarpkaya T. *Wave Forces on Offshore Structures.* Cambridge, UK: Cambridge University Press; 2010.
22. Patel MH. *Dynamics of Offshore Structures.* Oxford, UK: Butterworth-Heinemann; 2013.
23. Zhai W, Bai L, Zhou R, et al. Recent progress on wear-resistant materials: Designs, properties, and applications. *Adv Sci.* 2021;8:2003739.
24. Dhakal N. Tribology of newly developed EAL versus water in hydropower turbine bearings. *Tribol Int.* 2019;162:107101.
25. Freschi M, Paniz A, Cerqueni E, et al. The twelve principles of green tribology: Studies, research, and case studies—A brief anthology. *Lubricants.* 2022;10:129.
26. Singh H, Singh R, Kumar S, et al. A comprehensive review of tribological performance of coatings on automotive piston rings. *Manager's J Mech Eng.* 2024;14:48.
27. Mobley RK. *An Introduction to Predictive Maintenance.* Amsterdam, Netherlands: Elsevier; 2002.
28. Sharma V, Singh J, Hashmi AW. Quality control and regulatory compliance in biomedical manufacturing. *Adv Manuf Technol Biomed Sci.* 2025;18:332–350.
29. Singh AK. Industrial cases of microbial induced corrosion. In: Heitz E, Flemming HC, Sand W (Eds) *Microbially Induced Corrosion Its Mitigation.* Cham: Springer; 2020, pp. 81–106.
30. Batchelor AW, Lam LN, Chandrasekaran M. *Materials Degradation and Its Control by Surface Engineering.* Singapore: World Scientific; 2011.
31. Smogeli ØN. *Control of Marine Propellers: From Normal to Extreme Conditions*; PhD-Thesis; 2006.
32. Emtaubel S. Bearing misalignment and eccentric wear: A study on condition monitoring. *Int J Eng Res.* 2024;3:77–94.
33. Angelopoulos A, Michailidis ET, Nomikos N, et al. Tackling faults in the industry 4.0 era—A survey of machine-learning solutions and key aspects. *Sensors.* 2019;20:109.
34. Singh J, Kumar S, Kumar R. Support vector machine-based prediction of wear in Ni-20Cr_2O_3 thermal spray coating under bottom ash slurry conditions. *J Tribol.* 2023;39:84–105.
35. Singh J, Singh S. Support vector machine learning on slurry erosion characteristics analysis of Ni- and Co-alloy coatings. *Surf Rev Lett.* 2023;32:61 Available from: https://www.worldscientific.com/doi/10.1142/S0218625X23400061.
36. Wikle CK. Comparison of deep neural networks and deep hierarchical models for spatio-temporal data. *J Agric Biol Environ Stat.* 2019;24:175–203.
37. Singh J, Kumar S, Kumar R, et al. Machine learning regression tools for erosion prediction of WC-10Co4Cr thermal spray coating. *Results Surf Interf.* 2023;13:100156. Available from: https://www.sciencedirect.com/science/article/pii/S2666845923000636.
38. Singh J, Vasudev H, Chohan JS. Review on computational fluid dynamics based analysis of surface erosion in hydraulic machinery. *Int J Interact Des Manuf.* 2023;18: 5353–5380. Available from: https://link.springer.com/10.1007/s12008-023-01336-2.

39. Singh J, Sharma V, Singh A. Study on abrasive coal-slurry flow and erosion wear of NiCrBSiCoFeWC HVOF coating for coal-conveying slurry pumps using machine learning approach. *Int J Coal Prep Util.* 2025;2025:1–34.
40. Badihi H, Zhang Y, Jiang B, et al. A comprehensive review on signal-based and model-based condition monitoring of wind turbines: Fault diagnosis and lifetime prognosis. *Proc IEEE.* 2022;110:754–806.
41. Nsor M. Predictive maintenance using machine learning for engineering systems through real-time sensor data and anomaly detection models. *Int J Res Publ Rev.* 2024;5:5167–5183.
42. Liu R, Wang M, Wang H, et al. Recognition of NiCrAlY coating based on convolutional neural network. *Mater Degrad.* 2022;6:7.
43. Salem AM, Yakoot MS, Mahmoud O. Addressing diverse petroleum industry problems using machine learning techniques: Literary methodology— spotlight on predicting well integrity failures. *ACS Omega.* 2022;7:2504–2519.
44. Kuznietsov A, Gyevnar B, Wang C, et al. Explainable AI for safe and trustworthy autonomous driving: A systematic review. *IEEE Trans Intell Transp Syst.* 2024;25:19342–19364.
45. Sharma V, Singh J. Microstructural analysis on critical interfacial layers of FDM-based PLA and ABS polymers processed with laser machining. *Int J Polym Anal Charact.* 2025;2025:1–18.
46. Vikas S, Jashanpreet S, Chandmal SR, et al. Machine learning-based prediction and optimization of surface roughness in FDM-printed PLA using ANN and box-behnken design. *Mater Sci Technol.* 2025;41:1153–1173.
47. Hu Z, Hua L, Mei Z, et al. Corrosion fatigue life prediction model for 10CrNiCu shipbuilding steel under authentic marine environments. *Case Stud Constr Mater.* 2024;21:e03787.
48. Hua L, Hu Z, Wu W, et al. Investigation of marine corrosion characteristics of 10CrNiCu steel subjected to stress fields using an improved 3D cellular automata modeling. *Mater Des.* 2025;250:113606.

14 Challenges and Future Directions in AI and Tribology

Amrinder Mehta, Jashanpreet Singh, Brijesh Prasad, and Hitesh Vasudev

14.1 INTRODUCTION

To obtain an authoritative picture of the tribological response of materials, considerable numbers of atomistic, mesoscopic, and macroscopic computational frameworks, when read alongside experimental passive spectroscopy, direct force indentation, and extended-duty service trials, have become standard repositories in the discipline [1].

Artificial intelligence (AI) has begun to increase the rate of traditional tribological synthesis and analysis by supplying rapid approximations of signature features, generating quasi-exact meta-models of tribo-resistive behavior, and dynamically refitting system parametric surfaces. In the tribological domain, multilayer neural and semi-supervised ensemble architectures interrogate sparser high-dimensional tuples, including atomistically derived sliding, chemically normative spectroscopy-obtained coefficients of friction, graded moisture-removal energies, ep-bound and solvopathic dielectric micro-flows, and endurance fraction relations harvested through SCALPOL crystallization scans [2]. These architectures recover mappings that blind sampling irregularity thresholds through relevance regrouping and topological degrees of rank. Concurrent adaptive search regulators, bound to custom models of lubricant polar dielectric indexing and topographically regularized surface re-weld slopes, generate micro-brownized, interfacially isolated conditional distributions of heat generation and layer-rate dysfunction of the control state. By propagating knowledge hierarchies across spatial and chronal scales from aspheric shaft-abut mentally joint surfaces to service shed slit fins, the integrated curation environment guides experimental scaling, upgrades mid-term equipment uptime schedules, and triggers autonomous tracking of the equipment. The broad effect transforms mechanical load trains, geared and tributive bearings, and rotating icing membranes into more palatable carriers of persistent optimization, endurance, and reduced settling energy [3].

The hierarchical relationship between deep learning, machine learning, and AI is illustrated in Figure 14.1. The broadest definition of AI includes all computational techniques, such as expert systems, planning, search algorithms, and evolutionary algorithms, which allow machines to perform tasks that require human-like intelligence. Machine learning, a subset of AI, enables systems to recognize patterns in data without explicit programming. Classical algorithms, such as k-nearest neighbors,

 DOI: 10.1201/9781003624165-14

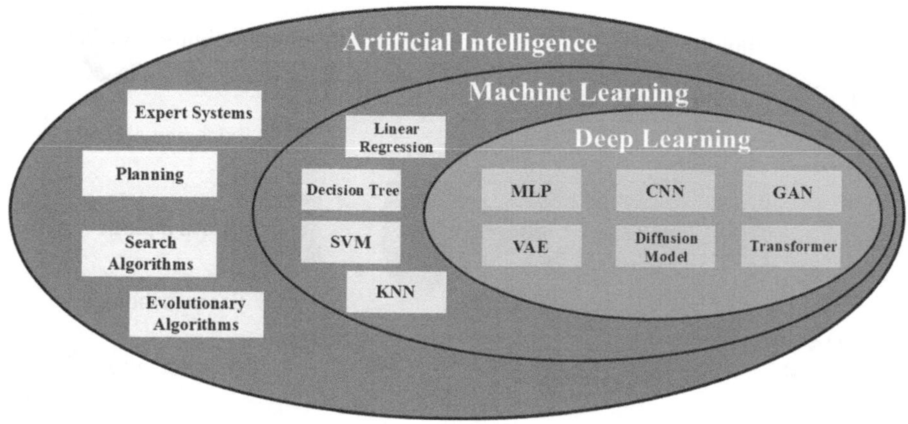

FIGURE 14.1 Diagram of the relationship between artificial intelligence, machine learning, and deep learning [4]. (Creative Commons Attribution 4.0.)

decision trees (DTs), support vector machines (SVMs), and linear regression, are examples of machine learning techniques. Deep learning (DL), which uses layered neural network architectures to automatically learn complex feature representations, has been further integrated into machine learning. Examples of DL models include multilayer perceptrons (MLPs) for general-purpose tasks, convolutional neural networks for spatial data, variational autoencoders for generative modeling, generative adversarial networks for realistic data synthesis, diffusion models for progressive denoising, and transformer networks for sequential and attention-driven tasks. This nesting demonstrates how the data-driven underpinnings of machine learning, which are found within the larger field of AI, are built upon DL [4].

The combination of AI with the study of friction and wearing surfaces yields a new field of "tribo-informatics," which combines tribology and data science to solve the complex puzzles of machine behavior and boost performance across the board. By applying smart data processing, machine learning, and clever statistics, tribo-informatics gathers knowledge from computer models, sensor webs, and laboratory experiments to paint a full picture of how surfaces slide, wear, and run smoothly [5]. The method automatically identifies key characteristics hidden inside huge datasets, unmasks subtle regularities within contact mechanics, and assembles models that predict future wear and lubrication needs based only on data. Operating across scales from the way atoms bump and slide past each other to how whole machine components hum along, tribo-informatics lets engineers tailor new materials, continuously watch and judge the condition of a part, and steer machinery in real time to work longer and use less energy [6].

Figure 14.2 presents the allocation of AI applications among the major subfields of tribological investigations. Research categorized as intelligent tribology occupies the largest share, comprising precisely 50% of the total documented AI endeavors. A further 27% is devoted to the fundamental theory of tribology, indicating a pronounced engagement of AI techniques in computational modeling and theoretical deduction. Basic component tribology comprises approximately 14% of research efforts and

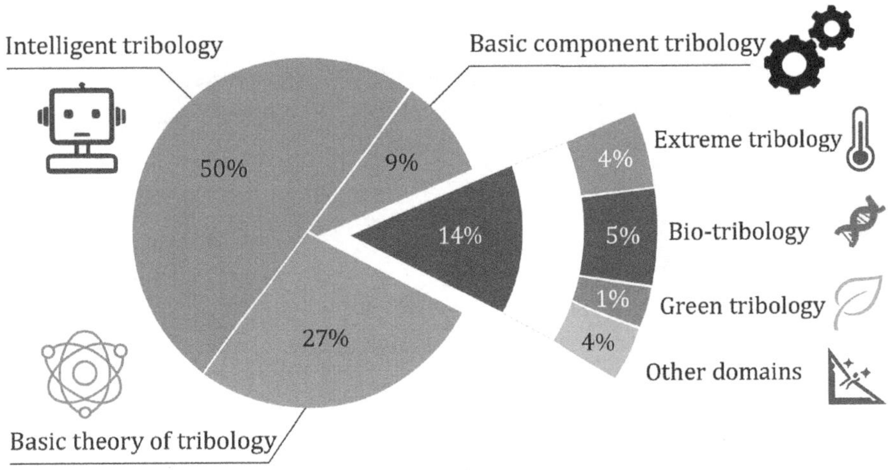

FIGURE 14.2 Proportion of AI technology in various research fields of tribology [7]. (Creative Commons Attribution 4.0.)

focuses on fundamental mechanical elements. The remaining areas include bio-tribology (5%), extreme tribology conditions (4%), other specialized domains (4%), and green tribology, representing the smallest segment at 1% [7].

Tribological science increasingly incorporates AI and machine learning across several complementary domains: theoretical foundations, intelligent component design, harsh-condition engineering, bio-systems, and environmentally sustainable applications. Within basic research, machine learning methods quantify the interplay among surface geometry, entrainment films, and effective friction, distilling empirical datasets into the governing laws. Intelligent design frameworks comprising deep neural networks (DNNs), ensemble DTs, and support vector classifiers systematically refine the topology of gear teeth, microstructure of bearing races, and lattice architectures of hybrid-additive parts [8]. Within extreme tribology, algorithmic predictors extrapolate wear metrics beyond the conventional limits of temperature and rotational speed. Parallel modeling of organic interfaces identifies the optimal tribological performance of orthopedic and dental implants. Finally, sustainable tribology redirects chemical genome-screener algorithms toward the discovery of non-hazardous lubricants, micro-coatings, and biopolymers, permitting the rapid design of next-generation tribological infrastructure. Collectively, these intertwined methodologies exploit heterogeneous data, augment *in situ* diagnostics, and accelerate empirical-to-parametric transitions across scales from molecular to kilometer and laboratory to application [9].

Figure 14.3 shows the underlying classification of machine learning, which is divided into three major branches on the right-hand side of the figure (the main branches being supervised learning, unsupervised learning, and reinforcement learning). Supervised learning involves classification tasks from labeled training sets and includes SVMs, DTs, rule-based systems, and a subcategory of inductive learning. The supervised, inductive k-nearest neighbors classifier also estimates

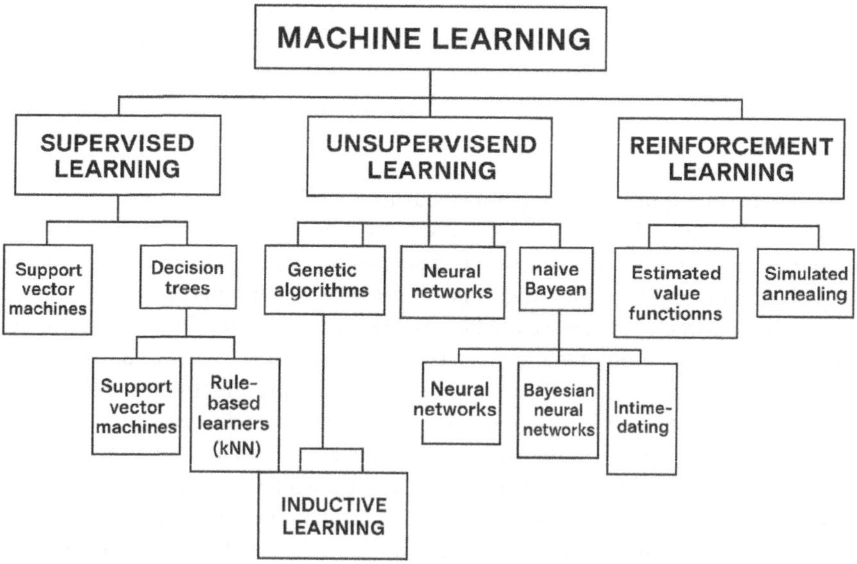

FIGURE 14.3 Structuring of machine learning techniques and algorithms [10]. (Creative Commons Attribution 4.0.)

an instance-based learner. Unsupervised learning contains algorithms that find structures within unstructured data, including genetic algorithms, artificial neural networks, and Bayesian neural networks. There are two main sub-areas in this category: networks and Bayesian systems, within which naive Bayes networks are specifically noted in the Bayesian subfield. The main branches are reinforced in the third section of the text, with learning through interaction and reward optimization, and estimated value functions and simulated annealing methods surround the reinforcement learning techniques. The structure above provides a complete picture of the pedagogical classification of machine learning methods: it is based on the learning types, data, and requirements and the techniques for rapidly advancing tasks from the different fields of computation [10].

Smart manufacturing automates the workflow by capturing real-time data and sensors, predicting malfunctions, and adjusting machine learning parameters while predicting and adjusting for planned and unplanned downtimes without losing control over the supply chain. Using the Internet of Things (IoT), smart manufacturing guides networked devices while producing goods and monitoring the production process, ensuring that the production system is both efficient and sustainable [11]. Digitally interconnected lines assist in assessing where real-time changes might be necessary by assessing changing demand rather than waiting for requests and clearing bottlenecks. Unrestricted by less intelligent systems, smart manufacturing uses data and digitized systems to mitigate waste while enhancing primary and secondary systems on the production floor. Automated systems inform the center systems of the location of blockages and control the reallocation of dependent systems [12].

Transparency through interconnected production lines eases knowledge flow, which augments strategic decision-making. Automated systems inform the central systems. The intelligent and adaptive ecosystem of smart manufacturing refreshes system workflows and operational boundaries with the help of AI. The interconnected system of agile decision-making maintains an intelligent production process and enables resilient operations at the equipment and enterprise levels [13].

As shown in Figure 14.4, machine learning can be divided into three central paradigms. In supervised learning, SVM and DT ascertain the optimal decision boundaries and hierarchical splits for classifying inputs and predicting outcomes from labeled data. SVM identifies and constructs a hyperplane that "maximizes the margin" between two classes, whereas DT iteratively divides the feature space into sub-spaces and computes the probabilities of the target values. In unsupervised learning techniques, such as K-means clustering and principal component analysis (PCA), there are no labels attached to patterns that the model attempts to "extract" or describe. K-means "clusters" datasets into groups according to their similarity, while PCA takes datasets with many dimensions and transforms them into their orthogonal "principal component" forms (PC1 and PC2) that describe the most variance in the data so that visualization and dimension reduction can be achieved. In the case of reinforcement learning, there is an agent–environment loop in which the learner operates. In this case, the agent chooses the actions to take and forms a policy, then

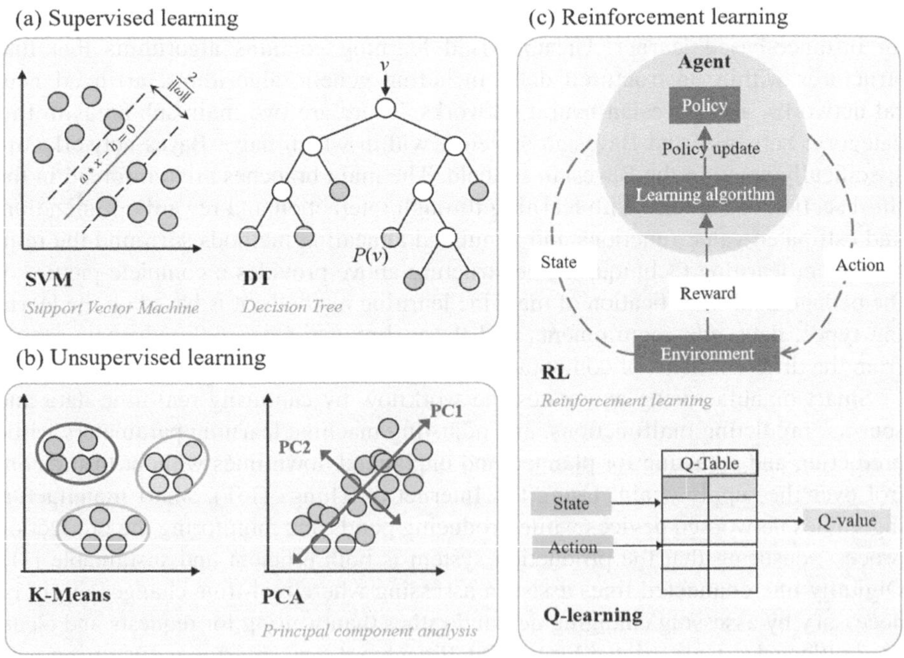

FIGURE 14.4 (a) The supervised learning models SVM and DT, (b) the unsupervised learning models K-means clustering and PCA, and (c) the framework of reinforcement learning and the Q learning mode [14].

gets rewarded by the environment, and subsequently modifies the policy according to the learning algorithm. The agent is guided by values to perform a specific action through a reward system. A value-based type of reinforcement learning is Q-learning, in which agents store state–action Q-values into a table and, through reward reinforcement, slowly change action values to converge on the optimal action policy that maximizes returns [14].

This review merges new studies on the problems and future scope of AI in tribology and offers a guide to scholars and professionals to help them tackle the intricacies of this newly evolving field.

14.2 CURRENT CHALLENGES IN AI-TRIBOLOGY

Traditional metasurface design begins with the assignment of phase and amplitude design targets to meta-atoms and the adjustment of their geometrical configurations using full-wave solicitation electromagnetic simulators. The rigidity and costly computations of this stepwise design approach make it very difficult to progress. Each step involves structural adjustments that require substantial amounts of money and time [15]. Hence, design modifications and adjustments constitute performance constraints of a meta-atom, which is irrelevant to volume calculations, making it even harder to prototype. These adjustments and prototypes lead to data complexity, which affects the wear and friction. These are often included in dynamic simulations or sensor stream monitoring. Slip and wear friction constitute multilevel interfaces of varying scales. Connecting data from computational and mechanical systems, all presented on a surface and hard drives, makes it a nightmare for data integrity and return on investment (ROI). Hence, advanced modeling strategies and design spaces exhibit disparate metasurfaces [16].

Figure 14.5 outlines an inverse design framework combining DNNs and genetic algorithms (GA) to generate metasurface meta-atoms that produce the desired wavefront transformations. Orbital Angular Momentum (OAM) field profiles and their corresponding phase amplitude requirements are used as design specifications, from which electromagnetic responses are targeted. These responses are defined as the input features of the DNN, which learns to model the complex geometries of the meta-atoms and their corresponding electromagnetic responses. In a GA (Galleschi–Amendola) loop, the initial set population of meta-atom topologies with predefined attributes evolved through a selection, crossover, and mutation process that the DNN uses to calculate the fitness of the population set. Each meta-atom's response was predicted and calculated by a DNN model without full-wave simulation, which is one of the most expensive predictive analytical processes; hence, design space exploration becomes very inexpensive. After population refinement, the GA proceeds to meta-atom shapes that obey the desired amplitude and phase profiles. On the right, the optimized metasurfaces, which consist of the inversely designed meta-atoms, are capable of controlling the incident waves to provide the desired OAM and lensing functions, which is a testament to the machine and evolutionary computation skills framework for metasurface design [17].

The deployment of opaque machine learning systems in contexts of significant societal consequences, such as healthcare delivery or criminal sentencing, has

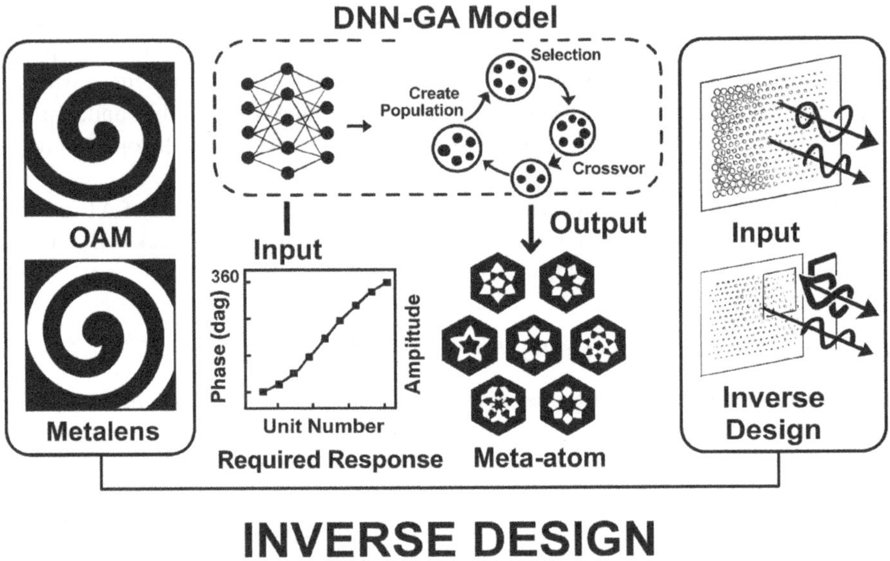

FIGURE 14.5 Schematic of the proposed inverse design [17].

become increasingly prevalent, and the resulting harms are already observable across multiple domains. The emerging scholarly consensus urges caution when pursuing explanatory techniques designed to render these black-box systems transparent ex-post [18]. More troublingly, retro programs or output attenuation, far from obviating risk, could furnish putative accountability with a flawed, yet superficially persuasive evidential veneer, thereby entrenching systemic bias and consolidating the operational tolerances within which that bias mutates over time [19].

Figure 14.6 illustrates a workflow aimed at bridging the clinical gap with DL for medical imaging through interpretability. The first step involves a black-box neural network that works on an image but does not provide reasoning. These black boxes are modified using different methods to provide reasoning through concept learning, case-based reasoning, counterfactual explanations, concept attribution, language-based concept descriptions, latent space visualization, attribution maps, anatomical priors, and internal network-based anatomical representations. These methods provide different levels of abstraction that describe the model at a rudimentary layer, highlight portions of the image, and then provide ancillary thinking with different interconnections of thoughts that help the model reach a conclusion. The strength and trustworthiness of these rationales are major focuses that undergo clinical validation, surface imaging biomarker correlation, adversarial robustness metrics, and quantifiable measurements of trustworthiness. The moment embedded interpretability is validated or proven under real-world conditions, DL approaches change from "black boxes" to fully functional clinical systems, which the clinician does not just "trust" but can also practically use for deployment in the clinical space for diagnosis and decision treatment planning [20].

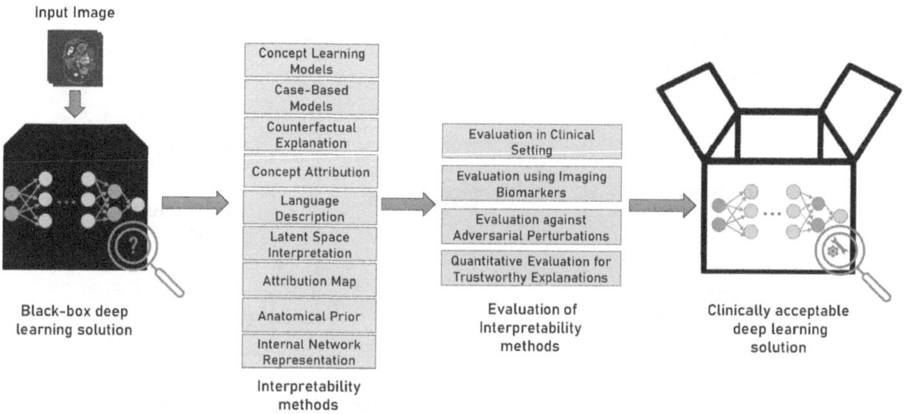

FIGURE 14.6 Overview of interpretability methods for DL solutions in medical image analysis [20]. (Creative Commons Attribution 4.0.)

The combination of the factors described above presents a multifaceted problem in evaluating and quantifying the quality of interpretability methods. The assessment of virtually intractable and real-world systems using unsupervised learning in tribology remains a largely unsolved problem [21].

14.3 APPLICATION DOMAINS

AI and DL are rapidly becoming recognized as pivotal engines for sustainable progress, and their evolving power aligns closely with the United Nations' sustainable development goals. By correlating and modeling the dimensions of satellite imagery and ground-sensor streams, these systems yield near-real-time observations and deployable early warnings, permitting organizations to mount quicker, more focused interventions against impending climate threats [22].

Figure 14.7 synthesizes the multifaceted deployment of AI and DL technologies in contemporary environmental health. At the center, a conflating node encapsulates these methods as an interconnected analytic backbone energizing discrete, sector-specific domains. To the left, health-centric pathways present the algorithm-driven determination of focal epilepsy, deployment frameworks for COVID-19 countermeasures, and ancillary biomedicine and clinical programs, collectively illustrating DL's augmentation of diagnostic clarity and pandemic readiness. Conversely, the right panel delineates environmental and safety trajectories, including environmental epidemiology frameworks that merge geospatial and health metrics, embedded surveillance systems for airborne and waterborne contaminants, workplace safety systems that prevent industrial hazards, and cross-continental studies of avian health that promote wildlife conservation and ecosystem vitality ratings. Through this topological survey, the diagram affirms the status of AI and DL as catalytic instruments for discrete intervention points in environmental health [23].

AI technologies are being systematically harnessed to govern critical inputs within agricultural systems, namely, water, fertilizers, and pesticides, while concurrently

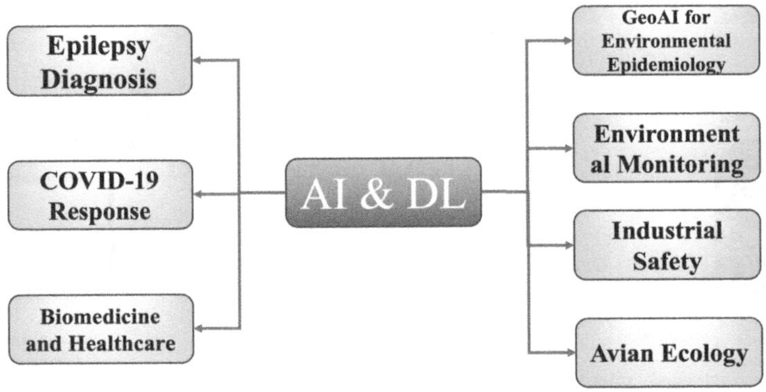

FIGURE 14.7 Applications of AI and DL in environmental health [23]. (Creative Commons Attribution 4.0.)

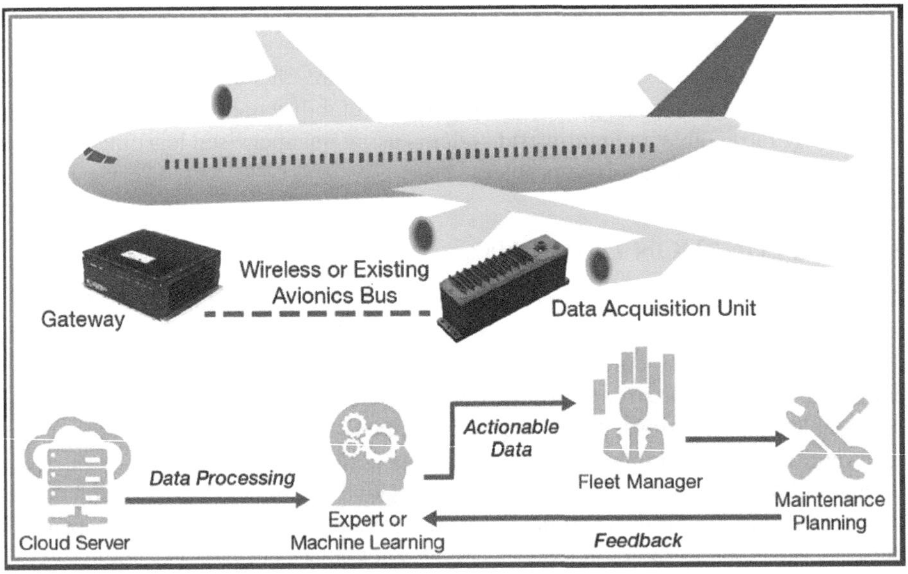

FIGURE 14.8 Shows the application of AI in predictive maintenance [26].

refining the planning, output, and stability of renewable energy installations and grid operations [24]. The architectonic fusion of machine learning, DL, and related algorithm families continues to permeate functions that stretch from the autonomous guidance of orbital vehicles to the real-time orchestration of global civil aviation, seeding unprecedented rises in system efficacy and safety margins while compressing design, testing, and operational cadences [25].

As you can see in Figure 14.8, this is an example of a predictive maintenance workflow applied to aircraft operation with the aid of AI. The primary onboard sensors transmit data to a data acquisition unit (DAU). The performance metrics, having been processed, cleansed, and aggregated, go through a series of pipelines.

Subsequently, the metrics are further analyzed using machine learning algorithms and/or a subject matter expert to produce assessments regarding a component's health and potential failures [26].

The increasingly sophisticated self-navigating satellites have advanced, and the control of intricate operations has become the domain of AI systems. The integration of AI into the aviation sector has also enabled automated advanced "smart" maintenance and control of air traffic. The primary objective of these functions and associated unmanned aerial vehicles has been to improve the fuel and safety efficiency [27].

14.4 FUTURE DIRECTIONS

A significant development within this field is the integration of informatics and data science with tribology, termed tribo-informatics. This encompasses the collection and management of data pertaining to various tribological systems and systems' data analytics processes. The integration of various systems will assist in the management of previously inaccessible data with respect to systems employing tribology [28].

Figure 14.9 presents a comprehensive analysis of the artificial neural network performance for wear prediction optimization across different architectural parameters. The bar charts demonstrate the average total errors when systematically varying the transfer functions, network types, neuron numbers, and layer configurations. Chart (a) shows a relatively consistent performance across different transfer functions, with errors of approximately ±60%. Chart (b) reveals a significant variation in the network types and neuron numbers, with some configurations achieving near-zero errors, while others exceed 60%. There are different results from different architectures, as shown in Chart (c), which shows the effect of different network types and the number of layers, as well as their performance. Chart (d) shows the impact of the number of neurons in multiple hidden layers on their performance across different

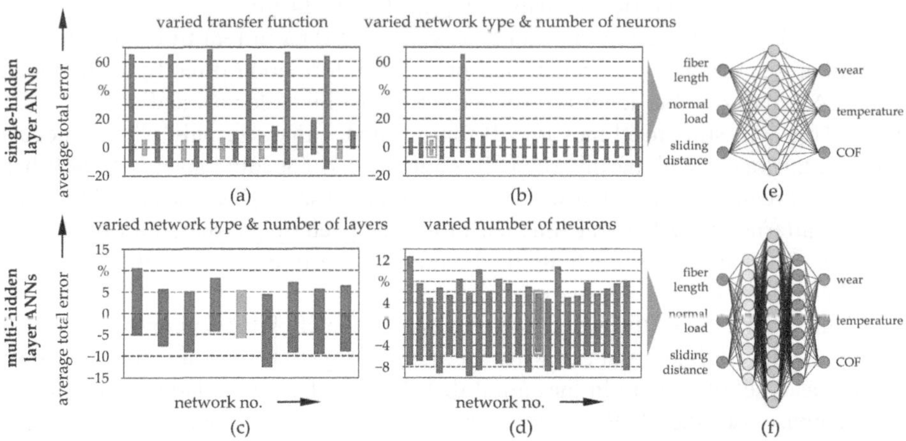

FIGURE 14.9 Average total errors for wear prediction using different ANN architectures when optimizing transfer function [29]. (Creative Commons Attribution 4.0.)

numbers of networks, where performance gaps are observed. Diagrams of neural networks (e) and (f) show the most efficient structures of single hidden and multiple hidden layers, respectively, where the inputs include fiber length, normal load, and sliding distance, and the outputs are the predicted values for wear, temperature, and the coefficient of friction, which predicts the complex interrelated structure that achieved the least values of error in the prediction [29].

Future AI and machine learning applications will progress beyond basic predictive analytics to embody full-field digital twins of high-fidelity, real-time virtual duplicates of tribological systems. Such models promise continuous observation, anticipatory maintenance, and fine-tuned control of frictional interfaces, minimizing both unplanned outages and overall energy consumption. Planned investigations will prioritize the automated characterization of micro- and nanoscale surface landscapes and the consequent synthesis of lubricated assemblies using intelligent design paradigms [30]. Elaborate computational algorithms will analyze large surface datasets to identify the microstructures and tribochemical coatings that provide the best tribological resistance, thus maintaining functional life and dependability [31].

Figure 14.9 shows the structure as well as the training process of generative adversarial networks (GANs), wherein two opposing neural frameworks engage in adversarial co-training and iteratively improve their functions. The generator receives as input a latent space vector from a noise distribution and converts it into synthetic samples, which try to approximate the distribution of empirical training samples. The generator, motivated to minimize generator loss, engineers increasingly credible samples to deceive the discriminator, whereas the discriminator, concerned with the discriminator loss, fortifies its ability to assign correct classifications. The interaction is conceptualized as a two-player minimax game, where conflicting objectives are derived from an implicit reward structure. Progressively, the generator becomes proficient in modeling data manifold boundaries, and the discriminator sharpens its detection boundaries. The equilibrium of this iterative scheme yields a generator that fabricates data that are indistinguishable according to the final discriminator from the statistics of the empirical distribution [32] (Figure 14.10).

With AI–IoT integration, we can start gathering data in real time from moving parts that need lubrication. Smart sensors send information directly to the cloud, where AI analyzes it to identify problems, forecast breakdowns, and suggest quick fixes. This will shift machines to self-tuning modes that use only what is needed and run far longer with hardly a twitch. To make eco-friendly choices, AI will help select the greenest greases and surface layers, thereby saving electricity and reducing waste. Patterns spotted by machine learning can speed up the search for biodegradable oils, foils, and films, pushing the dream of waste-free gears and the lean use of our planet's stock [33].

Figure 14.11 shows the workflow of a GA in which MLPs optimize multilayer power divider designs. It starts with encoding the geometry of a candidate power divider as a set of vectors in the initial distribution of the population. The forward design phase uses the MLP, which completes a device simulation, estimates the performance of each candidate, and assigns a defined fitness score. Competitors are termed as "parents." In the next design step, the GA "mutates" the parent encodings and generates "offspring" by means of a crossover. Using neural networks learned

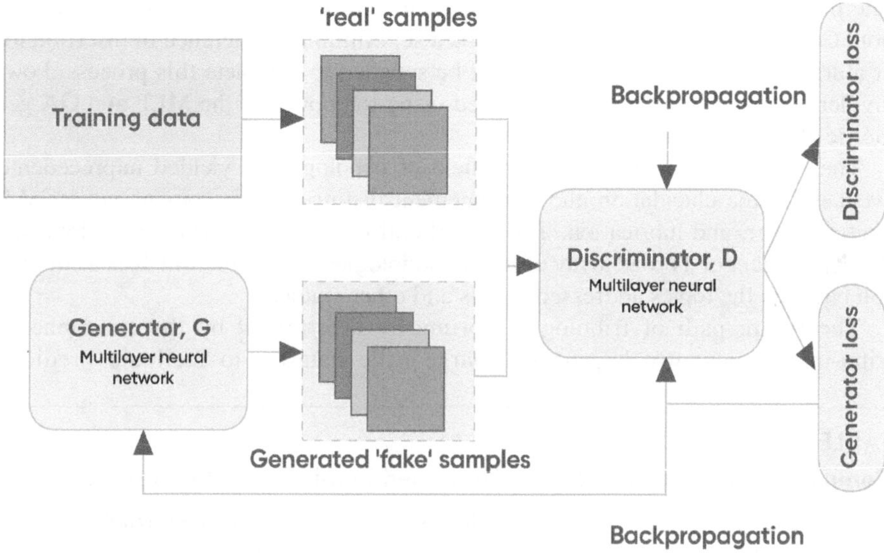

FIGURE 14.10 Generative adversarial network architecture [32]. (Creative Commons Attribution 4.0.)

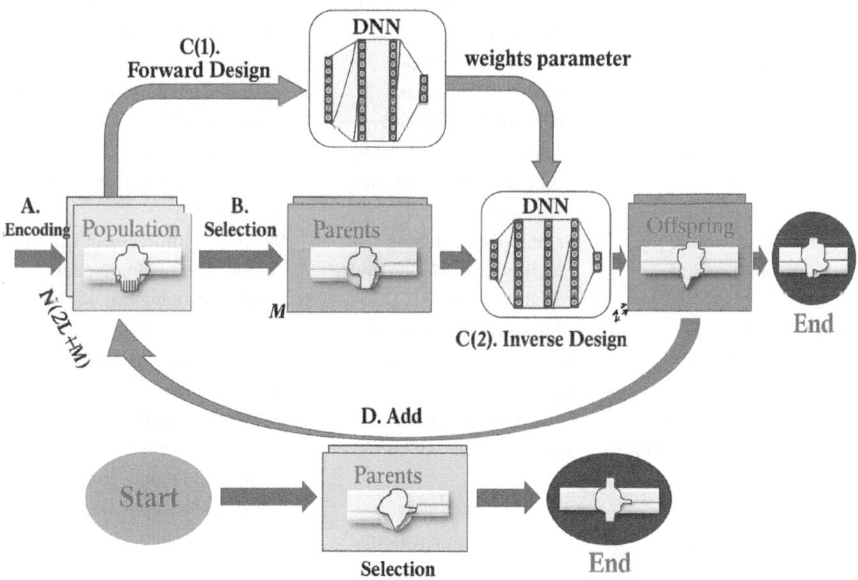

FIGURE 14.11 Optimizing power dividers based on multilayer perceptron [34]. (Creative Commons Attribution 4.0.)

during the simulation, the MLP rapidly predicts the performance without fully simulating them electromagnetically. Individuals that pass the fitness criteria and "survive" replace the updated population in place of poorly performing members.

The population "cycles" as unfit members are "swapped" in each successive iteration. Convergence criteria, which may be a set minimum difference in insertion loss or a desired power-splitting ratio, must be satisfied to complete this process. Power divider designs can be optimally refined using the power of the MLP and GA geometric characteristics [34].

The encroachment of AI into the field of tribology has yielded unprecedented avenues for the elucidation and enhancement of intricate phenomena governed by friction, wear, and lubrication. The consolidation of tribo-informatics is done with the deployment of a broad array of AI methodologies [35]. Table 14.1 lists a comparison between the topics addressed in this and other studies.

The future path of tribology is primarily determined by the emergence of tribo-informatics, which processes large-scale datasets to facilitate predictive

TABLE 14.1
Comparison between the Topics Addressed in this and Other Studies

AI Techniques	Challenges	Future Research Directions
Machine learning classification of MAO coating signals	High variability in coating porosity and thickness; noisy data	Develop robust feature extraction methods; integrate physics-based models into ML [36]
Supervised learning for corrosion prediction	Balancing conductivity with corrosion resistance; limited labeled data	Explore semi-supervised and transfer learning for scarce data; multi-objective optimization [37]
Nanocomposite electrodeposition modeling using ANNs	Uniform nanoparticle dispersion; complex co-deposition kinetics	Couple *in situ* monitoring data with adaptive learning; scale-up process modeling [38]
Deep learning for TiN coating quality assessment	Ensuring coating adhesion while preserving substrate properties	Investigate hybrid DL–mechanistic approaches; real-time feedback control [39]
Reinforcement learning for electrochromic cycling	Maintaining switching stability; degradation under repeated cycles	Integrate reinforcement learning with online degradation models; multiscale modeling [40]
Genetic algorithms in micro-arc oxidation optimization	High-dimensional parameter space; long experimental cycles	Use surrogate-based optimization; integrate autonomous experimentation [41]
Response surface-guided ANNs for plasma polishing	Surface finish variability; multiresponse trade-offs	Combine multifidelity simulations with machine learning; active learning to reduce experiments [42]
Convolutional neural networks for surface morphology analysis	Limited labeled images; complex texture patterns	Develop self-supervised vision models; expand datasets via synthetic augmentation [43]
Support vector regression for adhesion strength prediction	Modeling polymer–metal interfaces; nonlinear adhesion behavior	Explore kernel learning and graph-based models for interface complexity [44]
Ensemble learning for wear prediction	Transfer function selection; overfitting in single architectures	Investigate automated ML pipelines; meta-learning for architecture selection [45]

analytics and the creation of hyper-realistic digital twins, which portray the wear and lubricant dynamics under industrially accelerated conditions of mass abrasion. Auxiliary innovations, particularly those of algorithms devising the surfaces which are topologically micro-engineered to wear at predesigned rates, combined with the friction and thermal service loads during continuous operation, are progressively altering manufacturing in unprecedented ways. These innovations are primarily cantered on the economics of the reduction of wear and abrasion footprints and the stewardship of resources. This congruence of engineering innovations in tribology and manufacturing processes onto the principles of the circular economy is unprecedented. In achieving the targeted engineering innovations, tribology will enable these systems to operate at sustained hyperefficiency and extreme operational longevity with a significant reduction in the ecological footprint of the systems [46,47].

14.5 CONCLUSION

This review aims to offer a complete overview of tribology as well as tribological characterization. This overview review discusses challenges and future directions regarding AI in the field of tribology, focusing the emerging area, tribo-informatics, where tribological phenomena are studied through a new data-science approach. A systematic comparison of AI utilization across the major branches of the discipline indicates a predominant focus on intelligent-tribology, a point well documented in the studies. The following explanatory sections capture the field's unique challenges, namely, complex and hidden epistemology within data generative models, and some unsolved ethical issues. The overview describes AI applications in some dissimilar paradigms, particularly in the protection of public health, aerospace engineering, and predictive maintenance. Futuristics are then proposed, comprising the advanced development of digital twins, automated micro-etching of surface topographies, and custom formulation of surface antimicrobials, lubricants, and their antimicrobials. A comprehensive AI–IoT blend topology promises real-time data assimilation, on-the spot computation, and the evolution of self-optimizing machines within resource-efficient eco-positive limits. While progress on this front is inevitable, there still needs to be thorough and complex problem-solving before the field can fully leverage the siloed potential of the emerging intelligent tribological systems.

REFERENCES

1. Aydin F, Durgut R, Mustu M, Demir B. Prediction of wear performance of ZK60/CeO_2 composites using machine learning models. *Tribology International*. 2023;177:107945.
2. Baş H, Karabacak YE. Prediction of friction coefficient and torque in self-lubricating polymer radial bearings produced by additive manufacturing: A machine learning approach. *Proceedings of the Institution of Mechanical Engineers, Part J: Journal of Engineering Tribology*. 2023;237:2014–38.
3. Baş H, Karabacak YE. Machine learning-based prediction of friction torque and friction coefficient in statically loaded radial journal bearings. *Tribology International*. 2023;186:108592.

4. Si L, Niu R, Dang C, Bao X, Zhuang Y, Zhu W. Advances in artificial intelligence for artificial metamaterials. *APL Materials*. 2024;12:120602.
5. Biswas S, Satapathy A. A study on tribological behavior of alumina-filled glass–epoxy composites using Taguchi experimental design. *Tribology Transactions*. 2010;53:520–32.
6. Cai M, Zhang P, Xiong Q, Cai Z, Luo S, Gu L, et al. Finite element simulation of fretting wear behaviors under the ball-on-flat contact configuration. *Tribology International*. 2023;177:107930.
7. Yin N, Yang P, Liu S, Pan S, Zhang Z. AI for tribology: Present and future. *Friction*. 2024;12:1060–97.
8. Dai Y, Li K, Xiang Q, Ou M, Yang F, Liu J. Microstructure and tribology behaviors of WC coating fabricated by surface mechanical composite strengthening. *Applied Surface Science*. 2023;619:156759.
9. Dalli D, Buhagiar J, Mollicone P, Schembri Wismayer P. A novel hip joint prosthesis with uni-directional articulations for reduced wear. *Journal of the Mechanical Behavior of Biomedical Materials*. 2022;127:105072.
10. Wuest T, Weimer D, Irgens C, Thoben K-D. Machine learning in manufacturing: Advantages, challenges, and applications. *Production & Manufacturing Research*. 2016;4:23–45.
11. Deliwala AA, Dubey K, Yerramalli CS. Predicting the erosion rate of uni-directional glass fiber reinforced polymer composites using machine-learning algorithms. *Journal of Tribology*. 2022;144:091707.
12. Mahapatra SK, Satapathy A. Development of machine learning models for the prediction of erosion wear of hybrid composites. *Polymer Composites*. 2024;45:7950–66.
13. Fan F, Jalui S, Shahed K, Mullany B, Manogharan G. Mass finishing for additive manufacturing: Tribological analysis, surface topology, image processing, predictive model, and processing recommendations. *Tribology International*. 2025;204:110486.
14. Wang Q, Li Y, Li R. Integrating artificial intelligence in energy transition: A comprehensive review. *Energy Strategy Reviews*. 2025;57:101600.
15. Gallego M, Chávez SC, de la Roche J, Toro A. Effect of heat treatment on the wet abrasion resistance of WC-10Co4Cr coatings deposited onto stainless steel by HVOF. *Tribology-Materials, Surfaces Interfaces*. 2022;16:168–76.
16. Gou R, Luo X, Chen J, Wang X, Kang C, Lei Z. Study on the tribological properties of diamond and SiC interactions using atomic scale numerical simulations. *Tribology International*. 2023;178:108093.
17. Wu G, Si L, Xu H, Niu R, Zhuang Y, Sun H, et al. Phase-to-pattern inverse design for a fast realization of a functional metasurface by combining a deep neural network and a genetic algorithm. *Optics Express*. 2022;30:45612–23.
18. Gupta G, Satapathy A. Erosive wear characteristics of plasma-sprayed coatings of glass microspheres premixed with TiO_2 particles. *Tribology Transactions*. 2016;59:80–8.
19. Gupta G, Satapathy A. Erosion wear response of glass microsphere coatings: Parametric appraisal and prediction using taguchi method and neural computation. *Tribology Transactions*. 2014;57:899–907.
20. Salahuddin Z, Woodruff HC, Chatterjee A, Lambin P. Transparency of deep neural networks for medical image analysis: A review of interpretability methods. *Computers in Biology and Medicine*. 2022;140:105111.
21. Huang Y, Ding X, Yuan C-Q, Yu Z-K, Ding Z-X. Slurry erosion behaviour and mechanism of HVOF sprayed micro-nano structured WC-CoCr coatings in NaCl medium. *Tribology International*. 2020;148:106315.
22. Jatavallabhula JK, Shabana S, Pappula B. Development and evaluation of machine learning based predictive models for tribological properties of blended coatings at elevated temperature. *Journal of Bio-and Tribo-Corrosion*. 2025;11:25.

23. Fan Z, Yan Z, Wen S. Deep learning and artificial intelligence in sustainability: A review of SDGs, renewable energy, and environmental health. *Sustainability*. 2023;15:13493.
24. Kaushal S, Gupta D, Bhowmick H. Investigation of dry sliding wear behavior of Ni–SiC microwave cladding. *Journal of Tribology*. 2017;139:041603.
25. Kumar V, Verma R. Effect of GNP and laser-surface texturing on HVOF sprayed WC10Co4Cr coatings for high-wear resistance. *Tribology International*. 2023;178: 108057.
26. Shenwai PG, Choudhary A, Pokuri T, Basak A, Manikandan M, Singh B. On the role of artificial intelligence in aerospace engineering: Current state of the art and future trajectories. *The Aeronautical Journal*. 2025;2025:1–27.
27. Kushwaha AK, Rahman MH, Slater E, Patel R, Evangelista C, Austin E, et al. Powder bed fusion–based additive manufacturing: SLS, SLM, SHS, and DMLS. *Tribology of Additively Manufactured Materials*. 2022;2022:1–37.
28. Maalihan RD, Domalanta MRB, Ali MRR, Caldona EB. Kelvin probe force microscopy: Uncovering new frontiers in coating and corrosion studies. *Analytical Chemistry*. 2025;97:8133–46.
29. Egala R, Jagadeesh GV, Setti SG. Experimental investigation and prediction of tribological behavior of unidirectional short castor oil fiber reinforced epoxy composites. *Friction*. 2021;9:250–72.
30. Marian M, Almqvist A, Rosenkranz A, Fillon M. Numerical micro-texture optimization for lubricated contacts—A critical discussion. *Friction*. 2022;10:1772–809.
31. Marian M, Tremmel S. Physics-informed machine learning—An emerging trend in tribology. *Lubricants*. 2023;11:463.
32. Derrick M, David S, Frederic M, Tajebe B. A review on aerospace-AI, with ethics and implications. *Journal of Civil, Construction and Environmental Engineering*. 2025;10:60–74.
33. Neupane D, Bouadjenek MR, Dazeley R, Aryal S. Data-driven machinery fault detection: A comprehensive review. *Neurocomputing*. 2024;627:129588.
34. Ren Y, Zhang L, Wang W, Wang X, Lei Y, Xue Y, et al. Genetic-algorithm-based deep neural networks for highly efficient photonic device design. *Photonics Research*. 2021;9:B247–B52.
35. Ng WL, Goh GL, Goh GD, Ten JSJ, Yeong WY. Progress and opportunities for machine learning in materials and processes of additive manufacturing. *Advanced Materials*. 2024;36:2310006.
36. Mittal H, Kushwaha OS. Machine learning in commercialized coatings. *Functional coatings: Innovations and challenges*. 2024;2024:450–74.
37. Völker C, Kruschwitz S, Ebell G. A machine learning-based data fusion approach for improved corrosion testing. *Surveys in Geophysics*. 2020;41:531–48.
38. Abdesselam Y, Belloufi A, Rezgui I, Abdelkrim M, Catalin T, Chiriță B. Composite coatings based on nickel and Y_2O_3 nanoparticles: A comprehensive analysis of developments in electrodeposition and functional property optimization. *The International Journal of Advanced Manufacturing Technology*. 2025;137:3273–332.
39. Nyabadza A, Azoulay-Younes L, Vazquez M, Brabazon D. Machine learning for quality control of Tin-Copper electrodes. *Measurement*. 2025;250:117191.
40. Shalini S, Pratiksha J, Rahul M, Vinod Kumar S, Jayesh J, Jaiprabha y. Mechanisms of color switching in electrochromic materials: A comprehensive review of inorganic and organic systems. *Metallurgical and Materials Engineering*. 2025;31:821–32.
41. Jatavallabhula JK, Gaonnwe T, Nginda S, Rao Veeredhi V. Development and assessment of machine learning models for predicting fatigue response in AA2024. *Materials Research Express*. 2025;12:016505.

42. Zhang Z, Jia M, Wang L, Yu Y, Yang Z, Wang J, et al. Physical model-guided machine learning for accelerating laser induced plasma micro-machining process optimization. *Optics & Laser Technology*. 2025;183:112402.
43. Chen B, Zha J, Cai Z, Wu M. Predictive modelling of surface roughness in precision grinding based on hybrid algorithm. *CIRP Journal of Manufacturing Science and Technology*. 2025;59:1–17.
44. Fan C, Ding Y, Zheng Y. Bond strength and failure mode prediction model for recycled aggregate concrete based on intelligent algorithm optimized support vector machine. *Structures*. 2025;71:107999.
45. Kumaravel S, Suresh P. Ensemble machine learning for predicting and enhancing tribological performance of Al5083 alloy with HEA reinforcement. *Proceedings of the Institution of Mechanical Engineers, Part J: Journal of Engineering Tribology*. 2025;239:839–60.
46. Nosonovsky M, Bhushan B. *Green Tribology: Principles, Research Areas and Challenges*. London: The Royal Society Publishing; 2010. pp. 4677–94.
47. Ozkan D, Binal G, Ozgurluk Y, Odabas O, Yilmaz SE, Gok MS, et al. Characterization and tribology performance of refractory materials coatings with blast furnace slag (BFS) fabricated by thermal spraying. *JOM*. 2025;77:336–52.

Index

For Product Safety Concerns and Information please contact our EU representative GPSR@taylorandfrancis.com Taylor & Francis Verlag GmbH, Kaufingerstraße 24, 80331 München, Germany